UNIVERSITY OF BRISTOL
Food Refrigeration and
Process Engineering Research Centre
F R P E R C Churchill Building
Langford, Bristol. BS18 7DY
Tel: + 44 (0)117 928 9239   Fax: + 44 (0)117 928 9314

PHYSICAL, CHEMICAL AND BIOLOGICAL CHANGES IN FOOD CAUSED BY THERMAL PROCESSING

Proceedings of an international symposium sponsored by the International Union of Food Science and Technology, the Scandinavian Association of Agricultural Scientists and the Norwegian Agricultural Food Research Society, held in Oslo, Norway

PHYSICAL, CHEMICAL AND BIOLOGICAL CHANGES IN FOOD CAUSED BY THERMAL PROCESSING

Edited by

TORE HØYEM and OSKAR KVÅLE

The Norwegian Food Research Institute, Ås, Norway

APPLIED SCIENCE PUBLISHERS LIMITED
LONDON

APPLIED SCIENCE PUBLISHERS LTD
RIPPLE ROAD, BARKING, ESSEX, ENGLAND

ISBN: 0 85334 729 8

WITH 55 TABLES AND 166 ILLUSTRATIONS

© APPLIED SCIENCE PUBLISHERS LTD 1977

Printed in Great Britain by Galliard (Printers) Limited, Great Yarmouth

Contents

v

Session III: Applications Within Industry. Current Problems

Session IV: Reaching a Consensus

Design and Purpose of the Symposium

TORE HØYEM

The Norwegian Food Research Institute, P.O. Box 50, N-1432,
Ås-NLH, Norway

The idea of arranging this symposium originated in connection with the planning of the inauguration of the new building for the Norwegian Food Research Institute, which was planned to take place earlier this year. Our intention was to have the two events, the inauguration and this symposium, at the same time. For several reasons we had to run the inaugural ceremony separately on moving into the new premises in September 1975.

As this is the first international symposium for which our institute has been responsible, our first steps were somewhat reluctant, but both during the initial phase and later on we had the pleasure to co-operate with the Secretary-General of IUFOST, Professor Erik von Sydow. His experience and international contacts triggered the early preparations, and guided us through many details later on.

We also had the pleasure of meeting Professor Hawthorn and Dr Duckworth at the University of Strathclyde, Glasgow, who arranged a highly regarded symposium on 'Water relations of Foods' in 1974. To share their experience and obtain their advice has been of great value to us.

The design of international symposia of this kind is to some extent given by the 'Guidelines for International Symposia', worked out by IUFOST. The first point in these guidelines is as follows: 'The symposia should be as international as possible and of the highest possible quality. They should push the subject area forward and not just repeat what earlier meetings have covered'. The guidelines also say that there should be a 'Scientific Programme Committee' and a 'Local Organising Committee'. One difficult point in the guidelines is as follows: 'All speakers and active participants should be present by invitation only and the number of participants kept under strict control'. It is easy to understand how difficult it is to make a selection from all possible distinguished scientists in the world to join the symposium. According to past experience the number of participants in symposia of this kind should be kept low. The optimum number of participants seems to be somewhere between 70 and 100, and we have tried to meet this requirement.

We all know the communication problems that different languages create, even between scientists working in the same area. This everlasting problem is not ignored by us, yet we have decided to use English as the symposium language. We have not provided simultaneous translation in order to reduce the technical equipment and thereby the cost of the symposium. We deeply regret not having a multilingual service, which by the way would have been impossible at this place.

The theme we are going to treat at this symposium is extremely wide and extremely important. It is wide because any temperature condition has a bearing on *some* constituent of food, altering it in some way, for better or for worse. It is too easy to think only in terms of effects on the major components such as proteins, fats, and carbohydrates in this matter, because the smaller constituents like vitamins, amino acids, peptides, nucleic acid derivatives and many others should also be borne in mind—the food as it appears to the consumer is a very complex system, its eating quality being dependent on contributions from all the constituents mentioned and from their interactions.

The importance of the theme lies in the fact that the thermal processing of food is the most common way of preserving it, and making it edible. It is important to heat the food in such a way that it is satisfactorily preserved and still retains its expected taste, aroma and appearance, together with its functional properties. Moreover, the biological value of the food must be taken care of, so that throughout our handling of the food raw materials we always have its nutritional value in mind. We cannot allow ourselves, in the future, unintentionally and unnecessarily, to throw away part of our resources by improper handling.

The time available for all aspects of these important issues is less than five days. It is certainly difficult to draw up a suitable programme for a symposium over such a short time. This job has been done by the Scientific Programme Committee, which I mentioned earlier. It was of utmost importance to us to have broad-spectrum, experienced scientists on the Scientific Programme Committee, and I am happy to say that it was a pleasure to contact and work with the scientist we invited to chair the committee. Professors Bender, Hallström, Hamm and Pearson produced so many valuable suggestions and constructive ideas that the final task of putting together the programme was made rather easy for the Norwegian members of the committee, Director Kvåle and myself. It should also be mentioned that this committee has not had one meeting; all discussions have been carried out through the mail.

I feel that a major feature of the programme, namely the relationship between basic scientific research and its application in the food industry, is extremely important. The problems science is tackling in connection with the production and handling of food, and the solutions to these problems, should be made available to the food industry as soon as possible. There are many members of this audience who represent the food industry and related industries, and I hope that one of the results of this symposium will be that ties between the food industry and food research workers become stronger and more meaningful.

Apart from the scientific programme itself, all other arrangements have been taken care of by our organising committee. All the members of this committee are on the staff of the Norwegian Food Research Institute. As you will see from the programme, we have arranged a couple of activities to give you a break in the scientific programme.

Finally, at the beginning of the symposium, I want, on behalf of the Norwegian Food Research Institute, to thank everybody who has participated in creating this symposium both scientifically and technically. I also want to thank all of you who have made the trip to Oslo to help us give life to this meeting.

1

Food Research and the Food Industry

ANTON SKULBERG, M.P.

*The Norwegian Food Research Institute, P.O. Box 50, N–1432,
Ås–NLH, Norway*

Before turning to the main part of my topic I should like to point out some facts of life with which we have to contend, whether we are working in industry or in research. We who live in a rich part of the world take food for granted. We take it for granted to the extent that we want good food, but we do not want to pay much money for it. As many people as possible should be adequately fed with high-quality food products, but we must realise the fact that about one half of the world's population suffer from hunger or malnutrition. The number of people who each year die from hunger and illness caused by malnutrition is so high that it corresponds to a total eradication of the populations of Norway, Sweden, Denmark, Finland and the northern part of the British Isles.

Gross National Product (GNP) is the general unit used to measure the economic situation in different countries. The GNP *per capita* in the richest fifth of the world's population is 20 times higher than the corresponding figure in the poorest half of the world's population. In our part of the world we consider an economic growth of 3 % per year to be a rather modest increase. In order to maintain the *status quo*, developing countries would have to increase their economic growth by 60 %, and that is simply impossible. Therefore, as long as the rich countries increase their wealth, the gap between the rich and the poor countries is constantly widening. As you will understand, this gap therefore increases very rapidly.

This fact gives us a glimpse of our responsibilities as consumers as well as producers of food. The food industries in the rich countries face a variety of challenges and in particular to produce a variety of good food products at reasonable prices, to use the food resources in a responsible way, to utilise by-products and to prevent pollution of the environment by wastes from production.

Turning now to food research I would like to recall for you the classical picture of the scientist as an eccentric, absent-minded person, who pursues his own personal ideas without regard to the practical application of his results. In spite of that, research has been and is of the greatest importance to the development of industry and improvement of health and living standards. On the other hand, technical advance has caused unfortunate changes in the environment and in human living conditions. You may ask what James Watt might have done if he had been able to foresee the pollution and the difficult changes in social conditions brought about by the industrial revolution based on his invention of the steam engine. The food scientist will therefore never lack challenges and he certainly is faced by great responsibilities. Today the food scientist is not a lonely swallow doing what pleases him. He is normally employed by the food industry or attached to universities or public services.

Food industry and food research must share responsibilities and challenges in order to be able to meet the requirements and difficulties which face them. If we are to succeed, the two parties must know each other in order to create mutual understanding and to prevent misinterpretation of motives. Difficulties of this nature are only too apparent these days. Let us have a closer look at the mutual tasks of food science and food research, and let us start with the beginning.

RESOURCES AND RAW MATERIALS

The two main sources of raw materials for food are agriculture and fisheries. In addition, pure water is absolutely essential for food production. When we consider raw materials for food production we have to use a world-wide perspective.

As far as agriculture is concerned, the production of raw materials of animal origin generally takes place in the countries where industrial production is situated. Raw material of this nature is, however, to a large extent based on animal feedstuffs produced in other countries, and in particular in the developing countries. In a way this is a mutual advantage for the poor developing countries and the rich industrial countries. At the first glance it might seem very advantageous for the developing countries to sell raw materials for the production of feed concentrates to the rich countries. We should take care, however, not to tempt the developing countries into selling what they desperately need themselves, i.e. food for their own population. There are too many examples which show that developing countries have often concentrated more on agricultural production geared to selling products to the rich countries, whilst neglecting the food requirements of their own population.

It seems to me that planning for increased food production from agriculture on a world-wide perspective is a major task, and a field in which the food industry and food research must co-operate. This is indeed a responsibility for the politicians. They, however, badly need inspiration and ideas as well as concrete plans, and the food industry and food research may help them. We are all particularly concerned by the developing countries. Up to now we have tried to assist by helping them to apply modern agricultural machinery and methods which are suitable in industrial countries. We have done so with the best of intentions, but a closer look may show that this way may not be the best approach in countries where unemployment is one of the greatest problems. New, big and efficient machines need large areas of land, which have to be expropriated. The people living and working on this land are forced to leave their homes and they usually end up in the slums of the nearest city, where their miseries will inevitably increase.

We must by all means continue to help the developing countries, but the starting point must not be the concept of efficiency as practised in the rich donor countries. Nor must it be prestige projects which developing countries cannot maintain and continue themselves after the experts have left. As far as possible we must take time to investigate the real and urgent needs of the countries concerned and try to help them to help themselves. It is important that they use their own resources and capacities with the help of simple, but suitable, implements. I must say that in spite of the fact that the question of intermediate technology is a matter of controversy I believe that this way of approaching the problems is a good one in the initial phase of development aid.

If we really mean to help the poor countries to use their own food resources for their own population (and I hope we do) it will have a great impact on our own food production. Primarily it means that we will have to use and develop our own natural resources. In particular we will have to improve and extend our plant production and use our own resources for the production of animal products such as milk and meat. Grass is the second largest protein resource after fish, and except for extra-dry years, the extensive use of grass for feeding animals should particularly be looked into. Let us be sensible enough to realise that this is not an all-or-none situation. We shall always have to supplement each other in the world, and we must not go too far in any single direction, be it imports or self-sufficiency. The main point is that the food industry and food research are alert and that both are willing to face the situation in a realistic way and to do something about it.

I have talked about agriculture first because agriculture in a global context is the most important from the point of view of food production. In Norway, however, our main contribution to the world food supply is fish. Norwegian waters are spawning grounds and to some extent feeding areas for important fish species. This of course gives great opportunities to catch large quantities of fish, but at the same time we must clearly see the responsibilities that are associated with the possibilities.

For thousands of years we have exploited our fish resources by catching without replacement. We have talked a great deal about the protection of fish and other resources in the sea at international conferences and meetings. Some of these meetings are permanent entities where we meet and talk every year. The trouble is that we cannot easily agree, and the practical results are often rather poor. In the past we have seen examples of how we tried in vain to protect important natural resources by means of international agreements. I am referring to the eradication of whales, particularly in the Antarctic, in spite of the fact that international regulations were enforced. With shame I must confess that this country contributed substantially to this unfortunate episode.

At present there is a general consensus that fish resources are at stake mainly due to overfishing. The reason why we have been able to exploit the fish resources by catching for this long period of time without replacement is the reproductive ability of fish. In the past the quantities caught have been compensated for by reproduction. At present, however, we are attacking fish resources from two angles—by overfishing and by interfering with reproduction by pollution of the sea. Being one of the main spawning areas for cod, Norway must first of all realise the situation and try to take action to protect this important resource. Norway cannot do this alone. Collaboration with other nations is necessary, but from experience we know that new ways of collaboration must be applied, different from those of the past. In my opinion extended coastal fishery zones where the coastal state has both the right and the obligation to protect resources against overexploitation are necessary precautions in order to prevent eradication of important fish species. *But this is not enough.* Extended fishery zones are only one step in the right direction, and probably an intermediate step. In addition, it is necessary in fishing, as in agriculture, to plant as well as to harvest. We must not only produce by slaughtering, as we have done in the past. In the first instance, research, and later on practical approaches to this end are in my opinion necessary.

The first question to ask in a programme of this type would be if it is possible from a scientific point of view to increase the reproduction of cod. Should we in the future try to develop cod nurseries where the cod larvae by protection and artificial feeding might be helped to survive the most critical period in their life? Would this be an efficient reserve in case of extensive oil pollution?

About 110 years ago, the Norwegian marine biologist Professor G. O. Sars found that the mortality of the cod larvae after hatching and the survival of the young fry are the most critical problems in the life of the cod. Cod larvae must have food shortly after hatching, because their yolk-sac is very small. In fact the main reason why the cod spawn in Norwegian waters outside the Lofoten area is the simultaneous occurrence of large quantities of a certain type of zooplankton which is a suitable food for the larva. The survival of the larva is dependent on such a plankton drifting through the mouth of the larva within a short time after hatching. Lack of food may therefore cause a catastrophic mortality of the larvae, and a high death rate may also occur if the larvae are transported by the current to areas unfavourable for their further development. The start food is therefore an absolutely critical factor.

Each cod female produces 6–10 million eggs. Four to six months afterwards only 18 fry, weighing about 2 g, have survived, and after one year only 6 are left. Some scientists reckon that only 3 individuals reach the age of one year. The death rate from the age of one year to sexual maturity (3–4 years) is rather low, but only one or two individuals out of the 6–10 million reach sexual maturity. If we could reduce the dramatic death rate at the larval stage, great possibilities might be opened. First of all the quantity of fish might be substantially increased and secondly a genetic improvement of the cod species might be achieved by selection for important traits such as growth rate.

To cultivate fish instead of merely harvesting them is difficult because it means procuring food—and the right food for the fish. In the case of cod one would even have to provide the right baby food, to which special requirements are attached:

(a) The start food for cod must have the right particle size $\sim 40 \mu$m, ranging from 10–200μm.

(b) The density must be such that the particles are in suspension for some hours.

(c) Storage of the food for some time must be possible.

Extensive investigations indicate that the larva of *Calanus finmarchicus* is a suitable start food for cod larvae as well as a good food during the fry period. *Artemia salina* may be another suitable organism for this purpose. It should be underlined, however, that production of sufficient quantities of live food for feeding cod is very difficult. Extensive research to breed zooplankton is therefore necessary before putting these ideas into practice.

It should be mentioned that the production of zooplankton opens the possibility of recycling organic wastes, for instance human sewage. Wastes of this type may be utilised by algae which in turn may be used as food for zooplankton. In this way two goals may be pursued, namely prevention of pollution with organic waste, as well as creating a foundation for protein production by fish rearing. Development of non-living types of food for the cod larvae may also be another possibility which should be looked into by scientists.

In addition to the question of producing food the conditions for hatching and rearing larvae and fry must be investigated. The effects of parameters like temperature, light, depth and circulation of water and feeding intervals must be elucidated. Experience from start-feeding experiments with salmonids and plaice are expected to be of some value.

When the fry have reached a certain size they should be released into the sea. The time and conditions for their release will certainly be very important, but these questions must wait until the preceding problems have been solved. Scientifically, these problems might be overcome by reasonable research efforts on the part of the individual countries concerned and by scientific co-operation between them. Experiments of a similar nature are being carried out in some countries, such as Japan, USA, Great Britain and Germany.

The practical implementation of results which prove scientifically feasible may lead to a completely new situation as far as fish resources are concerned, provided a minimum of international co-operation in fish rearing and the establishment of fishing quotas may be achieved. International co-operation on rearing might for instance involve the breeding of different species by different countries according to natural conditions and possibilities. If we succeed, the problems created by extended territorial fishing zones may to a large extent be eliminated because of the increase of fish resources. In turn, the increased resources will create better possibilities for reaching international agreements. It should be underlined, however, that stronger efforts must be made simultaneously to prevent pollution of the sea. If pollution continues to increase, all other efforts will be in vain.

Developments along these lines are of great interest to the food industry because they will contribute to a substantial improvement out of the present difficulties of the fisheries. For the food scientists, such a programme would be a great challenge and indeed increased research efforts are essential in order to prove the feasibility of the ideas outlined above. Many research disciplines will have to contribute to research projects of this type.

FOOD PROCESSING

The processing of food is a traditional field for research activities within the food industry. Great achievements have been made by research within this field and contributions to industry and indeed to public health have been of great importance. An obvious example is the elaboration of procedures for safe processing of canned food which has made it possible to manufacture products which are both good and safe.

The main task of the food scientist in relation to food processing is to explain what is actually happening during the processing procedures, during storage and during distribution. The best way to overcome difficulties which are encountered is to find the reasons for the troubles. Once you know the reasons you will generally also find the cure. With the extremely complicated processing procedures and the vast number of different products found in industry today these tasks are enormous. It is therefore impossible for me to go into details. But it is worthwhile to mention that a comprehensive knowledge and experience of various fields of research and scientific and technical disciplines, for instance chemistry, biochemistry, microbiology, are necessary.

Packaging is a wide field of its own since one must know the properties of the packaging material as well as the food product. In addition, the interaction between product and packaging material is not easily foreseen. Too often packaging materials and methods are chosen on the basis of the immediate appearance of the finished product, and unfortunately the effect of the packaging on the keeping quality of the food itself may often be neglected or disregarded. A very simple example is washed carrots in plastic bags which look nice on the shelves in supermarkets, but the keeping qualities are much poorer than those of unwashed carrots stored and marketed without prepacking. There are many tasks to be performed by research within the fields of the processing, storage and distribution of food products, and in spite of the comprehensive research activities within these fields over the years I feel that they should continue and indeed be intensified.

FOOD HYGIENE

Food hygiene is getting more important as more complicated techniques are introduced and more chemical additives are applied. Research results are also throwing new light on old technological methods, traditional ways of preservation and the effect of food additives which have been used for a long time. Within this comprehensive subject, loaded with problems and uncertainties, research has one of its major fields of activity. In this area, as in most others, prevention is better than cure. The industry should therefore encourage research to aim at elucidating possible health hazards originating from processing methods and from additives. It is a great advantage to the food industry if they are the first to discover eventual health hazards or even prove suspicion of such problems. If the industry itself is alert and responsible in these respects it will gain confidence among consumers and official authorities, and that cannot be appreciated too much.

Similar considerations apply to a large extent to pollution problems. Here prevention is a most important keyword, and steps in this direction should be taken as soon as pollution risks are discovered. Precautions of this type are also very advantageous from an economic point of view. Adequate planning before starting production is one of the most important factors in this respect, and food scientists and scientists from other disciplines should be consulted during the initial stages of the planning of new plants.

As food scientists, we are confronted with considerable problems both *vis-à-vis* the food industry and the public in general. On one hand we may easily be deceived by the need to catch the public attention through the mass media and in other ways. This is easily done by frightening your fellow man by talking about all the health risks involved. The mass media are only too willing to exaggerate and to put what is said in what they think is a significant context. On the other hand we may be tempted not to give full details due to the repercussions that may be felt by our employers—the food industry. Scientific objectivity and responsibility must be our guidelines. The main thing, both for the industry and for us scientists, is to be believed by the public in general, and therefore there is no better way than objective honesty.

There may be some bodies who wish to dominate research and use it for their own purpose, such as industrial interests, consumers' associations and the mass media. The classical pattern is that the food industry on the one hand and the consumers' associations on the other try to prove their own points of view by different scientific results. There is also a tendency to apply science as a tool to promote particular interests. This is misuse of research and scientists must be aware that difficulties of this nature may increasingly be faced in the future. We must therefore watch our integrity and independence in order to maintain reliability. One of the main tasks of food research is to be a link between industry and official agencies. Scientists should be able to talk a language which both industry and government can understand. In this situation nothing is more precious than confidence, and the basis for confidence is truth and objectivity. Therefore, the best way in which we can serve both industry and society is to search for objectivity and to maintain scientific independence and integrity. Scientists who only try to please industry or consumers' associations or the authorities cannot help anyone.

I hope this symposium on the effect of heat treatment will be successful both to the food scientist and the food industry. Heat is an important way of preserving food. Preservation is an important stage in food economy because it forms the basis of the distribution of food. In many respects food distribution is more difficult than food production, and therefore progress in this field is of great importance.

2

Effect of Temperature–Moisture Content History during Processing on Food Quality

H. A. C. THIJSSEN AND P. J. A. M. KERKHOF

Eindhoven University of Technology, P.O. Box 513, Eindhoven, The Netherlands

INTRODUCTION

Food processing has five basic objectives: improvement of nutritional value, improvement of sensory acceptability, cost reduction, improvement of convenience for the consumer and increase in shelf life. Every operation has, however, side-effects which adversely influence product properties, albeit to a varying extent. This is so because during every operation many types of process may take place simultaneously—such as mechanical, physical, chemical, biochemical and microbial processes. The adverse effects can be classified as follows:

(a) Material losses
　　1. Losses of particulates (fines, dust etc.)
　　2. Losses of dissolved solids, e.g. leaching losses, drip losses
　　3. Losses of volatiles, e.g. aroma losses
(b) Deteriorative physical changes
　　1. Changes in structure, density, colour, permeability etc.
　　2. Changes in state of aggregation, such as melting, dissolution, crystallisation
(c) Physico-chemical destruction of cell membranes
　　1. Turgor losses
　　2. Solid losses (see (a) 2)
(d) Quality degrading chemical reactions
　　1. Enzymatic and non-enzymatic browning reactions
　　2. Vitamin destruction
　　3. Destruction of amino acids, such as lysine and methionine, resulting in loss of biological value
　　4. Oxidation of lipids, flavour components etc.

10

5. Hydrolysis of carbohydrates etc.
6. Destruction of pigments, such as chlorophyll, carotenoids and flavonoids
7. Development of undesirable flavours by reaction products
8. Destruction of carbohydrates and proteins
9. Retrogradation of starch
10. Cross-linking reactions of polymers, including proteins and carbohydrates, which adversely affect the water uptake of the product

Virtually all processes taking place during food processing are dependent on the temperature–water content history. Process kinetics, physical properties and physical and chemical equilibria are all influenced by temperature and water content. Every operation could in principle be optimised, if the effects of temperature and moisture content upon these phenomena and properties were known quantitatively.

EFFECT OF TEMPERATURE AND WATER CONTENT ON PROCESS RATES

To arrive at a general approach to the effects of water content and temperature during processing on food quality and process performance the various properties which change during the process or are liable to change, will be indicated by i and the value of that property i by P_i. The quantity P_i can be the water concentration, the enzyme concentration, the degree of oxidation of a certain constituent or the degree of discoloration, the number of micro-organisms per unit weight, the degree of mixing etc. A decrease or increase in the values of certain properties results in the desired effect of the process, for example reduction of water content or sterilisation. But a decrease or increase in the values of certain other properties causes adverse effects.

The rate of change of P_i in a given volume, specimen, or apparatus is indicated by r_i

$$\frac{dP_i}{dt} \equiv r_i \tag{1}$$

where t is the time. If the process is stationary and if the product in the apparatus is ideally mixed, Eqn. (1) reads

$$\frac{P_{i,1} - P_{1,2}}{\tau} = r_i \tag{2}$$

where the subscripts 1 and 2 indicate the feed to the apparatus and the product leaving the apparatus respectively and τ is the mean residence time in the apparatus. The rate r_i is in general a function of the values of many properties including P_i itself. The dependence of r_i on the value of its corresponding property P_i is indicated by

$$r_i = -k_i P_i^n \tag{3}$$

For physical processes such as drying, the value of n which indicates the order of the process generally lies between 0 and 1. The desorption and absorption rate, such as in drying or extraction, is during the regular regime first order in the concentration of the component to be desorbed or absorbed. But during the constant rate regime the concentration dependence is of the zero order. For chemical reactions in food, including degrading reactions and the destruction of micro-organisms, a first order dependence is assumed in general. Consequently, for $n = 0$

$$\frac{dP_i}{dt} = -k_i \tag{4}$$

and after integration

$$P_{i,0} - P_{i,t} = \int_0^t k_i \, dt \tag{4a}$$

and for $n = 1$

$$\frac{d\ln P_i}{dt} = -k_i \tag{5}$$

and

$$\ln P_{i,t}/P_{i,0} = -\int_0^t k_i \, dt \tag{5a}$$

Thus in the case of $n = 1$ a plot of the logarithm of the value of the property, which can be the concentration C_i of nutrient remaining, versus time yields a straight line.

Effect of Temperature

The temperature dependence of r and consequently also of k can usually be represented by an exponential function. For the inactivation of enzymes and micro-organisms Eqn. (6) holds in general

$$k_i = k_i^0 \exp{(a_i T)} \tag{6}$$

in which T is the absolute temperature, the constant a_i represents the dependence of k on temperature and k^0 is the value of k at $T = 0$. Slightly modifying Eqn. (6),

$$k_i = k_i^r \exp\left[a_i(T - T_r)\right] \tag{6a}$$

in which k^r is the value of k at the reference temperature T_r. Microbiologists use the value Z to indicate the temperature dependence and the decimal reduction time D instead of k. D is the time required to reduce the concentration of the reactant, or the number of spores, by one order of magnitude. Z is the required increase in temperature, expressed in degrees Fahrenheit, to reduce the value of D by one order of magnitude. It can be shown from the definitions that

$$D_i = \frac{2 \cdot 303}{k_i} \tag{7}$$

and

$$Z_i = \frac{4 \cdot 145}{a_i} \tag{8}$$

For most chemical reactions and physical processes the temperature effect is better described by the Arrhenius equation:

$$k_i = A_i \exp\left(- E_i/RT\right) \tag{9}$$

in which E is the activation energy, R the gas constant and A the frequency factor. Equation (9) reads with respect to a reference temperature T_r

$$k = k^r \exp\left\{- \frac{E}{R}\left(\frac{1}{T} - \frac{1}{T_r}\right)\right\} \tag{9a}$$

The relationship between Z, a and E, can be derived from Eqns. (6a) and (9a):

$$E_i = a_i T . T_r . R \tag{10}$$

If the temperature range of relevance to the process is not too large, the difference between Eqns. (6) and (9) is small. Table 1 presents the kinetic parameters of the principal physical properties, chemical and biochemical reactions, and destruction of nutrients, enzymes and micro-organisms.

Effect of Water Concentration

Most physical properties, including vapour pressure, viscosity, thermal conductivity, diffusivity, and relative volatility of aroma compounds, are greatly dependent on water concentration. It follows that the rate processes, such as rate of drying and rate of aroma loss, are also dependent on water

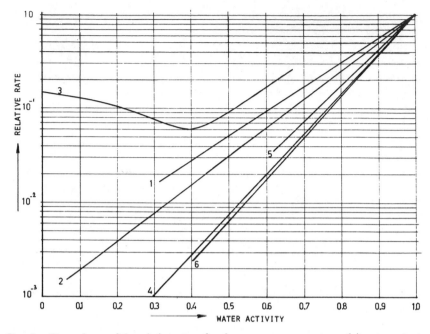

Fig. 1. Dependence of the relative rates of various processes on water activity at constant temperature. Curve 1, relative destruction rate of chlorophyll in spinach at 37 °C, La Jollo *et al.*[3]; 2, relative decay rate of ascorbic acid, Labuza[4]; 3, relative oxidation rate of potato chips stored at 37 °C, Karel[5]; 4, relative rate of inactivation of phosphatase in skim milk, Verhey[6]; 5, relative rate of inactivation of *C. botulinum*[7]; 6, drying rate of a slab of glucose at 30 °C slab temperature, Schoeber.[8]

concentration. Enzymatic processes, chemical processes and the destruction of enzymes and micro-organisms are likewise greatly affected by the water concentration. It is customary to use the water activity a_w rather than the water concentration itself in process calculations. This is because the water activity expresses the availability of water for the physical process or the reaction.

$$a_w = P_w^*/P_w^0 \tag{11}$$

where P_w^* is the equilibrium water vapour pressure of the food and P_w^0 is the vapour pressure of pure water.

 Sorption isotherms, representing the relationship between a_w and water concentration have been reviewed by Labuza[1] and by Loncin *et al.*[2] With the exception of Maillard reactions, which exhibit an optimum of the browning rate at an intermediate water activity between 0·5 and 0·7, and lipid oxidations, which have a minimum reaction rate at an intermediate level of water activity, all process rates decrease with decreasing water activity. The process rates relative to the rates at a water activity of unity are given for some chemical processes, the inactivation of spores or bacteria and the rate of drying in Fig. 1.

OPTIMISATION OF PROCESS CONDITIONS

Optimisation of Process Temperature

It is obvious that, when the various processes taking place during food processing, including the desired process such as drying and the quality degrading side-processes, all have different activation energies, the temperature of the product during processing will influence the quality. When the E-value of the property to be changed is higher or lower than the E-values of the quality degrading processes, and when there are no local temperature differences in the product, the quality will be fully preserved at extremely high or extremely low temperatures. The degree to which this can be achieved in practice depends on technical and economic considerations. When the E-value of the property which needs to be changed has an intermediate value, there is an optimum in the process temperature. In this case the overall quality degrading effect should be minimised. Because physical rate processes have in general considerably lower E-values than chemical processes (see Table 1) the quality of the product will be most effectively preserved at the lowest possible temperature. An example is the concentration or drying of heat-labile foods. The E-value of the drying rate is 10 kcal/mol or lower, whereas the E-values of the degrading chemical reactions are somewhere between 15 and 30 kcal/mol. The increase in drying time that will be the consequence of a lower temperature is fully offset by much lower rates for the chemical reactions. In sterilisation the reverse is true, the E-values of the destruction processes of micro-organisms are considerably higher than those of physical and chemical degrading processes and reactions. For products which can be heated more or less homogeneously this led to the development of the High Temperature Ultra-Short Time (HTUST) sterilisation process.

Optimisation of Water Content

In processes in which one is free to select the water content the quality is

TABLE 1

Activation energies at high water activities of physical properties, rates of physical processes and rates of various groups of reactions

Property or process	$k_{250\,F}$ (s^{-1})	$D_{250\,F}$ (s)	E (kcal/mol)	Z (°F)
Physical properties	n.a.[a]	n.a.	0·5–50	
Water vapour pressure			10	
Water diffusion coefficient			2–10	
Heat transfer coefficient			0·5–7	
Viscosity water (20 °C)			0·002	
Viscosity glucose (25 °C)			48	
Physical process rates	w.v.[b]	w.v.	2–10	
Drying rate				
Crystallisation rate				
Extraction rate, etc.				
Enzyme reactions			4–15	
Chemical reactions			15–120	
Hydrolysis	10^{-1}		15–26	
Thiamine destruction	10^{-4}–10^{-3}		27	
Chlorophyll destruction	10^{-4}–10^{-3}		7–20	
Maillard browning	10^{-4}–10^{-1}		25–50	
Protein denaturation			80–120	
Destruction micro-organisms			50–160	
Yeasts		6 (55 °C)		—
Moulds (ascospores)	10^{1}–10^{2}	10^{-2}–10^{-1}		10–23
Vegetative bacteria	10^{6}–10^{13}	10^{-13}–10^{-6}		7–25
Spores	10^{-2}–10^{0}	6–600		12–18

[a] Not applicable.
[b] Widely varying, very dependent on process conditions.

best preserved at that water concentration at which the process rate of the desired change in the food property is highest compared with the r-values of the quality degrading processes and reactions. Consequently if dr_i/da_w of the desired process is higher or lower than those of the degrading processes, the quality will be preserved most effectively at the highest or lowest possible water activity. Again if dr_i/da_w of the desired process has an intermediate value, the optimum will be at an intermediate moisture level. The relative value of dr_i/da_w can for some processes be obtained from Fig. 1. When the water concentration varies during the process, such as in drying when water removal is the objective of the process, moisture ranges which are favourable for adverse processes should be passed as quickly as possible.

EXAMPLES OF PROCESS OPTIMISATION WITH RESPECT TO TEMPERATURE AND MOISTURE CONTENT HISTORY DURING PROCESSING

There are innumerable processes in which moisture, temperature or both affect the quality of the product. By way of illustration one example will be given in the field of drying and one example in the field of sterilisation.

Quality Preservation During Spray-Drying

When heat-labile, aroma-containing liquid foods such as coffee extract are spray-dried, the volatile aromas should be retained as fully as possible and the heat-labile constituents should suffer minimum thermal damage. The shorter is the time required to lower the water content at the surface of the droplet to an activity smaller than 0·9, the higher will be the aroma retention. This is because at water activities below 0·9 the surface of the droplets becomes selectively permeable for water and starts to retain the larger aroma molecules. Consequently it is only during the drying time elapsing between the release of the droplets at the nozzle and the moment that the surface water activity of the droplets reaches the critical value that the droplets will lose aromas through their surface. This period of aroma loss decreases with increasing initial dissolved solids content in the feed to the spray-drier and with increasing air temperature. It has been shown[9, 10] that even very volatile aromas can be retained almost completely. However, optimum process conditions for aroma retention may result in severe thermal degradation of the product. At high initial concentrations of dissolved solids and high inlet air temperatures of the dryer the period of constant rate evaporation is very short indeed. This means that before a significant fraction of the water has been evaporated, the water activity at the surface of the droplet starts to decrease and the droplet temperature starts to rise rapidly from the wet-bulb temperature to the temperature of the air in the dryer. In other words, the droplet temperature rises to values well above 100 °C while the water content in most of the droplet is still high. Because most heat-labile compounds are very sensitive to heat at high water activities this causes severe decomposition of these compounds. The effect of the length of the constant rate period on the retention of aroma and the retention of the heat-labile compounds is presented schematically in Fig. 2. It is obvious that there will be an optimum in the process conditions if the heat-labile compounds contribute with the aroma to the quality of the dried product.

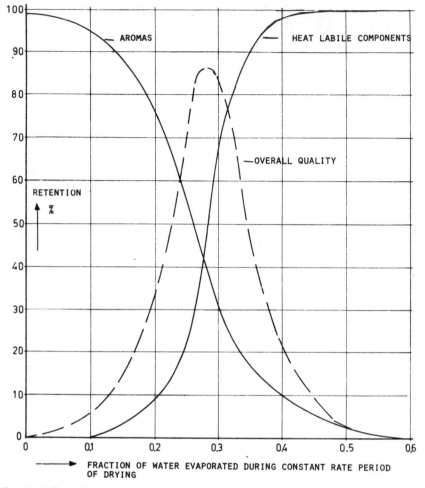

FIG. 2.   Effect of fraction of water evaporated during the constant rate period of drying on the retention of volatile aromas, on the retention of heat-labile non-volatiles and on the overall quality of the product after drying.

Calculation of the Lethal Effect of Heat Sterilisation of Canned Foods and Optimisation of Nutrient Retention

Introduction

It is well known that the HTUST method cannot be applied to canned foods. Very high temperatures will have caused severe thermal degradation of the food near the container wall, long before the food at the centre of the container has risen in temperature. The variation of the temperature in the can centre with a dimensionless time Fo for various values of the dimensionless time Fo_h at which the cans are cooled again after the onset of heating is given in Fig. 3. A relatively low retort temperature, however, will also cause great loss of quality by adverse chemical reactions because of the very long time it will take to attain sterility. Ball and Olson,[11] Teixeira *et al.*,[12] Hayakawa,[13] and Jen *et al.*[14] have presented methods of estimating nutrient degradation in canned food during thermal processing. The method of Ball and Olson requires major computational effort. The finite difference method of Teixeira also requires complex computerisation in application. The method of Jen is in this respect a significant step forwards. All the same the method is rather time-consuming and requires tedious interpolations of the Stumbo tables.[15] The method introduced in this paper can be used without consulting any tables. The calculation procedure quickly produces the process conditions that give the desired degree of sterilisation at minimum loss of quality.

Basic program

The temperature distribution in a can as a function of time is calculated analytically. Subsequently the conversion of heat-labile compounds, including the inactivation of micro-organisms, is calculated numerically and integrated over the total can volume. This integrated value of the reduction of the heat-labile constituents is a function of a number of key variables.

$$\ln \frac{\bar{C}_{i,t}}{\bar{C}_{i,0}} = f(Fo_h, N_{i,1}, N_{i,2}) \tag{12}$$

in which

$$Fo_h = \frac{at_h}{R^2}\left(\frac{1}{R^2} = \frac{1\cdot2}{h^2} + \frac{2\cdot8}{d^2}\right)$$

$$N_{i,1} = \frac{R^2 k_i^r}{a}$$

$$N_{i,2} = a_i(T_h - T_0)$$

where:

$\bar{C}_{i,t}$ = mean concentration of compound *i* in the can after time *t* at which heating has been started.

$\bar{C}_{i,0}$ = mean concentration of component *i* at time $t = 0$.

a = effective thermal diffusivity of the food in the can. (In the case of

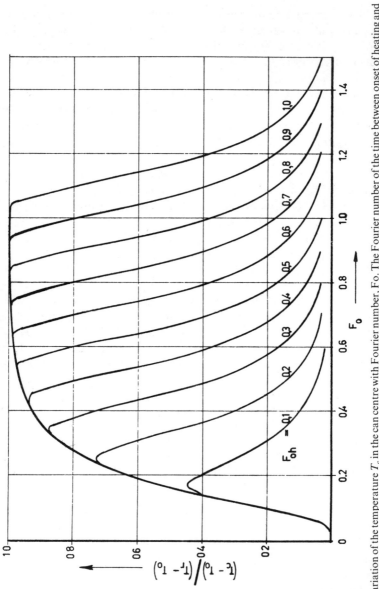

FIG. 3.   Variation of the temperature T_c in the can centre with Fourier number, Fo. The Fourier number of the time between onset of heating and onset of cooling of the can is taken as parameter.

pure conductive heating a is the thermal diffusivity; for fluids the *effective* diffusivity is considerably higher.)

t_h = time elapsed between the onset of the heating of the can and the onset of the cooling of the can.

h = can height.

d = can diameter.

T_h = temperature of heating medium.

T_0 = temperature of food before heating.

In the calculations the cooling water temperature is taken equal to the temperature of the food before heating. The numerically obtained results are in close agreement with the values of the reduction for the heat-labile constituents calculated by the Stumbo method. The numerical results are subsequently correlated with the key variables of Eqn. (12).

Correlation of the results

For the correlation two new combinations of key variables are introduced:

$$Y = \frac{- \ln \bar{C}_{i,t_h}/\bar{C}_{i,0}}{U} \ (Fo_h = 1) \tag{13}$$

in which $U = N_{i,1} . \exp(N_{1,2})$ and C_{i,t_h} indicates the mean concentration of the heat-labile constituent for a heating time t_h and at the instant that the can temperature has been brought back to the initial temperature T_0, so for $t = \infty$

and

$$V = \frac{- \ln \bar{C}_{i,t_h}/\bar{C}_{i,0}}{U} + 1 - (Fo_h + Y) \tag{14}$$

The following correlations have been obtained:

(a) The correlation of $N_{i,1}$ with $N_{i,2}$ at $Fo_h = 1$ and for various constant values of the reduction $\bar{C}_{i,0}/\bar{C}_{i,t_h}$. It appears that at constant reduction of the heat-labile constituents the relationship between $\ln N_{i,1}$ and $N_{i,2}$ is almost linear. The correlation is presented in Fig. 4.

(b) The correlation of Y with U and $N_{i,2}$

$$Y = 1 \cdot 088 - 0 \cdot 132 . \ln N_{i,2} - 0 \cdot 0088 \, [\ln(U + 1)]^2$$
$$\times \exp\{- 0 \cdot 0075 \, [\ln(U + 1)]^{2 \cdot 66}\} \tag{15}$$

(c) The relationship between V and $Fo_h + Y$ with D_{250} (in seconds) as parameter. The results are represented graphically in Fig. 5. It is

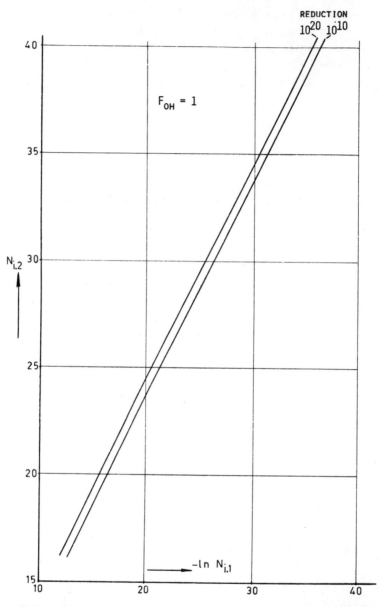

FIG. 4. Relationship between $N_{i,1}$ and $N_{i,2}$ for constant reduction of the heat-labile constituents at $Fo_h = 1$. The reduction factor $\bar{C}_{i,0}/\bar{C}_{i,t_h}$ is taken as parameter.

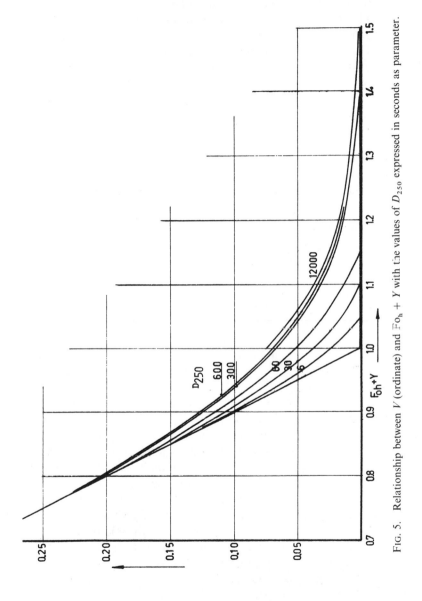

Fɪɢ. 5. Relationship between V (ordinate) and $F_{0_h} + Y$ with the values of D_{250} expressed in seconds as parameter.

interesting to note that all values of V versus $Fo_h + Y$ are on the same curve for identical values of D_{250}.

(d)   The relationship between $dN_{i,2}/dFo_h$ and $Fo_h + Y$ at constant conversion. The relationship is given in Fig. 6.

(e)   The relationship between the ratio of a_i of the destruction of the micro-organisms and a_i of the quality component (chemical reaction) and the value of Fo_h at which the desired reduction of the micro-organisms causes minimum loss of the heat-labile quality component (see Fig. 7).

Calculation procedure

The successive steps of the calculation of the optimum process conditions that give the desired sterilisation at minimum quality loss will be given.

(a)   Calculate k_i^r at the initial food temperature and a_i from the literature values of Z and D_{250} for both the micro-organisms to be destroyed and the heat-labile quality component. Calculate subsequently $N_{i,1}$ for micro-organisms and heat-labile components.

(b)   Choose the desired reduction of the micro-organisms; this gives the value of $\bar{C}_{i,0}/\bar{C}_{i,t_h}$ and consequently of $\ln \bar{C}_{i,t_h}/\bar{C}_{i,0}$ for the micro-organisms.

(c)   Read from Fig. 4 the desired value of $\bar{C}_{i,0}/\bar{C}_{i,t_h}$ for the micro-organisms and its value of $N_{i,1}$, the corresponding value of $N_{i,2}$ at $Fo_h = 1$.

(d)   Calculate Y for the heat-labile compound at $Fo_h = 1$

(e)   Calculate with the aid of Fig. 6 the value of Fo_h at which the ratio of $dN_{i,2}/dFo_h$ at constant reduction of micro-organisms and $dN_{i,2}/dFo_h$ at constant reduction of the heat-labile component is equal to the ratio of the value of a_i of the micro-organisms and the value of a_i of the heat-labile component. At this Fo_h value the quality is optimally retained at the required sterilisation. The optimum value of Fo_h can also be directly obtained from Fig. 7.

(f)   With the optimum value of Fo_h and Y for the micro-organisms, V is obtained from Fig. 5. From the value of V the value of $N_{i,2}$ and consequently of the retort temperature T_h can be calculated. Subsequently the value of Fo_h gives the heating time t_h.

(g)   Calculate with the aid of Fig. 5 the loss of the heat-labile quality component.

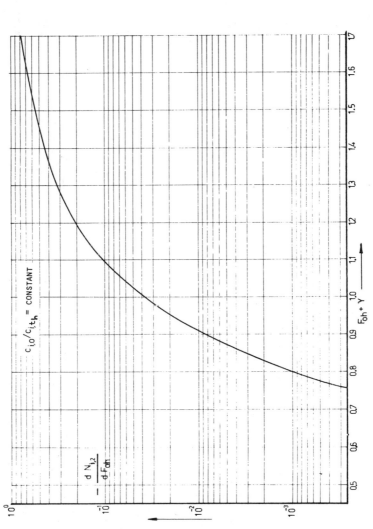

FIG. 6. Relationship between the first derivative of $N_{i,2}$ with respect to Fo_h at constant reduction of the heat-labile constituent and $Fo_h + Y$

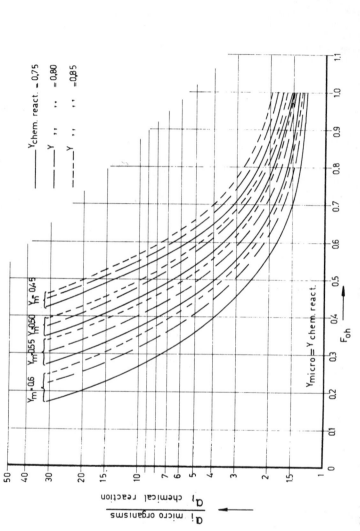

FIG. 7. Relationship between the ratio of the q_t values of the micro-organisms and of the heat-labile quality component and the optimum value of F_{O_h} at which the desired sterilisation effect causes minimum loss of quality. The Y_m values of the micro-organisms and the Y_{chem} values of the heat-labile quality component are taken as parameter.

Results

The results of the calculation procedure are given for two can sizes in Table 2 for the destruction of *Clostridium botulinum* and optimum retention of thiamine. As would be expected the optimum temperature is higher for smaller can sizes. The temperature will approach infinity for a can size of which the value of R^2/a approaches zero. It is also obvious that the quality is better preserved in smaller cans. The variation of the retention

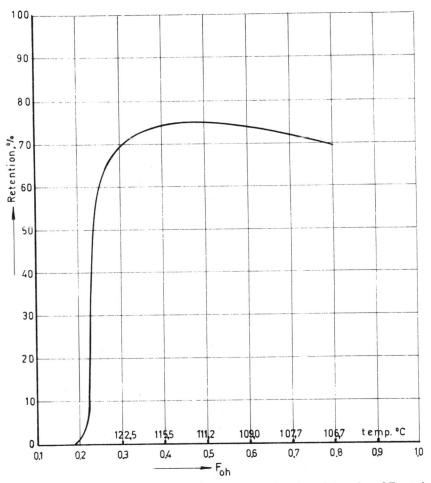

FIG. 8.   Retention of thiamine in $10 \times 10\,cm$ cans as a function of the value of F_{oh} and the corresponding retort temperature. Reduction of *C. botulinum* 10^{10}; *C. botulinum*: $D_{250} = 12\,s$, $Z = 16\,°F$; thiamine: $D_{250} = 9000\,s$, $Z = 45\,°F$; $R^2/a = 1040\,s$.

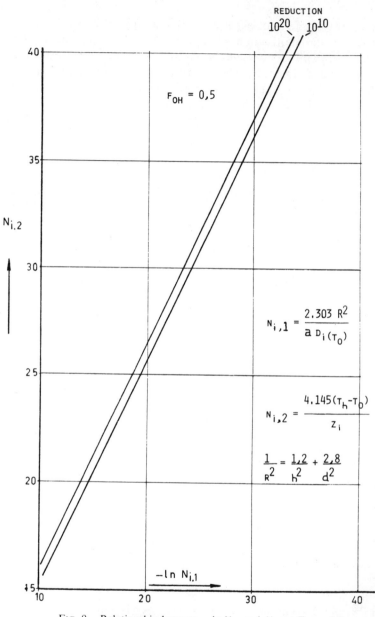

FIG. 9.   Relationship between $-\ln N_{i,1}$ and $N_{i,2}$ at $Fo_h = 0.5$.

TABLE 2
Optimum process conditions for the retention of thiamine[a]

Reduction of spores of *C. botulinum* (10^{10})		
Can size (cm):	10×10	5×5
Fo_h	0·465	0·475
Retort temperature (°C)	112·3	117·6
Heating time t_h (s)	$4·84 \times 10^3$	$1·24 \times 10^3$
Thiamine retention (%)	75	89

[a] *C. botulinum:* $D_{250} = 12$ s, $Z = 16$ °F; *thiamine:* $D_{250} = 9000$ s, $Z = 45$ °F

of thiamine with Fo_h is illustrated in Fig. 8. Clearly the quality retention is mainly dependent on Fo_h and consequently on the retort temperature at very low values of Fo_h.

'Short-cut' method

The energy of activation of most quality-degrading reactions lie between 15 and 40 kcal/mol. For this range the optimum in the sterilisation conditions lies between $Fo_h = 0·45$ and $Fo_h = 0·55$. Because the optimum is rather flat $Fo_h = 0·5$ can be taken as a good average value. Consequently the calculation can be short-cut by using directly the relationship between $-\ln N_{i,1}$ and $N_{i,2}$ for $Fo_h = 0·5$. The relationship is given in Fig. 9.

REFERENCES

1. Labuza, T. P. (1968). Sorption phenomena in food. *Food Technol.*, **22**, 263.
2. Loncin, M., Bimbenet, J. J. and Lenger, J. (1968). Influence of the activity of water on the spoilage of foodstuffs. *J. Food Technol.*, **3**, 131.
3. La Jollo, F., Tannenbaum, S. R. and Labuza, T. P. (1971). Reaction at limited water concentration. *J. Food Sci.*, **36**, 850.
4. Labuza, T. P. (1972). Nutrient losses during drying and storage of dehydrated foods. *CRC critical reviews in Food Technology*, Chemical Rubber Co., Cleveland, Sept., p. 217.
5. Karel, M. (1975). *Principles of Food Science, Part II: Physical Principles of Food Preservation*, Marcel Dekker, New York and Basel, p. 259.
6. Verhey, J. G. P. (1973). Vacuole formation in spray powder particles. *Neth. Milk Dairy J.*, **27**, 3.
7. Karel, M. (1975). *Principles of Food Science: Part II, Physical Principles of Food Preservation*, Marcel Dekker, New York and Basel. p. 48.
8. Schoeber, W.. (1976). Regular regimes in sorption processes. Thesis. Eindhoven University of Technology.

9. Thijssen, H. A. C., (1972). Prevention of aroma losses during drying of liquid foods. 4th European Symposium 'Food', Frankfurt, 25–28 October 1971, *Dechema Monographien*, Vol. 70, pp. 353–67.
10. Kerkhof, P. J. A. M. and Thijssen, H. A. C. (1975). Quantitative study of the effects of process variables on aroma retention during the drying of liquid foods. AIChE meeting on 'Dehydration and concentration of Foods', Los Angeles, Calif., USA., Nov. 16–20.
11. Ball, C. D. and Olson, F. C. W. (1957). Sterilisation. In *Food Technology*, McGraw-Hill, New York, Toronto and London.
12. Teixeira, A. A., Dixon, J. R., *et al.* (1969). Computer optimisation of nutrient retention in the thermal processing of conduction heated foods. *Food Technol.*, **23**, 845.
13. Hayakawa, Kan-ichi (1969). New parameters for calculating mass-average sterilising values to estimate nutrient retention in thermally conductive food. *Canad. Inst. Food Technol. J.*, **2**, 167.
14. Jen, Y., Manson, J. E. *et al.* (1971). A procedure for estimating sterilisation of and quality factor degradation in thermally processed foods, *J. Food Sci.*, **36**, 692.
15. Stumbo, C. R. (1973). *Thermobacteriology in Food Processing*, 2nd edn., Academic Press, New York and London.

3

Heat Preservation Involving Liquid Food in Continuous-Flow Pasteurisation and UHT

BENGT HALLSTRÖM

Chemical Centre, Food Engineering, Lund University, P.O. Box 50, S-25053 Alnarp, Sweden

Thermal preservation techniques involve heating the product to a certain temperature, holding the product at this temperature for a certain time and then cooling it to storage temperature. The maximum temperature level and the holding time are chosen to give a specified thermal destruction of bacteria. The necessary heating and cooling treatment may be carried out either in a continuous-flow heat-exchanger or with the product packed in its consumer package. Which method to be used is a matter of the type of product and type of package, but also a matter of quality demand and technical development.

One incentive in this development work is the fact that an increased temperature combined with a decreased holding time, but having the same lethal effect on micro-organisms, means a lower ratio for other chemical reactions, e.g. those denaturing the product. Further, in a continuous-flow process the product temperature may be raised and lowered much faster than in a batch process, and, furthermore, the time–temperature distribution for the different parts of the product is more even. Finally, the development of aseptic filling has made continuous-flow sterilisation (UHT) possible.

Liquids containing particulate material mean special problems which have not yet been fully solved. In the continuous-flow heat-exchanger the centre of the particles still needs a longer time than the liquid itself, thus resulting in overheating of the liquid. Further, aseptic filling of such products is not yet commercial.

Since during heating of a product the time–temperature curve is convex upwards, and during cooling concave, the heating process has a greater influence than the cooling process on physical, chemical and biological changes and is therefore more interesting to study. This emphasis may also

31

Bengt Hallström

TABLE 1

Some examples of heat treatment, pasteurisation and sterilisation

Heat treatment	Pasteurisation	Sterilisation
Heating and cooling inside the package	Tunnel pasteurisation of beer	Autoclave and hydrostatic sterilisation of canned food
Heating in continuous flow, cooling in the package	Pasteurisation and hot filling of juice	'Flash 18'
Heating and cooling in continuous flow	HTST^a-pasteurisation of milk	UHT-sterilisation of milk products

^{a} HTST = high temperature, short time.

be observed in the historical development of the topic. Table 1 gives some examples of the development steps for some types of products.

For heat treatment before filling, e.g. in continuous flow, there are several technical alternatives available:

(a) Indirect heat exchange; the product and the heating/cooling fluid are separated by a wall—the heating surface.
1. Tubular heat-exchanger
2. Plate heat-exchanger
3. Scraped surface heat-exchanger
(b) Direct heat exchange; the steam is condensed into the product thus involving dilution of the product. Cooling is done by means of evaporation (flash cooling) whereby the process is so controlled that the same amount of water as was previously added as steam condensate is removed.
1. Steam injection ('steam into milk')
2. Steam infusion ('milk into steam')
Further, product to be heated/cooled may be mixed into an already heated/cooled product which is recirculated at a high rate.
(c) Electrical methods.
1. Resistance heating
2. Inductance heating
3. Dielectric heating
(d) Mechanical friction heat; heat is generated in the product itself while it is flowing through a very narrow slit where the surfaces are moving at high speed in opposite directions.

These methods of heating/cooling have been used in building up different process systems on the market both for pasteurisation (HTST) and continuous-flow sterilisation (UHT). For pasteurisation the same type of heat-exchanger is normally used throughout the process. Plate and tubular heat-exchanger systems are well established on the market. Other systems are those which make use of direct exchange with steam; in these cases preheating is normally carried out indirectly and only the final heating is by direct heat exchange. For viscous products, systems are available utilising scraped surface heat-exchangers. Electrical heating so far has few applications. A French manufacturer is bringing out a plant with resistance heating, mainly used for fruit juices. According to the patent literature, research pertaining to dielectric heating is going on but so far no commercial equipment is available.

Pasteurisation is a well-established and well-known process. In the following I will therefore concentrate on the UHT process, which is still under development and thus more interesting. Much of what is discussed below is, however, also applicable to pasteurisation in general.

UHT PROCESSES

Table 2 shows a list of manufacturers of the UHT systems most used today. A more detailed list and description of these systems is given by, for instance, Burton.[1]

TIME–TEMPERATURE RELATIONSHIPS

Earlier, when different UHT processes were evaluated and compared both with regard to the sterilisation effect and product denaturation, only the sterilisation temperature and the average holding time were considered. However, the processes listed in Table 2 differ very much in, for instance, heating-up time (from almost zero for the direct methods to more than 120 s for some of the indirect methods) and further development demanded more accurate analyses. It should be mentioned that these problems are similar for the fermentation industry and a great part of the theoretical work is done in this field.

The time–temperature curves for the average particle in a Newtonian liquid treated in a heat-exchanger can be easily deduced. Similar investigations have also been done by Deindoerfer and Humphrey.[2] The

TABLE 2
UHT systems and their manufacturers

Indirect heating	
Tubular heat-exchangers	Cherry–Burrell Corp., USA
	Stork–Amsterdam BV, Netherlands
Plate heat-exchangers	Ahlborn AG, W. Germany
	Alfa–Laval AB, Sweden
	APV Co., England
	Sordi, Italy
Scraped surface heat-exchangers	Cherry–Burrell Corp., USA
Direct heating	
Steam injection	Ahlborn AG, W. Germany
	Alfa–Laval AB, Sweden
	APV Co., England
	Cherry–Burrell Corp., USA
	Rossi & Catelli, Italy
Steam infusion	Breil & Martel, France
	Ets. Laguilharre, France
	DaSi Industries, Inc., USA
	Paasch & Silkeborg, Denmark
Electrical methods	—
Mechanical friction	SEFFAC, France

following equation applies to the general case and gives the temperature T as a function of the time τ:

$$T = T_{s0}[1 - b \exp(-K\tau)] \tag{1}$$

where

$$b = \frac{T_{s1} - T_0}{T_{s0}} \tag{2}$$

and

$$K = \frac{UA}{Wc}\left(1 - \frac{wc}{w_s c_s}\right) \tag{3}$$

In these equations T_{s0} and T_{s1} are the inlet and outlet temperatures of the heating fluid; $T_0 =$ inlet temperature of the product; $U =$ overall heat transfer coefficient; $A =$ heating surface; w and w_s are the mass flow rates of product and heating fluid; c and c_s are the specific heats of the product and the heating fluid; and $W =$ mass of the flowing product in contact with surface A.

For an isothermal heat source, for instance condensing steam, $T_{s0} = T_{s1}$ = T_s and $wc/w_s c_s$ becomes zero, giving

$$T = T_s \left[1 - \frac{T_s - T_0}{T_s} \exp\left(-\frac{UA}{Wc} \tau \right) \right] \qquad (4)$$

For a countercurrent heat source of equal flow rate and equal heat capacity, $wc = w_s c_s$, and the heating and cooling curves reduce to straight lines:

$$T = T_0 \left(1 + \frac{T_{s1} - T_0}{T_0} \cdot \frac{UA}{Wc} \cdot \tau \right) \qquad (5)$$

For direct heat exchange U is very high as there is no wall separating the heat-exchanging fluids. This gives a very rapid heating up; no figures are available and normally the time is regarded as less than one second. There is a problem if the steam is not completely condensed, which disturbs the flow and the temperature distributions. So for instance Stroup et al.[3] have measured temperature variations after the steam injector up to $\pm 15\,^\circ$C.

These equations are accurate only for average conditions or so-called plug flow. However, there are two types of irregularities which disturb this model. There is a distribution of flow velocity over the width of the channel, depending on the type of flow (laminar or turbulent), the flow of heat (heating or cooling) and the physical properties of the product (Newtonian or non-Newtonian liquid). Further, there is a temperature profile over the channel the shape of which is a function of the same parameters.

Flow Velocity and Temperature Distribution

For laminar flow the velocity profile is parabolic and the mean velocity is half the value of the maximum velocity. For turbulent flow the velocity profile follows a seventh-order equation giving a mean velocity of 82% of the maximum velocity. For a non-Newtonian fluid and laminar flow

$$U_{mean} = \frac{n + 1}{3n + 1} \cdot U_{max} \qquad (6)$$

where n is the power-law (pseudoplastic) index. A Newtonian liquid is one in which the stress–strain relationship is linear and defined by a single viscosity parameter. A non-Newtonian liquid does not obey this condition. A normal model is the power-law equation according to which the shear stress is proportional to the radial velocity gradient to the power of the index n.

This relationship between U_{mean} and U_{max} means that the fastest particle of the fluid moves through the plant in a shorter time than the previous equations express. The velocity distribution is also influenced by the flow of heat. For a fluid to be heated the U_{mean}/U_{max} ratio becomes higher than the figures given and for a fluid to be cooled, lower than these figures. This is a result of viscosity changes due to temperature distribution and is therefore normally negligible for low-viscosity liquids.

In a liquid to be heated the part of the product closest to the wall of the heating surface has a higher temperature than that on the centre. The ratio of the 'over-temperature' depends on flow conditions.

A combination of the effects of velocity distribution and temperature profile results in a non-uniform treatment of the different elements of the liquid. The centre receives the least thermal treatment. A complete and general solution of the mass-fraction–time–temperature distribution for a liquid passing a heat-exchanger has so far not been presented. This distribution formula should be combined with a kinetic model for physical, chemical and biological changes to give the final results of the changes of the product.

Simpson and Williams[4] have investigated laminar flow of non-Newtonian fluids in a tube for heating, holding and cooling. Both the sterilisation effect and chemical changes are calculated. Holdsworth[5] has given equations and guidelines for the calculation procedure. He also presents a diagram giving the percentage of 'over-sterilisation' in isothermal flow (holding). Lin[6] has also worked with non-Newtonian liquids and laminar flow and theoretically applied this in the destruction of *C. botulinum* and denaturation of vitamin B_{12}.

Particulate material

For liquids containing particulate material the relationships become still more complicated and these problems are not yet completely solved. Ruyter and Brunet[7] have, however, developed a computer program for certain presumptions; this results in the degree of 'over-sterilisation' for the liquid, e.g. the ratio

$$\frac{F_{liquid}}{F_{particle\ centre}} \tag{7}$$

Mathematical Models

By combining mathematical kinetic models for biological, chemical and physical changes in the product with the time–temperature formula for the process, different process systems can be compared and evaluated. Various kinetic models have so far been used. First the Arrhenius equation which

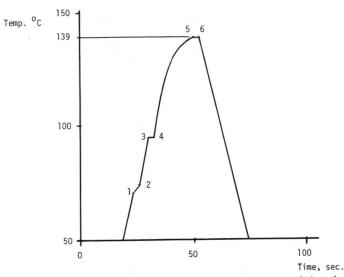

FIG. 1. Time–temperature relationship for indirect heated UHT system. 0–1, preheater; 1–2, homogeniser, 71–74 °C, 25 s; 2–3, recuperative heater, 74–95 °C, 4·5 s; 95 °C, 2 s; 4–5, steam heater, 95–139 °C, 17·5 s, $T = 142·3[1 - 0·3324 \exp(-0·1524\tau)]$; 5–6, holding tube, 139 °C, 2·4 s; 6–7, recuperative cooler, 139–50 °C, 21·6 s.

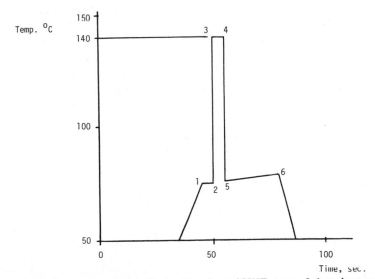

FIG. 2. Time–temperature relationship for direct heated UHT system. 0–1, preheater; 1–2, 75 °C, 5·1 s; 2–3, steam injector, 75–140 °C, ∼0 s; 3–4, holding tube, 140 °C, 5·1 s; 4–5, flash cooler, 140–76 °C, ∼0 s; 5–6, homogeniser, 76–79 °C, 24·1 s; 6–7, cooler, 79–20 °C, 15 s.

indicates that the degree of change is proportional to

$$A = A' \int \exp\left(-\frac{E}{RT}\right) d\tau \tag{8}$$

where $A' = $ constant, $E = $ the activation energy of the reaction and $R = $ the gas constant. This equation is a little too complicated to use and the following expression has been used for a lot of changes

$$F_Z, C_Z = \int 10\left(\frac{T - T_0}{Z}\right) d\tau \tag{9}$$

where $Z = $ the slope index of thermal death–time curve of micro-organisms or nutrition and T_0 a reference temperature. This expression is well known for calculating the lethal effect (F) and also for calculating enzymatic and sensory changes ($C = $ cook-value).

A simple model is used by Kiermeier,[8] who regards the surface under the time–temperature curve (the 'sum of heat') as an indication of sensory changes in UHT-treated milk. This gives the equation

$$K = \int (T - T_0) d\tau \tag{10}$$

Example: evaluation of milk sterilisers

In the following example I have tried to apply Eqn. (10) to two commercial UHT systems; one of the most common direct systems and one of the most used indirect systems. Temperature profiles have kindly been supplied by the companies and are illustrated in Figs. 1 and 2. In these figures are also indicated the equations for the time–temperature relationships used, in accordance with the flowsheet for the process. In the

TABLE 3
Values of A/A', F and c calculated for direct and indirect heating systems

Process	Parameter	Direct heating	Indirect heating
Destruction of micro-organisms	$F_Z = 8\cdot5$	14·6	16·7
	$F_Z = 10$	6·8	8·7
Browning of milk	$C_Z = 21$	7·2	14·5
Average chemical changes	$C_Z = 25$	3·5	7·9
Destruction of *C. botulinum* ($E/R = 37\,300$)	A/A'	$0\cdot31 \cdot 10^{-38}$	$0\cdot40 \cdot 10^{-38}$
Denaturation of thiamine ($E/R = 11\,800$)	A/A'	$0\cdot20 \cdot 10^{-11}$	$0\cdot54 \cdot 10^{-11}$

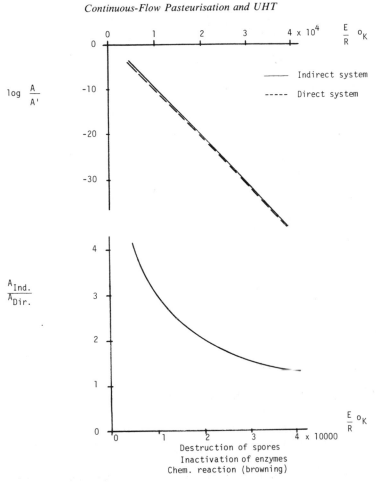

FIG. 3. Ratio of changes for indirect/direct UHT systems according to the Arrhenius equation.

following considerations only mean velocities are regarded. For these time–temperature relationships the equations given above have been calculated with different values of the constants in the models; *i.e.* the activation energy E in the Arrhenius equation (8) and the Z-value and the reference temperature T_0 in Eqn. (9). In the calculation of F_Z, $T_0 = 121\,°C$ has been used and for C_Z, $T_0 = 100\,°C$.

The first diagram (Fig. 3) shows the ratio of changes according to the Arrhenius equation and gives A/A' for some interesting values of E/R. In

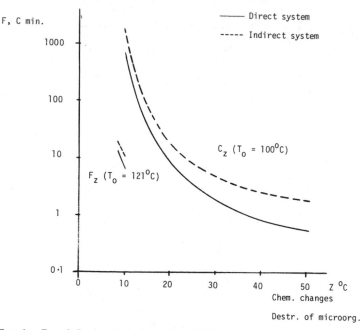

FIG. 4.   *F*- and *C*-values in indirect/direct UHT systems according to Eqn. (9).

order to illustrate an eventual difference between the two systems, $A_{direct}/A_{indirect}$ has also been plotted.

In Fig. 4 Eqn. (9) is used, thus giving the corresponding values for some values of Z and T_0 of interest. A/A', F and C for certain values are also given in Table 3.

For browning of milk $Z = 21$ ($Q_{10} = 3$) has been used according to Burton[9] and for average chemical changes of milk $Z = 25$ ($Q_{10} = 2\cdot5$) according to Pien.[10] In the Arrhenius equation $E/R = 37\,300$ and $11\,800$ is taken from Simpson and Williams.[4]

REFERENCES

1. Burton, H. (1972). *UHT Processing systems for milk and milk products.* IDF Monograph on UHT milk.
2. Deindoerfer, F. H. and Humphrey, A. E. (1959). Analytical methods for calculating heat sterilisation times. *Appl. Microbiol.,* **7**, 256–63.
3. Stroup, W. H., Parker, R. W., Dickerson, R. W. and Read, R. B. (1972).

Temperature variations in the holding tube during ultra-high-temperature pasteurisation by steam-injection. *J. Dairy Sci.*, **55**, 177–82.
4. Simpson, S. G. and Williams, M. C. (1974). An analysis of high-temperature/short time sterilisation during laminar flow. *J. Food Sci.*, **39**, 1047–54.
5. Holdsworth, S. D. (1970). Continuous sterilisation of foods. *Process Biochemistry*, March, 57–62.
6. Lin, S. H. (1976). Continuous high-temperature/short-time sterilisation of liquid foods with steam injection heating. *Chem. Eng. Sci.*, **31**, 77–82.
7. de Ruyter, P. V. and Brunet, R. (1973). Estimation of process conditions for continuous sterilisation of foods containing particulates. *Food Technol.*, July, 44–51.
8. Kiermeier, F. (1972). Ultrahocherhitzung und ihr Einfluss bei Milch und Milchprodukten. *Deutsche Milchwirtschaft*, **23**, 836–44.
9. Burton, H. (1975). Some factors influencing the quality of liquids processed by the ultra-high-temperature system with aseptic filling. *Proc. 6th European Symposium—Food*, Cambridge, Sept. 8–10.
10. Pien, J. (1972). *Chemical and Physico-chemical Aspects, Laboratory Control.* IDF Monograph on UHT milk.

4

Heat Preservation involving Liquids and Solid Foods in Batch Autoclaves and Hydrostatic Sterilisers

W. F. Hermans

*Research and Development Department, Stork Amsterdam BV,
Staringstraat 46, Nieuw Vennep, The Netherlands*

The heat preservation of liquid and solid foods in batch or in continuous processing equipment includes preservation inside the container. The product to be sterilised is previously prepared, precooked if necessary and sometimes even presterilised. Then it is put into a container under hygienic but non-aseptic conditions, hermetically sealed and transported to the actual sterilising equipment. Here we have to distinguish between batch-type and continuous-flow equipment.

Batch-type equipment is characterised by a pressure vessel in which the whole process takes place. The containers are stored on pallets, in crates or on lorries, and brought into the pressure vessel, which is closed afterwards. Then the sterilising process can start. The older types were manually operated, the more recent ones are equipped with a programme-control unit, so that the whole operation, including de-aeration, filling with steam or hot water, pressure- and temperature-control, holding time and cooling programme is done automatically, thus avoiding the possibility of human failures. The automatic programme-controller also enables the operator to supervise several batch sterilisers at the same time. Batch-type equipment is basically simple and extremely suitable for small charges and a wide variety of types of container. Special features, like rotation, can be built in relatively cheaply. The main concern, however, with this equipment is to ensure equal temperature distribution during sterilisation. Thoroughly mixing or agitating the heating medium during this phase of the process is absolutely essential. Another point worth mentioning about batch equipment is the relatively high consumption of steam and cooling water. This can be suppressed by the combined running of several batch autoclaves and the installation of big storage tanks for hot and cold water, but this makes the whole set-up more complicated and requires careful operation.

Also, there still remains the large amount of labour to load and unload the autoclaves.

Continuous sterilisers are designed for higher production capacities (300–1500 cpm, or hourly capacities of 18 000–90 000, and even more) and a limited variety of types of container. They are characterised by two or more mechanical or hydrostatic pressure locks on each end of the sterilising compartment. The containers are transported via the locks through the whole installation. If mechanical locks are applied, the pressure is changed instantaneously while passing the lock. Since many containers cannot stand such a sudden pressure drop, more than one lock is sometimes required, to build up the outside pressure to equilibrate with the pressure inside the container. In hydrostatic locks the outside pressure is slowly changed, while at the same time heat exchange with the container can take place, thus influencing the pressure inside the container, and keeping a balance between changes of pressure and temperature. The gentle change in temperature while passing the hydrostatic lock also makes this type extremely suitable for handling glass containers, thus avoiding thermal shock. Since hydrostatic locks need a certain height, the rest of the installation is built up in towers, in this way giving the characteristic shape of the well-known 'hydromatic'. Special features, like rotation, whether it is 'end-over-end' or axial, is also done in hydromatics, but this requires a higher investment. However, the constant moving and half-rotation every now and then gives a certain amount of agitation.

One of the main advantages of continuous sterilisers from a technological point of view is the fact that temperature differences in the actual sterilisation chamber, within certain limits, are small. Every container follows the same track. Therefore, it is sufficient to maintain strict temperature control in a few spots. In fact every section of the continuous steriliser is working as a heat-exchanger in steady state: a product flow enters and leaves and the heating or cooling medium provides a countercurrent flow.

Since the continuous steriliser, although basically simple, is in fact a more complicated installation than the batch autoclave, the advantages of the continuous steriliser only pay off with production capacities higher than say 200 cpm or 12 000 per hour. Furthermore, the carriers have to be designed for the containers to be handled, which restricts the application to a limited size and form of container. The processing time, however, can be varied widely, simply by changing the transport speed. Another very important part of the continuous steriliser is the feed and discharge system which has to offset the labour costs required for loading and unloading the batch autoclaves. For instance, one man can guard the feed and discharge system of a 60 000 container per hour continuous steriliser, while several employees are kept very busy to load and unload the equivalent number of batch autoclaves in order to reach the same hourly capacity. Also, heat recuperation in continuous sterilisers is an important advantage, thus cutting down the consumption of steam and cooling water by more than 50% compared to batch autoclave systems.

A comparison between batch and continuous processes depends on the product to be sterilised and the container in which the product is treated. As far as the product is concerned the following types can be distinguished:

(a)   Whole, low-viscosity liquids (milk, pharmaceutical liquids).
(b)   Whole, high-viscosity liquids (baby foods from milk or other bases, with additives but still homogeneous).
(c)   Low-viscosity liquids with solid particles (peas, haricot beans, carrots in brine).
(d)   Viscous liquids with solid particles (concentrated soups, pieces of meat in gravy).
(e)   Solid foods (stews, corned beef).

As far as the container is concerned, there is a tremendous variety, but in general the following divisions can be made:

(a)   Not sensitive to pressure or temperature (small rigid metal cans).
(b)   Not sensitive to pressure, sensitive to temperature (crown-corked bottles).
(c)   Sensitive to pressure, not sensitive to temperature (big, rigid metal cans).
(d)   Sensitive to both pressure and temperature (plastic containers, pouches, vacuum-closed glass jars).

Physically, the heat preservation of foods, mainly containing water, whether it is free or bound, has two aspects:

(a)   Heat transfer from the outside media, via the container wall, to the most slowly heated point of the product.
(b)   Pressure changes, accompanying the temperature changes, which cause pressure differences over the container wall.

As regards chemical and biological aspects, which can be equated with product quality, there are two additional aspects:

(a)   Sufficient 'lethality'.

(*b*)   Minimal chemical changes, including losses on vitamins and nutritive value.

Since the rate of chemical change increases much more slowly with increasing temperature than the lethal rate increases, heat preservation is carried out at higher temperature levels, thus minimising the processing time. Heat transfer from the medium to the wall is mostly very good. Heat transfer coefficients up to 2000 kcal/m/h/°C are no exception. Heat transfer from the container wall to the product however is much less effective, depending on the type of product. Low-viscosity products, thoroughly shaken, can reach heat transfer coefficient values of 500–600 kcal/m^2/h/°C, but solid foods, e.g. stews, have a low coefficient of heat conduction (~ 0.5 kcal/m^2/h/°C) and therefore the heat transfer coefficient is limited to 50 or 80 kcal/m^2/h/°C.

Another complication in heat transfer is the type of product which is solid at lower temperatures, but is a viscous fluid at higher temperatures, e.g. concentrated soups, or products in a fat gravy. This is especially difficult below the temperature where these products set hard, thus forming a jacket and inhibiting heat transfer. These products can cause problems during the cooling phase.

One of the characteristics of the batch autoclave is that, immediately after closing the door, the sterilising temperature and pressure can be set. On the other hand, there is always some time required to ventilate the autoclave to get rid of enclosed air. This is the reason that in practice it takes at least several minutes before the required sterilising conditions are achieved inside the autoclave.

Another important point with batch autoclaves is that, even with thorough mixing of the sterilising medium, whether it concerns saturated steam, steam–air mixtures or water under pressure, it is unavoidable that during the first minutes after reaching the sterilising conditions, considerable temperature differences are still found between various locations inside the autoclave. This is the more serious when products with high internal heat transfer coefficients are processed, and these are just the products which need to be sterilised at the highest possible temperature in the shortest possible period. Therefore rotation of the product inside the autoclave might be worth while for this type of product.

When considering autoclaves a third point must not be forgotten, and that is the waiting time for the first containers during loading the autoclave before the last containers are in, and the door can be closed. For instance, a 300 cpm filling line, feeding several 10 000-can autoclaves requires at least half an hour waiting time for the first cans to be loaded until the sterilising process can start. This sometimes is a disadvantage, especially for hot-filled cans.

Continuous sterilisers of the hydromatic-type are characterised by hydrostatic legs on both sides of the sterilising compartment. For these it also takes several minutes to transport the containers from the entrance to the top and subsequently down into the hydrostatic leg before they are exposed to sterilising conditions. But as soon as the container is positioned in the hydrostatic leg, heat transfer can start. This almost makes up for the coming-up time in an autoclave. Although the sterilising conditions have to be controlled carefully, unavoidable temperature differences throughout the sterilising compartment are not as critical as they are in autoclaves, since every container is following the same track, and is thus being exposed to the same conditions. Even continuous sterilisers, not being provided with rotating carriers, have an advantage over the still-cook autoclaves in this respect, in that the container is always moving and once every few minutes rotates over the top and bottom shafts. This certainly does not make up for a constantly and carefully chosen rotation, as far as whole, low-viscosity products are concerned, but it does so to a great extent for all other types of product. Finally, since the containers are continuously fed to the steriliser, the waiting time is the same for each container and equal to the transport time from the filling machine to the steriliser entry. Normally this is not more than a few minutes.

The modern sterilisation process aims to destroy microbiological life and to stop enzymatic activity in the product with the least possible effect on the chemical and nutritive value of the product. This tends to increase processing temperatures and reduce processing times. The obstacles to obtaining this certainly do not lead to a straight choice between the batch autoclave and the continuous steriliser, but are instead due to limitations of the heat transfer coefficient inside the container. And this is even more complicated for products containing solid particles, because the total heat transfer depends on the heat transfer coefficient from the container wall to the liquid and from the liquid to the solid particle. This latter heat transfer coefficient depends on the velocity differences between the liquid and the solid particle, and therefore depends on several factors:

(a)   Agitation of the container.
(b)   Percentage of headspace in the can.
(c)   Viscosity of the surrounding liquid.
(d)   Differences in density between the liquid and the solid particles.
(e)   The ability of free movement for the solid particles in the liquid.

Considering these factors it is possible to estimate the optimal processing conditions for every type of product in a certain container. Throughout the years much research has been performed, most of it concentrated on products in metal cans. The influence of decreasing processing times and increasing processing temperatures on improved product quality is obvious. However, in only very rare cases is it really possible to obtain all the benefits of this knowledge under industrial circumstances. One individual container well-defined in shape, headspace and content might be exposed to the optimal sterilisation process, but a great number of those cans have to be processed according to a compromise, due to unavoidable variations in, for example, headspace; the rate of the liquid–solid particle fraction; size of the solid particles, etc. In this respect I refer to a paper presented in 1964 by Hersom[2] mentioning the problems which may arise if these factors are not considered.

A mathematical study by Bimbinet and Duquernoy[3] specifically concerned with containers containing solid particles in a low-viscosity liquid leads them to the conclusion that the gap between the lethal values in the liquid and the solid particles is bigger in HTST processes than it is in processes performed at somewhat lower temperatures.

For practical reasons, agitation of containers during a heat preservation process, is performed by rotation, whether end-over-end, or axial. In batch autoclaves this is achieved by rotating the baskets; in continuous sterilisers, by rotating the carrier. The effect of the rotation depends on the revolutions per minute (R) and on the radius (N), according to the formula RN^2. Since in batch autoclaves the radius for a container in the centre of the basket is different from the radius at the edge, this must also lead to a compromise in choosing the processing conditions. Furthermore, unless the heating and cooling media can be ideally mixed throughout the whole batch autoclave, heat transfer to containers at the edge of a basket is better than to containers in the centre. In this respect the continuous steriliser with rotating carriers has certain advantages.

With containers other than the well-known cans, such as glass jars with ventilating caps, plastic containers and pouches the possibility of agitation is quite restricted. These containers also introduce other problems.

Glass jars with ventilating caps (the Omnia cap) have to be kept in a vertical position during the whole process. These jars (0·5 or 1 litre content) are mainly used for vegetables in brine. The thin metal cap, with a diameter of about 10 cm is pressed over a rim on the neck of the glass jar. During heating up the liquid and the gas–vapour mixture in the headspace expand. The excess above the liquid can eventually disappear via the leaking cap. As soon as the cooling phase starts the internal pressure decreases, which causes a vacuum-closing of the cap. Since the cap cannot stand more than one atmosphere pressure difference between the outside and the inside, a well-controlled balance has to be maintained between temperature and pressure during the cooling phase. Processing these containers in batch autoclaves with controlled pressure during the cooling is already done. But it is also possible in continuous sterilisers of the hydromatic type—taking into account that due to short stops of several minutes caused by disturbances in the line after the steriliser, the cooling of the product goes on—while the outside pressure remains constant. Considering the heat penetration, this type of container is far from ideal, and the processing conditions are not influenced by the choice between 'still-cook' batch autoclaves and continuous sterilisers provided with pending (or 'paternoster') carriers.

Plastic containers are used in the dairy industry for milk and milk products. In flow sterilisation milk is bottled under non-aseptic conditions into a polyethylene bottle, hermetically heatsealed with an aluminium cap and then mildly after-sterilised for about 14 minutes at 116°C. Since polyethylene at temperatures above 100°C becomes very weak and unable to stand pressure differences between the inside and the outside of more than a few metres water column, a thorough pressure control in this temperature region is required. Furthermore, the headspace in the bottle has to be kept within certain limits to avoid blowing up. These bottles are mainly processed in continuous hydromatics, using additional air pressure in the sterilising compartment.

Since plastic materials such as polyethylene are not completely gastight, they are not suitable for sterilised products which are supposed to have a long shelf life. This is why flexible containers for preserves of vegetables or meat products are provided with a gastight layer. These pouches are mostly made from a laminate of different materials, including aluminium to protect against light and PVDC to obtain sufficient gastightness.[7]

The pouch is extremely suitable for HTST processes. The extended surface in relation to the content permits a fast heat penetration, which is much better than in cylindrical containers. However, several aspects have to be considered carefully in order to achieve these promising advantages. Except for pouch-filling machines provided with vacuum-closing, it is unavoidable that, together with the product, some air is enclosed. During the residence period under sterilising conditions the outside pressure should be sufficient to avoid expansion of this enclosed air, thus preventing the formation of an insulating air-layer between the pouch-wall and product.

According to Dalton's law the total pressure inside the pouch is the sum of the partial pressures of air and vapour. The vapour pressure depends on the temperature of the product, so the air pressure in the sterilising medium has to be such that expansion of the enclosed air is avoided.

The pouch should preferably be processed in a horizontal position, so that the product can spread out over the maximal given surface, thus keeping its thickness to a minimum.

When air is enclosed it is advisable to process the pouch in a steam–air mixture or in a water spray under pressure, rather than in a waterbath. This is for two reasons:

(*a*)   The difference in density between the usual products and water is only very small. If for any reason, e.g. the streaming conditions of the surrounding water, differences in thickness of the pouch occur, the forces which are available to flatten out the pouch contents are very small. Therefore, once differences of thickness are introduced, equalisation to the flat condition is inhibited.

(*b*)   Being immersed in water, the air will try to rise. Since there is a small difference in density between the product and water, a small amount of enclosed air will cause the pouch to float, thus retaining its own air-balloon and leading to considerable variations in thickness.

Sometimes plates or racks must be used to keep the thickness of the pouch within certain limits. Such plates however must not introduce any internal pressure inside the pouch because this will certainly rupture the aluminium layer, or damage the seals. Therefore it is necessary for the pouch always to be able to move freely within the given space. Changes in volume of the pouch should be limited by applying additional air pressure on the outside. The processing conditions for pouches should be based on the following factors:

(*a*)   Maximal thickness of the pouch.

(*b*)   Amount and position of enclosed air.

(*c*)   The size of the solid particles.

Most pouch material does not permit thorough agitation, because this leads to friction between the pouch and the carrier, which causes damage to the outside coating and print. Therefore the heat penetration has to be based mainly on conduction.

Enclosed air must preferably be kept to a minimum, otherwise this can greatly influence the heat penetration, especially in horizontally positioned pouches. If a layer of air is able to separate the upper pouch-wall from the product, the heat can mainly be transported from the lower wall, and the distance to the slowest heating point is in that case highly increased. This can be avoided to a great extent, by increasing the outside air pressure, thus keeping at least a partial contact between the upper wall and the product.

Although in theory a very attractive container, enabling fast heat transport due to the heat transfer from both sides of a thin layer, the safety margins in processing should be slightly larger than with conventional tin containers. Sometimes these safety margins are considered a disadvantage, because part of the product should be oversterilised. But in fact this is only true to a small extent, at least to a smaller extent than is the case in cans, where the outer product ring is exposed for much longer to the sterilising temperature, as is the slowest heating point.

The processing of pouches in continuous sterilisers, although more difficult in terms of handling, has certain advantages over batch autoclaves, due to the light agitation, first of all by the half-rotation around the top and bottom shaft, but also by the slight trembling of the moving carrier. The carrier in the continuous steriliser has to be designed in order to give a minimum of damage during movement, and a visor is required to avoid floating out of the carrier.

The pouch is still not legally permitted in every country. The USA is quite reluctant, although recently it seems that the FDA may approve in the near future.[5] Some disadvantages include:

(a)   Costs of the pouch material, including carton overwrap.[6]
(b)   Problems associated with handling, causing a much higher percentage of damaged pouches than is usual in the canning industry.
(c)   Difficulties in controlling leakage after processing.
(d)   Low filling speeds (40–60 pouches per minute) on modern filling machines.
(e)   The pouch content is much smaller than the content of, say, metal cans, and therefore the throughput of pouches per hour has to be much higher than the number of cans per hour in order to obtain the same production capacity.

Although much literature is published in this field, it is rather questionable whether pouch-processing will take over a considerable part of the market, which is now mainly supplied with cans and glass jars. It certainly will for special products, as has already been shown in Italy and in Japan, where sauces containing vegetables and pieces of meat are widely available.

The heat preservation of products with retention of quality means that the heat transfer from the medium to the inside, including the solid particles, must be regulated, and where possible improved. Heat preservation inside containers is highly dependent on heat transfer from the container wall to the actual product. This heat transfer is reduced by the limited means of agitating the contents, and this is particularly important for the more viscous products. In this respect aseptic processing seems to anticipate on the requirements for HTST preservation. In the dairy industry this technique is now widespread since several types of aseptic filling machines are on the market. Using scraped surface heat-exchangers, operating under aseptic conditions, a similar technique could be applied to highly viscous products, and on products containing solid particles, thus increasing the heat transfer.[9] But up to now there has hardly been an aseptic filler on the market which is reliable for such foods as preserves and meat products.

This is for several reasons:

(a) The aseptic containers used in the dairy industry are specially designed to enable aseptic filling. Aseptic fillers, based on filling the conventional glass bottle, have failed so far.

(b) Spoilt milk is easy to recognise from its smell and taste. This is not the case with preserves and meat products and this makes the canning industry rather reluctant.

(c) The capacity of aseptic fillers in the dairy industry is low (60–80 cpm) compared to the usual rate of filling in the canning industry.

(d) In the dairy industry a spoilage of 0·1 % on aseptic fillers is generally accepted. This figure is much lower for the conventional in-container processes.

(e) Aseptic fillers in the dairy industry are designed such that some regular failures in operation do not necessitate a renewed presterilisation of the filler in order to continue the filling operation.

Although several attempts have been made to adapt conventional fillers to aseptic conditions, the real solution of this barrier is probably to be found in a complete new filler design and container material. Also an interim step towards aseptic processing, which was made in the dairy industry by introducing the 'two-phase' sterilisation system, is not very likely in the canning industry. The 'two-phase' system is based on HTST presterilisation of a product, by which the commercial sterility is already obtained. This is followed by filling under non-aseptic but hygienic circumstances. Subsequently a mild after-sterilisation is performed, usually with an F_0-value of 2 to 3, to compensate for possible contamination during the filling and closing operation. For whole, low-viscosity products, like milk, with a good knowledge of the possible bacterial contamination between the two phases, this process has proved to be very feasible. Many of these lines are installed in the dairy industry.

To apply the same system to vegetable and meat products would require a thorough investigation, and due to the slow heat penetration in more viscous products would perhaps only be applicable to low-viscosity products.

SUMMARY

Summarising, the heat preservation of liquids and solid foods in batch autoclaves and in hydrostatic sterilisers only enables certain general conclusions to be made.

(a) Considering the properties of the product and the container, an optimal process can be designed as well in batch autoclaves as in continuous sterilisers, as long as modern techniques of agitation and temperature and pressure control are applied. Due to the fact that temperature variation in continuous sterilisers is less since every container is following the same track, and the waiting time between filling and processing is very short, the continuous steriliser has a slight advantage over the batch autoclave.

(b) Batch autoclaves are more suitable for a small capacity and a great variety in container size, since the necessary adaptions to handle these variations can be obtained from relatively low investment costs.

(c) Hydrostatic sterilisers are more economic in terms of energy consumption and labour costs than batch autoclaves. The inherent versatility is restricted under practical conditions to a smaller variety of container types. Therefore the continuous steriliser is more suitable for high production capacities.

REFERENCES

1. Clifcorn, L. E. *et al.* (1950). A new principle for agitating in processing of canned foods. Continental Can Company, Inc., Bulletin No. 20.
2. Hersom, A. C. (1964). Sterility problems associated with rotary cooker coolers. Paper presented at the International Food Industries Congress.
3. Bimbinet, J. J. and Duquenoy, A. (1974). Simulation mathématique de phénomènes interessant les industries alimentaires. *Industries Alimentaires et Agricoles*, **91**, No. 4.

4. Thorpe, R. H. and Atherton, D. (1972). Sterilised foods in flexible packages. Chipping Campden Research Association, Technical Bulletin No. 21, Jan.
5. Stockman, S. A. (1976). Approval of retortable pouch nears. *Food Engineering International*, March, page Int. 15.
6. Eisner, M. (1972). Die Rotationssterilisation in der modernen Konserven-produktion. *Die Industrielle Obst- und Gemüseverwertung*, **57**, 97–103.
7. Nieboer, S. F. T. (1976). Retortable pouches respond to higher costs. *Food Manufacture*, September, 37–8.
8. Guedez, O. and Bates, R. P. (1975). Laboratory procedure for the pressure processing of pouches. *J. Food Sci.*, **40**, 724–7.
9. de Ruyter, P. W. and Brunet, R. (1973). Estimation of process conditions for continuous sterilisation of foods containing particulates. *Food Technol.*, July, 44–50.

5

Concentration by Evaporation

H. A. LENIGER

Food Science Department, Agricultural University, 10 *Englaan, Wageningen, The Netherlands*

INTRODUCTION

Concentration can be defined as the withdrawal of part of a solvent from a solution. In the food industry, the solvent that we want to withdraw is almost always water, so in our further discussion we shall confine ourselves to this subject. The solution we want to concentrate need not necessarily be clear—solids or fluids can be dispersed within it.

Concentration cuts the cost of transport, handling, and storage. It reduces the water activity, too, so it is also a method of preservation. Concentration often constitutes a pretreatment to drying, when, and in so far as, the cost of concentration favourably counterbalances the savings in the cost of drying. Occasionally, concentration forms the pretreatment for crystallisation.

It is obvious that concentration belongs to the group of physical separation methods. It will be equally obvious that, for the partial withdrawal of water, one should utilise differences in properties that exist between water and solutes. And in the process of evaporation, one does take advantage of the difference in volatility. There are, however, other means, and one must carefully weigh the pros and cons to find the most appropriate method in a particular situation. One can, for instance, concentrate foods by freezing out, whereby a favourable equilibrium is utilised between an aqueous solution and almost pure ice. Sometimes, one takes advantage of the equilibrium between a gas hydrate and a dilute solution. There are also various membrane processes, based on a difference in transport rate (rate of diffusion). In dialysis, water diffuses from one solution to another that has a higher osmotic pressure. In pervaporation, water diffuses through a membrane, after which it is evaporated and removed as vapour. In reversed osmosis, water diffuses through a membrane and is then drained off as such. We shall not give any further thought to these methods, however, and shall focus our attention solely on concentration by evaporation.

In principle, what one attempts to do when concentrating foods by evaporation is to remove water in such a way that the composition of the concentrated solution that remains does not change, except for the water content, of course. In other words, one wants to withdraw water without impairing the quality of the product. The ideal one aims for in this process is to produce a product which, after being reconstituted to its original concentration by the addition of water, cannot be distinguished from the original product. The word 'ideal' is used here because in reality certain changes will always occur in the composition of the product and thus affect its quality.

The interesting thing about evaporation is that one can choose from a wide range of different types of evaporators and that evaporation can take place under strongly different processing conditions. One can select conditions that are so favourable that hardly any harm at all is done to the quality of the product. Or, if one chooses, one can produce obvious changes in the product. Now it is a fact of life in food processing that a method that does little harm to the original quality of the product usually costs more than a method that does greater harm. A higher quality concentrate, in other words, is more expensive to produce than one of poorer quality. This means that, when choosing an evaporator and deciding the conditions under which the product will be processed, a compromise will usually have to be made between quality and costs; or if you prefer to put it this way, quality and costs will be optimised.

A thought that occurs to me is that if one could succeed in evaporating a product without any effect on its quality, one might speak of an operation; but because in reality some changes will always occur, it is better to speak of a process.

QUALITY

When we ask ourselves what can happen during evaporation, we come to the conclusion that the most important events are chemical reactions between the components (including water), which are usually present in great numbers and in widely varying concentrations, and thermal degradation. These reactions can affect a product's appearance, taste, nutritional value and so on, and depend on concentrations, time, and temperature. As well, one must be prepared for enzymatic and even microbiological changes in which, apart from time and temperature, water activity comes into play. Of more importance in many cases is that evaporation causes a loss of volatiles, and thus of aroma; during evaporation, components with a relative volatility in relation to water greater than one enter, for a smaller or larger part, the vapour phase and are removed along with it. Apart from the properties of the aroma compounds, the extent of the loss is determined by the amount of water that is evaporated and by the processing conditions under which this takes place.

One can state that as a rule the effect of evaporation on quality is an unfavourable one. But there are exceptional cases where changes might be regarded more as an advantage than a disadvantage. An interesting possibility would be to combine evaporation with intentional chemical reactions, with the evaporator then functioning as a reactor.

The importance attached to the extent to which the quality of a product deteriorates during evaporation depends on a great many factors. In the first place, some products are naturally more heat sensitive than others. With some products, the retention of its taste may be regarded as all-important, whereas with other products taste is a quality factor of secondary importance. The same applies to aroma compounds: in some products aroma is a vital hallmark of quality, in others it does not matter so much. Further the sensory perception of changes varies widely; one can imagine considerable chemical changes that are scarcely noticeable, although the reverse is also possible. With some aroma compounds, a major part can be lost before anyone really notices it; with others even a slight loss means a drop in quality. Of great significance is the attitude of the consumer to the changes caused by evaporation. This attitude varies from one product to another and from one consumer to another. Unfortunately, time does not allow us to delve further into this most interesting subject.

What finally settles the question of the extent to which a product may deteriorate is whether the consumer is prepared to pay more for a better quality product. Generally speaking one should not entertain too many illusions on this point. Naturally there will always be a few consumers who are prepared to pay for high quality products. But if they are asked to pay disproportionately high prices for only a slight improvement in quality, their interest will quickly wane. So we can conclude that it is the quantitative aspects of quality and costs that are significant. However, as will be shown, many difficulties are met in quantifying the results of the evaporation process.

Perhaps we ought to point out that evaporation in itself is not an isolated thing; it is part of a flow sheet, and one must also give thought to what precedes evaporation and what follows. It is quite possible that the pretreatment and aftertreatment of a product, and here we include storage, cause so much deterioration of the quality that it will make little sense to devote much care to the quality of the product during evaporation. This is something that is all too often lost sight of.

TIME AND TEMPERATURE

The combination of residence time and temperature during the evaporation process deserves careful attention. Unfortunately, it is extremely difficult, when one thinks about it, to make any intelligent remarks about chemical changes that occur in an evaporator. We are faced with two groups of problems. In the first place, one ought to have some knowledge of the reactions between the many components (present in widely varying concentrations) as a function of temperature and time. This problem is discussed by Dr Hermans in the previous chapter. There is another matter, however, about which something must be said here, and that is to consider the temperatures and times that are relevant.

Roughly one can reason as follows. Evaporation can be done under an extremely wide range of temperature and time combinations. If the water could be withdrawn at a very low temperature and in a very short time, one would have the ideal situation. Obviously, however, quite apart from the costs, there are physical limitations. Theoretically, the lowest possible temperature is just above freezing point, but in practice the lower limit must be taken at room temperature. It is impossible to say what residence time is needed to withdraw part of the water at such temperature. This will depend both on the type of evaporator and on the properties of the product. It is a matter of the time which is necessary to supply the required heat of vaporisation to the product.

To simplify the problem we shall first consider a one-stage evaporation process. In such a process—and naturally remaining within the realm of physical possibilities—we find two extreme cases, between which there is a whole range of transitional forms. There are evaporators with a small fluid content and there are evaporators with a large fluid content. Film evaporators contain only a small amount of solution. They come in many different types, but an example chosen at random is shown in Fig. 1. Shown in Fig. 2 is an example of an evaporator with a large fluid content. A film evaporator has a relatively large heating surface, or in other words, the

H. A. Leniger

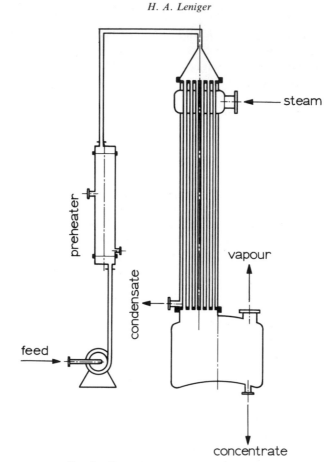

FIG. 1.   Example of falling-film evaporator.

surface that supplies the heat of vaporisation is large in proportion to the fluid content. In contrast, an evaporator with a large fluid content has a relatively small heating surface. Now if, for the sake of simplicity, we assume that in both types of evaporator, the same heat transfer coefficient from the heating medium to the solution can be realised, the capacity of the evaporator will be determined by its heating surface and can be expressed in kg evaporated water per unit of time. To enable a fair comparison between the two types of evaporator, they must of course have the same capacity. So now we have the same number of m² heating surface, but the film evaporator has a much smaller fluid content than the type in Fig. 2. This means that under the same processing conditions, the average residence

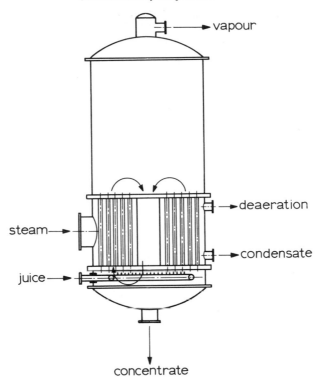

vapour

deaeration

steam

condensate

juice

concentrate

FIG. 2.   Example of evaporator with large liquid content.

time of the film evaporator is much shorter than in the other type; when the fluid in the evaporator is thoroughly mixed, the average residence time of the solutes will be equal to the fluid content in m^3 divided by the quantity of fluid leaving the evaporator in m^3 per unit of time. Evaporators with a small fluid content are, in general, more expensive than those with a large content and the same evaporation capacity.

We are now in a position to draw an important conclusion. If the same processing conditions apply, an evaporator with a small fluid content will produce a better quality product because of the shorter average residence time than will one with a larger fluid content. The choice of which evaporator to use will then depend on the difference in costs incurred in producing the products and whether the higher costs of a better product can be regained by demanding a higher price.

With a certain type of evaporator, it is possible to vary the processing conditions. Still speaking of a single stage evaporation, one can select the evaporating temperature by choosing the pressure in the evaporator. Apart from this, one can of course select the temperature of the heating medium, which is usually steam. This means that one can vary not only the evaporation temperature but also the driving force of evaporation, which is the temperature difference. One would now suppose that the capacity of an evaporator is independent of the evaporation temperature and is proportional to the temperature difference. After all, one can express the capacity of an evaporator as

$$\text{water evaporation (kg/s)} = \frac{\text{heat flux (kJ/s)}}{\text{heat of vaporisation (kJ/kg)}}$$

while the heat flux (kW or kJ/s) is equal to

$$\text{heat transfer coefficient } k\,(\text{kW/m}^2/^\circ\text{C}) \times \text{heating surface (m}^2)$$
$$\times \text{ temperature difference } (\Delta\theta)$$

But it is not as simple as that. If one considers a tube evaporator with natural circulation, one finds that at the same boiling temperature the overall heat transfer coefficient increases with increasing temperature difference. This implies that the evaporating capacity—which is proportional to the product of temperature difference and heat transfer coefficient—increases more than proportionally with the temperature difference. This relation is illustrated in Fig. 3. From this figure one can also see that the overall heat transfer coefficient at constant temperature difference rises considerably when the evaporation temperature is increased.

Two remarks need to be made. In the first place, the overall heat transfer coefficient is used in Fig. 3 because it determines the evaporation capacity. In an evaporator, however, this k-value is almost entirely determined by the high thermal resistance on the side of the product, thus the low individual heat transfer coefficient from wall to boiling liquid, in respect to which the resistances in the wall and on the side of the heating medium—as a rule steam—is negligible. The effects mentioned are mainly due to an improvement in the heat transfer on the side of the product when the viscosity becomes lower because of a rise in temperature and when the circulation becomes stronger because of a greater temperature difference. The second remark on this subject is that data as given in Fig.3 should not be generalised. They are applicable only for an evaporator of certain dimensions. If accurate data are required, they should be determined by experiments for each individual type of evaporator. The effects of boiling

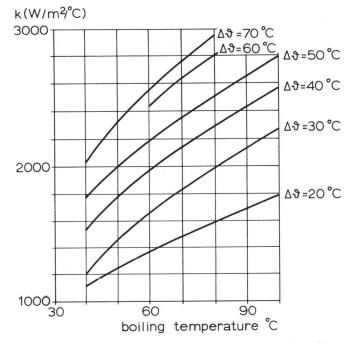

FIG. 3.   Influence of temperature difference and boiling temperature on the total heat transfer coefficient k.

temperature difference in a film evaporator will obviously be somewhat different, but here too viscosity plays an important role.

One can read from Fig. 3 that, for instance with the same evaporation temperature, a 50% higher heat transfer coefficient can be obtained by making the temperature difference about twice as large; this means an evaporation capacity three times greater and a residence time three times shorter. From the point of view of the quality of the product, such a measure seems favourable; the much shorter residence time will probably have more effect than the higher wall temperature with which the product comes into contact. In general, the combination that will be applied will consist of increasing the temperature difference and simultaneously raising the evaporation temperature. Referring once again to Fig.3, we see that a × 3 larger capacity and a × 3 shorter residence time can also be obtained by, for instance

$$\text{raising } \Delta\theta \text{ from} \sim 20\,°C \text{ to} \sim 35\,°C$$

$$\text{raising } \theta \text{ from} \sim 46\,°C \text{ to} \sim 77\,°C$$

so that k increases from ~ 1200 to $\sim 2100 \, \text{W/m}^2/\,^\circ\text{C}$. These figures show that a high evaporation temperature means a short residence time, and vice versa. This is actually the normal case. In principle, one can always choose between evaporating at a high temperature for a short time or a lower temperature for a longer time. But a simple answer to the question of which combination is best from the point of view of quality is hard to give. This will depend on the type of evaporator used and on the particular product. Considering the analogy between the influence of the temperature on the water vapour pressure and that on the rate of reactions, one might possibly compare different time/temperature combinations with the aid of the product of time and water vapour pressure.

Once it has been empirically established how large the heating surface should be for two evaporators of the same type and equal capacity, one of which works with a high evaporation temperature and a short average residence time and the other with a low temperature and longer time, it is possible to state with certainty that, irrespective of the influence of the processing conditions on the quality, the product from the low temperature evaporator has a considerably higher cost price. The reason for this is in the first place the much larger heating surface required by a lower evaporation temperature, but it is also due to the volume of the water vapour, which rises rapidly as pressure is reduced, and to the low temperature of the water vapour. The large volume demands expensive pipes and pumps, while the low temperature makes the condensation of the vapour difficult and costly. So, speaking generally, evaporation at a low temperature is an exceedingly expensive business, and its advantages, to say the least, are doubtful.

It remains to be said that the problem of optimising the dimensions and processing conditions of an evaporator with a given capacity is very complicated indeed. This is because there are such a vast number of possible combinations of evaporation temperatures, temperature differences, and heating surfaces, all of which can lead to the desired result. When one goes more deeply into this problem, one faces the question of which temperature and which residence time one must speak of when considering the influence of the evaporation process on quality. This is not so easy to answer. The simplest situation is found in the simplest film evaporator. Here, a film of the product that is to be concentrated flows downwards over a heated wall. The flow is laminar up to a Reynolds number far below 2300; if there were no temperature differences, the flow profile would be a half-parabola, but because the product is heated on the wall, the profile assumes a slightly flatter shape. This is shown schematically in Fig. 4. Heat is transferred through the film by conduction and evaporation takes place at the surface;

there is thus no formation of vapour bubbles. The temperature profile approximates to a straight line; at the evaporating surface, the temperature is determined by the pressure prevailing above it and by the local concentration. Complications are that because water and volatiles evaporate, the concentration of non-volatile compounds at the surface increases while that of volatile components decreases. The evaporation of

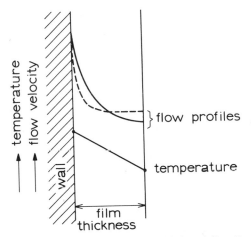

FIG. 4.   Temperature and flow profile in a falling film.

water means that the average concentration increases in a downward direction; because of this, there is a tendency for the thickness of the film to decrease; the increasing viscosity, however, produces a reverse effect. Only in an actual situation will one be able to analyse this further. Sufficient knowledge is available about the residence time distribution with laminar flow and one could say something about it in non-isothermal cases. However, when one considers that in the vicinity of the wall the product is exposed to a relatively high temperature while at the same time its residence time near the wall is relatively long, and considering further that the temperature coefficient of chemical reactions is high, one comes to the conclusion that most of the changes in the product occur in a thin layer along the wall and that the reactions further from the wall are comparatively negligible. Thus, one can state that the conversions that take place in a film evaporator will depend on the wall temperature and will be proportional to the area of the heating surface.

In some film evaporators the product is stirred. There are also types whose stirring mechanism continuously scrapes clean the heating surface. In such an apparatus less harm is done to the product, on the one hand because the heat transfer coefficient is better and hence the residence time can be shorter, and on the other hand because a smaller part of the product is exposed to the wall temperature. Obviously, such film evaporators are much more expensive than simpler types. This also applies to an evaporator in which a very thin film is spread over the heating surface under the influence of centrifugal force; this enables the residence time to be very short. It is no easy task to make a comparison between the various types of film evaporator. Strictly speaking, one would have to optimise each type and then compare the installations thus obtained, all with the same capacity in m³ evaporated water per hour. A complication is that one must take into account conversions in the product in feed and outlet pipes.

A more complicated case is that of an evaporator through which the solution that is to be concentrated circulates. From the point of view of residence time one has in theory two borderline cases: either plug flow or ideal mixing. Plug flow can be obtained either with a series of circulation evaporators or with very long tubes through which the product flows once only, thus with no recirculation. Evaporators with short pipes can bring about a vigorous mixing, which can be augmented by stirrers or by a circulation pump. One can now assume that the conversion of undesirable reactions is only very slight, so that the concentration of the reactants scarcely changes. In that case, the conversion is proportional to the average residence time of the reactants, or in other words, of the solutes. One can now calculate that the average residence time of the solutes in a plug flow evaporator is shorter than in an evaporator with ideal mixing. Under otherwise comparable conditions, a plug flow evaporator is therefore the better choice. When intensive mixing occurs, one can assume the reaction temperature to be average between wall temperature and boiling temperature and take the average residence time as reaction time. The residence time is a matter of how long the solutes, which react with one another or decompose, remain in the evaporator. With less intensive mixing, much depends on the type of flow in the evaporator and thus on the product properties. It would then be wise to take into account a somewhat higher temperature than the average and a somewhat longer than average residence time.

There are still other types of evaporators that we could consider, but we shall confine ourselves to only one other type. This is what is known as a flash-evaporator. Here, the product to be concentrated is heated, without boiling, in a heat-exchanger. It is then atomised under vacuum and is thus partially evaporated in an extremely short time. Any changes in a product treated in this way will occur exclusively in the heat-exchanger; with the help of theory from chemical reaction engineering, we can calculate what happens. The difference between this and a chemical reactor is that in an evaporator it is a matter of only very limited conversion.

COSTS OF EVAPORATION

The cost of a product concentrated by evaporation is determined by such factors as the price of raw materials and packing, the cost of operations which precede evaporation and those that follow. As far as the actual evaporation itself is concerned, the degree of concentration, the initial expenses of the evaporating equipment—with its depreciation, interest, and such like—and finally, the energy consumption are important. The ratio of all these cost factors will vary from case to case. We shall not go any further into the price of raw materials and packing, nor into the cost of pretreatments or aftertreatments. The cost of evaporation will obviously be strongly influenced by the degree to which the product is concentrated. One will have to weigh the pros and cons in each individual case of whether it is worthwhile concentrating the product to a higher degree of dry matter content, while not losing sight of the fact that the withdrawal of each subsequent quantity of water will mean a disproportionate increase in costs. Our further review of costs will be confined to those involved in the purchase of the evaporating installation and of its energy consumption.

There are many different types of single stage evaporators and within each type one can find differences in capacity and in the materials used in their manufacture. If one considers evaporators of the same capacity and manufactured from the same materials, one will soon realise that any additional measure intended to improve the quality of the product means a higher purchase price. These higher costs are then supposed to be recovered by a higher price for a better quality product. The improvements that can be made devolve, in principle, on lowering the evaporating temperature, shortening the residence time, or a combination of the two. As has already been explained, a lower evaporating temperature demands a disproportionately larger heating surface because the heat transfer coefficient from the wall to the product declines with the lower temperatures. At the same time, the evaporator with its accessories becomes more expensive because the larger volumes of water vapour produced demand all kinds of costly technical measures. These disadvantages can be partially offset by taking measures to improve the heat transfer to the product. In the first place, one could consider increasing the difference in temperature between the heating medium and the evaporating solution. In general the combination of a low evaporating temperature and a great temperature difference will be an attractive one, but one has to consider the possible consequences of the higher wall temperature to which the product is exposed. This is not just a matter of heat damage, but also one of scale formation. By increasing the temperature difference, the heat transfer is improved because the flow and the circulation in the evaporator become stronger. But this effect can also be obtained through all kinds of constructional measures, and then one does not need to increase the wall temperature. Evaporators with very long vertical pipes work well because the fluid in them is forced rapidly upwards by the steadily increasing quantity of vapour. Of course mechanical aids can also be used to improve the flow; sometimes this is even essential, because of unfavourable product properties. One has to consider, however, that these aids not only increase the purchase price of the evaporator, but that a pump for forced circulation requires energy, and that the energy consumption increases by about the cube of the fluid velocity in the pipes. A construction such as this is therefore only used in special circumstances. We have already pointed out that one can equip film evaporators with stirrers or scrapers, or even let the heating surface rotate. It need hardly be said that one must always carefully consider whether such expensive fittings are justified. It also speaks for itself that rotating parts can shorten the lifetime of the equipment and also increase its maintenance costs.

Improving the heat transfer to a product at a constant evaporating temperature means that, for a given capacity, a smaller heating surface will suffice. Unfortunately, few data are available as to the extent to which heat transfer can be improved by the measures mentioned above. More data are also urgently required about the relation between the heating surface and the purchase price—and this for various kinds of evaporators—to enable a comparison to be made between them. More research must be conducted on these subjects.

With the desired capacity, a given heat transfer coefficient and temperature difference, and a certain evaporating temperature, it is possible to shorten the average residence time, with the same heating surface, by giving the evaporator a smaller fluid content. Such a measure should, in principle, have a favourable effect on the quality of the product. Yet little information is available about the extent to which the price of an evaporator rises by reducing the residence time under otherwise similar processing conditions. One wonders whether it is worthwhile spending so much more on an evaporator merely to obtain a shorter residence time?

A problem that is very similar to the one just treated is that of scaling-up the evaporation process. To some extent, one can of course achieve a greater capacity by such measures as increasing the temperature difference and the evaporation temperature. If one wants to increase capacity to a greater degree, however, while retaining the same processing conditions, one will have to enlarge the heating surface. In principle, this is an attractive measure because an evaporator with a heating surface ten times as large, thus with a capacity ten times as large, will only cost five times as much. But unfortunately there are usually other factors than the initial costs of the evaporator that are decisive for the scale of the evaporating process. In those initial costs, the price of the construction materials is an important item; it is wise to ask oneself whether one really needs expensive materials like stainless steel; quite often, substantial cuts in initial costs can be made by using cheaper materials.

With some products the aroma is an all-important quality factor. This applies particularly to fruit juices. An extremely expensive measure has long been commonly applied when fruit juices are being concentrated; the expense refers not only to the initial costs but also to the energy consumption. This measure consists of the recovery and concentration of the aroma that has been removed in the evaporation process. We shall come back to this subject a little later on.

It is hard to say precisely how far the initial costs of an evaporator and its accessories affect the total cost of a product. But what can be said, and quite emphatically, is that the initial costs of the equipment play a much less important role than they used to. The reason for this is that the cost of energy, considered high even in earlier days, has gradually become the decisive factor in evaporation and nowadays constitutes the largest cost factor of the entire process. In recent years energy costs have increased more than any of the other price indexes; and it is quite possible that in the future they will account for an even higher percentage of evaporation costs than they do at present. This turn of events will most certainly exert its influence on the optimisation between costs and quality, and even on the choice between a thermal concentration process and other processes that use less energy. Any measures by which the consumption of energy could be reduced have always been important in evaporation, but they now demand more attention than ever.

As an illustration of the changing circumstances, we might point out that the price of gas and oil is now four times higher than it was five years ago, whereas over the same period, the general price index has risen by a factor of two. To generate heat in the form of steam, the energy costs account for at least 70 % of the total costs. Nowadays one must expect to pay US$10 per ton for steam. Significant in all this is that many food plants are fairly small and that they have to pay a relatively high price for their energy.

In the one-stage evaporation to which we have confined our discussion so far, the withdrawal of a kilogram of water requires, in theory, a kilogram of steam; in practice this is slightly more. In the water vapour that is formed at a relatively low temperature, there is a large amount of heat that has to be

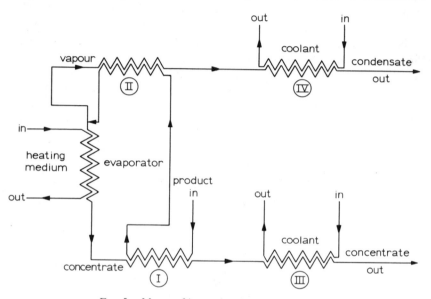

Fig. 5.   Means of improving thermal economy.

removed in a condenser with a cooling medium. It is obvious that this heat could be utilised, and this has indeed been happening for quite some time. One can, for instance, follow an arrangement as shown in Fig. 5; here, the product that is to be evaporated is preheated by using it to cool the concentrate that is emerging from the evaporator and then by using it to condense the vapour from the evaporator. The heat of vaporisation, however, has to be supplied to the evaporator, while concentrate and condensate require further cooling. One must further consider that a heat-exchanger becomes very large if one wants to improve the efficiency, and that a heat-exchanger must be dimensioned in such a way that as much heat as possible is transferred at the lowest possible flow resistance. This is a subject in itself and we cannot go into it any more deeply here.

It has long been the custom to use the water vapour from an evaporator as a heating medium in a following evaporator; of course both temperature and pressure are lower in the second evaporator, than in the first. Here, one speaks of a double stage evaporator and in general of multiple stage evaporation. There are various ways of arranging the evaporators with respect to the direction of flow of the solution. The most common is the

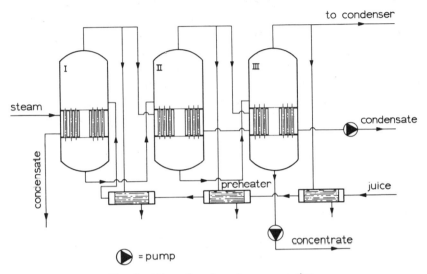

FIG. 6. Direct-flow three-stage evaporation.

direct-flow design illustrated in Fig. 6. Here, the solution that is to be concentrated flows in the same direction as the heat flow. No pumps are needed between the evaporators and the feed can be preheated stepwise as shown in Fig. 6. The most highly concentrated solution is evaporated at the lowest temperature. This may be an advantage for heat sensitive materials, but if the solution becomes too viscous at the lowest temperature, it may be necessary to use a counterflow installation or a mixed installation. The residence time of the solute in the evaporator operating at the highest temperature is then longer than in the case of direct flow.

In a multiple stage installation, a kilogram of steam evaporates a kilogram of water in all stages, so with a four-stage system, one would need 0·25 kg steam per kg evaporated water. In reality, due to heat losses, the steam consumption is slightly higher. To be able to understand the operation and the consequences of an installation such as this, one has to apply the following reasoning. With evaporation in more than one step the total temperature difference between the fresh steam and the cooling medium in the condenser is distributed over more than one evaporator. Heat flows from the heating medium to the cooling medium and on its way, is used for transport of water vapour more than once. If we leave out of consideration the influence of boiling point elevation, hydrostatic pressure, and flow resistance between the evaporators, and the influence of the temperature difference on the heat transfer coefficient, we find that a single stage evaporator evaporates n-times as much as each equally large evaporator of an n-stage battery, but also uses n-times as much as steam. The one-stage evaporator, after all, works with a temperature difference n-times as great. The choice of n depends on the permissible temperature limits: the temperature of the fresh steam is as a rule determined by the heat sensitivity of the product; the lowest boiling temperature by the attainable vacuum, the viscosity of the product, the technical problems involved in handling large quantities of water vapour at low pressure, and so on. Furthermore, n depends on the cost of the apparatus in relation to the cost of the steam. A multiple stage system does not therefore, have a greater capacity, but merely a lower energy consumption. The cost of the evaporators with their accessories are approximately proportional to their number. At a certain number of stages a point will be reached where any further increase in initial costs will not be compensated for by any reduction in energy cost. Where the minimum in total costs lies will depend mainly on the energy cost; if these are higher it will pay to evaporate in more stages. This is shown schematically and simply in Fig. 7.

In practice, all n-evaporators of a multiple stage installation must have a larger heating surface than a one-stage evaporator of the same capacity. The reasons for this are that the heat transfer coefficient with a temperature difference n times less becomes less favourable, that more of the totally available temperature difference is lost through the repeated occurrence of hydrostatic pressure, that there are n flow resistances instead of one and that these correspond with n temperature differences, and finally that the influence of boiling point elevation presents itself n times instead of just once (the boiling point elevations naturally differ in the various stages). A multiple stage installation is less attractive for use with extremely heat sensitive products. There are two reasons for this. First, to be able to work with a large number of stages, the highest temperature selected must be rather high indeed and this can affect product quality. Secondly, where the lowest temperature is concerned, one is greatly restricted because, as explained earlier, evaporation at a very low temperature is exceptionally expensive. When evaporating a sugar solution to produce sugar, one goes so

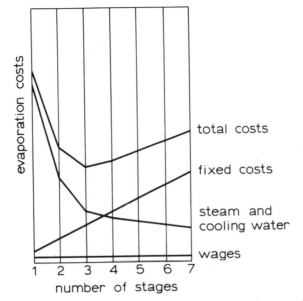

FIG. 7.   Influence of the number of stages on the cost of evaporation.

far that thermal decomposition occurs at the highest temperature, which causes a loss of yield. This, however, is amply compensated for by a lower steam consumption. A multiple stage installation is also unfavourable, however, because a product's average residence time in the installation is longer than in a one-stage evaporator (not exactly n times longer because, with a direct flow arrangement, the concentration in the first evaporator(s) is low and therefore the residence time(s) of the solutes short). Multiple stage evaporation can therefore only be used if the effect of the longer residence time on the product quality is merely slight.

In a variant of the multiple stage principle flash-evaporation is exclusively applied in each stage. Figure 8 illustrates a three-stage installation of this type. Fresh steam is only applied as heating medium in the heater. A great many evaporation stages can be attained by working with only very slight temperature differences in each stage. In this way steam consumption is kept very low; when fresh water is obtained from sea water through evaporation in a large number of stages, one can, for instance, evaporate 7 kg water per kg steam. The disadvantages of this system for application in the food industry are about the same as those encountered in the traditional multiple stage installation.

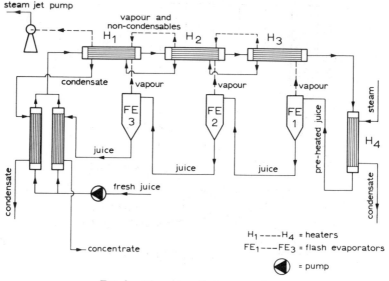

FIG. 8.   Three-stage flash-evaporation.

One can also make good use of the heat of condensation of the vapour by applying what is known as thermocompression. The vapour is adiabatically compressed to superheated steam of higher pressure, which is used again in the same evaporator as heating medium. Some aspects of this process are shown schematically in Fig. 9. In this figure consideration is given to the use of a mechanical compressor. Because the efficiency of the cycle is less than 100%, a little surplus energy is supplied by the compressor and a corresponding amount of vapour removed. Instead of a mechanical compressor, a steam-jet compressor can also be used (see Fig. 10). Here too a small amount of vapour, roughly corresponding with the amount of high pressure steam used, has to be removed. A one-stage evaporator equipped with a steam-jet compressor uses about the same amount of steam as a two-stage evaporator without a compressor. The initial costs of the first installation are lower. Mechanical compressors only come up for consideration when the price of electrical energy is low. Moreover, thermocompression by mechanical means can only be applied cheaply when evaporating at fairly high pressures because otherwise the volume of vapour is too great.

Thermocompression is actually a variant of a method whereby the heat of condensation is brought to a higher temperature level with the aid of a heat pump, and is then once again supplied to the evaporator to provide the

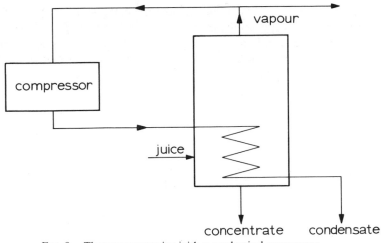

FIG. 9. Thermocompression with a mechanical compressor.

required heat of vaporisation. One can, for instance, condense a refrigerant such as ammonia or freon and then supply the heat of condensation to the evaporating solution in the evaporator. The liquid cooling medium can then be allowed to re-evaporate by withdrawing the required heat of vaporisation as heat of condensation from the water vapour produced during the evaporation process. The now gaseous refrigerant is compressed and the

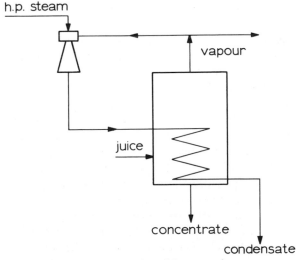

FIG. 10. Thermocompression with a steam-jet compressor.

cycle can begin anew. This is shown schematically in Fig. 11. It offers
the opportunity of evaporating products at a very low temperature because
the thermocompression is not applied to the very large volume of water
vapour but to a much smaller volume of a medium with a higher vapour
pressure. Considering all the foregoing factors, however, it is rather
doubtful whether such a method is justified.

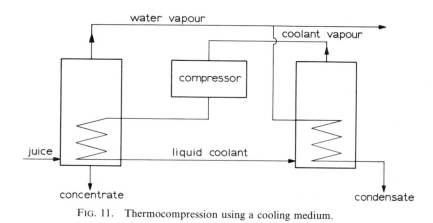

FIG. 11.   Thermocompression using a cooling medium.

With regard to the residence time and temperature of vaporisation, an
evaporator working with thermocompression is by the nature of things a
one-stage evaporator. Thus in this respect the application of
thermocompression is favourable from the point of view of quality. A
further reduction in steam consumption can be obtained by combining
thermocompression with multiple stage installations. In this way,
evaporation can be done in two or three stages, a compressor being used in
the last stage.

When energy costs exert, or are expected to exert, a·predominating
influence, one will have to consider what concessions can be made as regards
quality and what other measures can possibly be taken to maintain quality
as far as possible. Multiple stage evaporation will probably be needed in
many cases. In the evaporation of milk a limit has possibly been reached as
far as thermal economy and total product costs are concerned. If one goes
beyond this limit and applies any higher temperatures the taste of the
product will probably change so much that it will not be tolerated. This, of
course, will differ with each individual product. With a fruit juice one can
conceive that when a substantial proportion of its aroma has been

withdrawn, there would be no objection to proceeding further with its concentration at a relatively high temperature. In such a case, one could use a multiple stage installation arranged in such a way that part of the solution is evaporated at a relatively low temperature; the vapour thus formed could be withdrawn to be concentrated, after which, for the sake of a low energy consumption, higher temperatures could be applied without any problem.

The following example may illustrate the effect of the increase in steam prices. A product with a dry matter content of 10 % is concentrated to 25 % and 50% dry matter in a single stage evaporator. With steam prices at, respectively, US$2·5 and US$10 energy costs per ton concentrate are:

10–25%: 1500 kg water to be evaporated, costs US$3·75–US$15

10–50%: 4000 kg water to be evaporated, costs US$10–US$40

This example also demonstrates the need for the degree of concentration to be optimised in relation to transport and storage costs.

A different situation exists if a plant can generate its own power and can use exhaust steam for heating purposes. One finds such a situation in sugar factories, where heat is used with great efficiency. In smaller plants, however, this is out of the question.

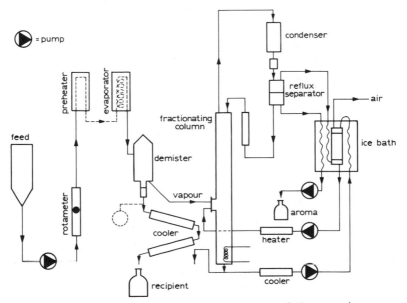

Fig. 12. Aroma recovery and concentration after flash-evaporation.

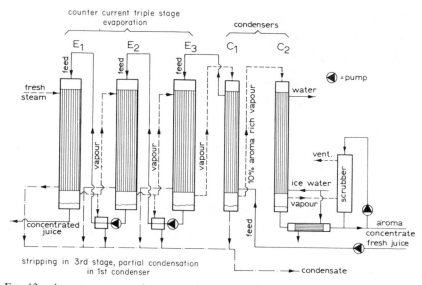

FIG. 13. Aroma recovery and concentration by partial condensation in combination with three-stage evaporation.

RECOVERY AND CONCENTRATION OF AROMA

A few final remarks now about the combination of evaporation and the recovery and concentration of volatiles. This combination is common practice with fruit juices. The most usual method is the flash-evaporation of, say, 10 or 20 % of the preheated juice, whereby a substantial portion of the volatiles is vaporised. As shown in Fig. 12, the vapour enters a distillation column and the stripped juice can be further concentrated in an evaporator. The bottom product of the fractionating column is almost pure water and the aroma concentrate can be removed at the top; this can be stored separately and later added again to the concentrated juice. It is also possible to use a multiple stage installation and to use the vapour of the first stage as a heating medium in the second; there, this vapour, rich in aroma, is condensed and can afterwards be fed into a distillation column. If the prevailing temperature in the first stage is too high, another arrangement can be used so that the volatiles are withdrawn at a lower temperature. Finally, it is possible to condense the vapour partially. The scheme shown in Fig. 13 will then be created.

6

Time–Temperature Relationships in Industrial Cooking and Frying

MAGNUS DAGERSKOG

SIK, The Swedish Food Institute, Fack, S-40021, Göteborg, Sweden

INTRODUCTION

With the rapidly growing market for industrially prepared foods, demand is increasing for equipment suitable for large-scale cooking and frying. A fair evaluation of such methods requires good knowledge of the relationships between processing conditions, resulting time–temperature exposure and the related quality changes. This kind of knowledge is necessary also for optimising existing methods and for the development of new methods or combinations of methods.

With solid foods, to which this paper will be limited, heat transfer and temperature development with time is of essential interest, since it will determine not only the necessary processing time, but also temperature distribution and the accumulated outcome of all time–temperature dependent physical and chemical changes that occur. The aim of the heating process is to obtain a palatable and digestible food product with a minimum of quality loss.

What separates cooking and frying is mainly the surface heating conditions. Cooking occurs in a wet or moist environment with food surface temperatures equal to or below the actual boiling temperature of water. Frying, however, is characterised by a combination of surface dehydration, crust formation and browning in a dry environment, with surface temperatures reaching considerably above 100 °C.

In the present paper time–temperature relationships during cooking and frying processes will be discussed primarily on the basis of our own experience from experimental work and computer simulations in this area.

THE COOKING PROCESS

During the cooking process the food material is affected in many ways including changes in the protein, fat and carbohydrate portion with both positive and negative results in quality and nutritional status, which is discussed in more detail during another session. As distinct from frying, the foods used for cooking often require a higher time–temperature exposure to become palatable, such as for low grade meat and for roots and beans.

Conventional Cooking Methods

The common attire of institutional methods for food cooking, such as batchwise kettles and steam cooking cabinets, are used to a considerable extent by the food industry, while high-pressure cookers are finding increasing use, mainly in the catering field. However, continuous methods of heating wrapped or unwrapped foods in water, steam and steam–air mixture are also in use. Common to these methods is that heat is transferred to the surface and conducted into the central parts. Because of the high heat transfer capacity of boiling water or condensing steam ($\alpha = 1500$–$10\,000\ \mathrm{W/m^2/^\circ C}$) the surface temperature reaches the cooking temperature very fast, while the temperature rise at the centre depends primarily on the dimension of the food product and the thermal diffusivity. This is further illustrated in Fig. 1 where the calculated centre temperatures for three simulated food spheres of different radius, exposed to four different temperatures are shown. The thermal diffusivity used in this example ($1\cdot55 \times 10^{-7}\ \mathrm{m^2/s}$) is applicable to products with 75–80 % water content, as in most vegetables and meats.

Heat Treatment Calculations

To be able to evaluate the influence on quality of different cooking methods, it is necessary to calculate the combined influence of time and temperature on rheological, chemical and sensory properties. Leonard *et al.*[1] proposed a cook value (*C*-value) which is calculated in analogue to the sterilisation value (*F*-value) as follows:

$$C = \int_0^t \frac{T(t) - 100}{10^{Z_c}}\,\mathrm{d}t$$

where C is the cook value in equivalent minutes at $100\,^\circ\mathrm{C}$ and Z_c is the necessary temperature rise in $^\circ\mathrm{C}$ needed for a 10-fold increase in reaction rate for the chemical and sensory changes.

Mansfield[2] mentions a Z_c-value of $21\,^\circ\mathrm{C}$ while Leonard *et al.*[1] and

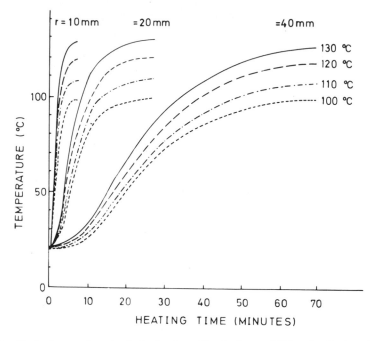

FIG. 1. Centre temperature rise during heating of food spheres of different radius in water at different temperatures.

Reichert[3] used a value of 33 °C as an approximation for chemical changes. In fact, very few experimental measurements of C- and Z_c-values for different foods have been reported in the literature. At our institute, we have started to determine these data for rheological and sensory changes in different foods. For white potatoes we have found, from cooking experiments with small cylinders at different cooking times and temperatures, that a C-value of 6 equivalent 100 °C minutes was sufficient for finish cooking (Fig. 2). We also found a Z_c-value of about 17 °C for all three varieties that were tested. In Table 1 some of the corresponding results with other foods are listed.

By the use of a computer program for the numerical calculation of temperature distribution and integration of C-values during cooking of potatoes in the form of 'rotation ellipsoids' it is now possible to study the resulting 'texture gradient' throughout the potato. Figure 3 shows the results from cooking of potatoes, at 100 °C constant temperature to a final C-value of 6 in the central parts. The cooking period ends after 21 min

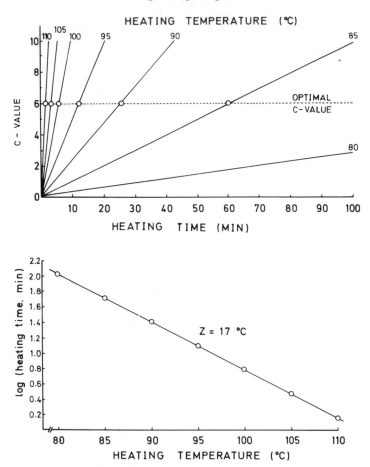

FIG. 2.   Calculation of the Z_c-value for texture changes in potatoes, by heating small potato cylinders in water at different temperatures. Optimal *C*-values correspond to easy cooking.

heating, causing the *C*-value accumulation at the surface to stop, thereby the texture gradient evens out to some extent.

For the spheres of Fig. 1, with a radius of 20 mm, the *C*-value gradient ($C_{surface} - C_{centre}$) during cooking at different temperatures has been calculated and plotted against *C*-value of the centre in Fig. 4. From the figure it is evident that a higher cooking temperature results in higher surface overheating but shorter cooking time. At higher final *C*-values, as required for foods more difficult to cook, a higher cooking temperature can be allowed for the same size of *C*-value gradient.

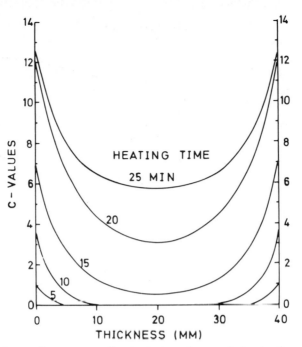

Fig. 3. *C*-value profiles corresponding to texture changes during heating of potatoes (40 × 60 mm) at 100 °C in water.

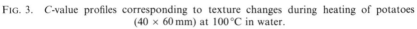

TABLE 1

Experimentally determined temperature dependence of C-values

Food factor	$Z_c(°C)$
Sterilised minced fish	
Taste	24
Consistency	28
Appearance	23
Off-taste	29
Off-odour	29
Off-colour	23
Lightness	25
Sterilised green peas	
Fresh appearance	35
Fresh taste	24
Hardness	15
Consistency	16
Cooked potatoes	
Hardness	17
Cooked, low-grade meat	
Tenderness	24

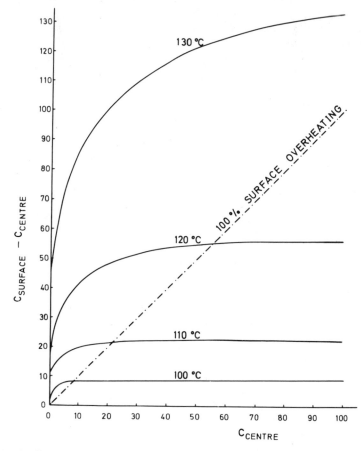

FIG. 4.   Surface overheating as a function of the C-value of the centre during heating of food spheres ($r = 20$ mm) at different temperatures. A Z_c-value of 25 °C was used.

If the temperature levelling effect after cooking is also calculated, the final C-value difference will be lower than shown in Fig. 4, which is indicated in Fig. 5 for a 120 °C cooking temperature.

Process Simulation

With the necessary thermal and other physical data at hand, and with the help of systematic experimental studies of time–temperature changes and related quality changes, mathematical models can be developed for most cooking processes. The combined effect of a number of processing

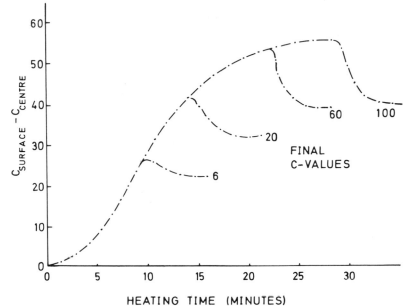

HEATING TIME (MINUTES)

FIG. 5. Surface overheating during heating and cooling of food spheres ($r = 20\,mm$) at 120 °C to four final C-values.

variables, such as thickness, starting temperature, cooking temperature programme and cooking medium can be studied and optimised by computer simulations with regard to the temperature distribution and related time–temperature dependent quality changes.

Figure 6 illustrates computer simulation of three different pressure cooking processes for potatoes. A decreasing cooking temperature results in lower surface overheating in comparison to the other processes. The final C_{100}-value at the centre after cooking is 6 minutes in all cases.

Unconventional Cooking Processes

To overcome the difficulties associated with conventional cooking methods, where the heat transfer to the food interior is limited by its thermal conductivity and the resulting high temperature and quality gradients, direct in-depth heating with electromagnetic energy sources has developed over the last decade. For cooking and reheating of foods in catering and in the home microwave ovens are becoming standard equipment in several countries. In Fig. 7 the resulting temperature gradient after heating a 30-mm slab of meat by microwaves and in a conventional

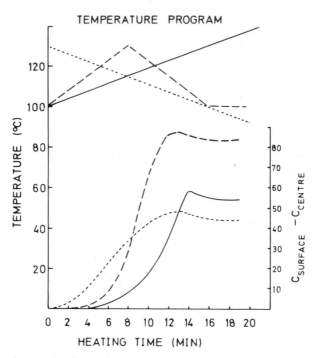

FIG. 6.   Surface overheating during cooking of potatoes (40 × 60 mm) with three different temperature programmes. Cooling is initiated to give a final C-value of 6 min at the centre after the cooling period.

oven are compared. It is evident that faster heating with less overheating of the outer parts is obtained, which should result in higher yield and final quality. Cooking by microwave heating alone for industrial purposes is usually considered uneconomical, unless very positive effects can be obtained in terms of yield and quality. As a rule a combination with some conventional heat source such as steam, hot air, infrared heating, etc. will be required.[4]

Industrially precooked foods represent a rapidly growing market. At least two large plants for rapid continuous microwave cooking of chicken parts have been reported by May[5] and Smith.[6] It is claimed that processing time could be reduced by about 75 % depending on the degree of cooking desired and the material, with 5–10 % reduction in weight loss. Promising results with microwave cooking of wieners have been obtained in Canada,[7] and with combinations of microwave heating and various browning methods for meat patties in our laboratories.[8]

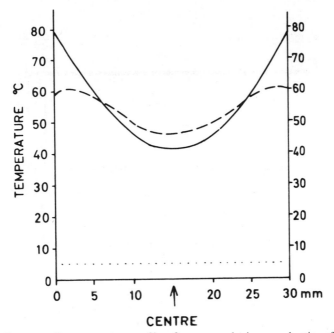

FIG. 7.   Corresponding temperature profiles after oven and microwave heating of a 30-mm thick slab of meat.[9,10] ——— Oven heating (17 min). ———— Microwave heating (5 min).

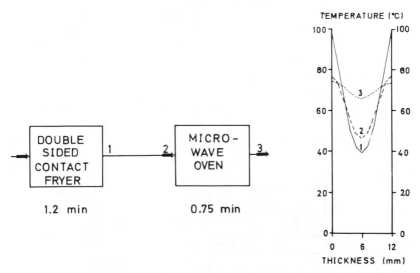

FIG. 8.   Temperature profiles during finish cooking of prebrowned meat patties by microwave heating.[9,10]

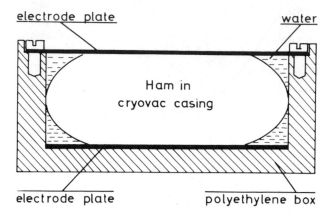

FIG. 9.   Cross-section of mould-electrode arrangements for capacitive dielectric heating at 60 MHz.[17]

An industrial application which combines surface browning and microwave heating was recently introduced in a Swedish food plant, the so-called 'INPRO-process'. Important advantages in comparison to deep-fat frying are claimed, both in process control and economy, and in sensory and nutritional quality. In Fig. 8 the temperature profiles after pan frying and before and after microwave heating have been calculated.[9,10]

For foods with large dimensions, like ham, capacitive dielectric heating at 27–60 MHz may be useful for the cooking process. As found by Bengtsson and Green[11] such a process resulted in acceptable temperature distribution and substantially reduced heat treatment times and juice losses with indications of an advantage also in sensory quality. In Fig. 9 the mould-electrode arrangement is shown.

As an alternative to microwave heating, electrical resistance heating has been suggested for some applications. For boiling potatoes this method has been developed into a practical process by a Swedish company, the so-called

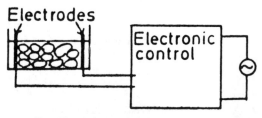

FIG. 10.   Principles of the OSCO method.[12]

OSCO method.[12] The principle of the method is to apply alternating electrical current at 50 Hz between two electrodes in water or in a salt solution. The boiling process is controlled electronically (Fig. 10). The experimental results show that the time needed for boiling can be reduced from 25–30 min to 5–7 min, depending primarily on the salt concentration of the water.

THE FRYING PROCESS

Possibly the most important and difficult operation in industrial or institutional food preparation is the frying operation. The process should give an attractive and tasty surface crust, combined with rapid in-depth heating to the desired degree of cooking and with as small losses in yield, nutritional value and sensory quality as possible. Until fairly recently, deep-fat frying was the only method suitable for large-scale industrial processing.

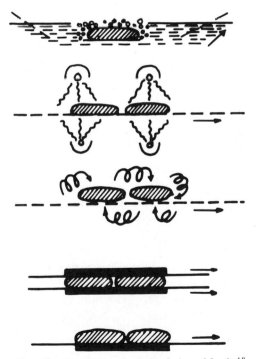

FIG. 11.   Possible methods for frying of foods.[18]

Inherent problems are a lack in flexibility between different products, large working volumes of fat and risks for fat degradation and fat exchange between bath and sample. For this reason deep-fat frying is much criticised and people are looking for alternatives. In principle, there are basically three other possible methods, based on different heat transfer mechanisms. As shown in Fig. 11, these are convection heating, contact heating and infrared radiation.

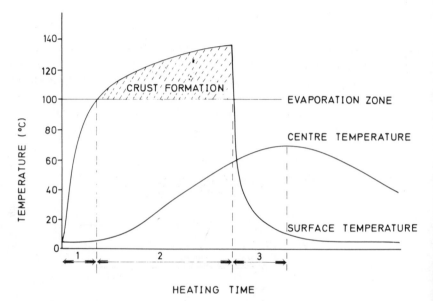

FIG. 12.   Characteristic time–temperature lapse during frying of foods.[19]

For all frying methods, the heat transfer and temperature rise is characterised by a time–temperature diagram as in Fig. 12 with three clearly distinguishable periods. The length of the second and third period is primarily related to the sample dimensions, while the length of the first period and the degree of crust formation depends on the actual frying method and heat transfer capacity. The frying time (second period) is almost independent of the frying method because of the 100 °C evaporation zone formed inside the crust acting as a temperature driving force during heating of the central parts.

Contact Frying

During recent years contact frying of foods has gained increasing interest for industrial purposes as an alternative to deep-fat frying. In the case of single-sided contact frying the product has to be turned over during the process in order to treat both sides. Temperature measurements and computer simulations at SIK[13] of the resulting temperature profiles during the frying of a 30 mm thick slab of meat (Fig. 13) show that the top side cools down very fast, causing prolonged heating. The measurements also point

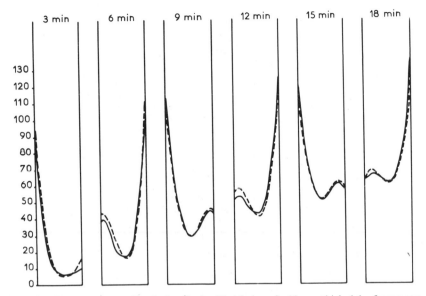

FIG. 13.   Temperature profiles during single-sided frying of a 30-mm thick slab of meat at a pan temperature of 175 °C.[13]

out that a high 'turning over' frequency reduces the necessary frying time. For industrial purposes a Norwegian company has developed a continuous single-sided frying process with a moving Teflon conveyor heated by electrical resistance platens.[14] Other available processes work with some kind of conveyors carrying the food over a conventional frying table.

By the use of simultaneous double-sided contact frying the necessary frying time can be reduced to less than 50 % compared to single-sided frying of a product of normal thickness. For meat patties Dagerskog and Bengtsson[15] studied the relationship between crust colour formation, yield, frying time and pan temperatures for different recipes. Reproducible results were obtained which could be presented in the form of nomograms, such as that shown in Fig. 14, useful for process optimisation. As an example such a

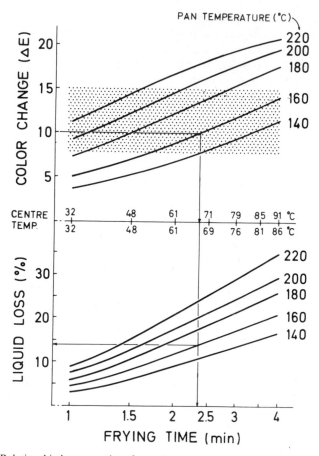

FIG. 14.   Relationship between colour formation, frying time, centre temperature and yield
during double-sided contact frying of meat patties of 11-mm thickness:[15]

diagram can be utilised, starting out from a desired crust colour change of
$\Delta E = 10$ (colour difference in NBS units) and a pan temperature of 160 °C.
A necessary frying time of 2·3 min and a resulting centre temperature of
67 °C and weight loss of 14% is obtained. However, it is important to
remember that the diagram is valid only for one recipe, sample thickness
and initial sample temperature.

The surface and centre temperatures measured during the frying of meat
patties at different pan temperatures (see Fig. 15) show that the rate of heat
transfer is almost independent of pan temperature during the first 2 min. An

exception is at a pan temperature of 120 °C where the surface temperature does not reach 100 °C at once. With increasing frying time the evaporation zone recedes inwards and centre temperatures begin to deviate, depending on the pan temperature used. The temperature development of the surface crust is not regular, but demonstrates a minimum after about 2–2·5 min

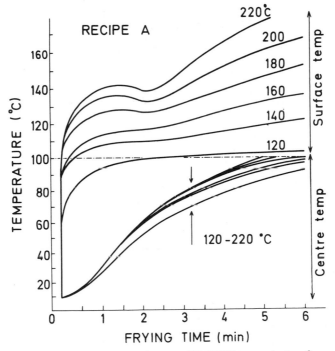

Fig. 15. Influence of pan temperatures between 120–220 °C on product surface and centre temperature during double-sided contact frying of meat patties of 11-mm thickness.[15]

frying time, approximately corresponding to centre temperatures in the fibre shrinkage range of 60–70 °C. This cooling effect may be explained by liquid being pressed out during onset of shrinkage. Temperature profiles in Fig. 16 show that the evaporation zone penetrates in the meat surface about 2 mm on each side during double-sided contact frying.

In Table 2 the results from two single-sided frying experiments are summarised, together with data calculated for double-sided pan frying to reach the same centre temperatures and colour changes. This comparison shows that double-sided frying requires lower pan temperature and less than half the time at equivalent yield, centre temperature and colour.

Magnus Dagerskog

TABLE 2
Comparison of double-sided and one-sided pan frying

	One-sided frying					Double-sided frying		
Frying time (min)	Pan temperature (°C)	Liquid loss (%)	Centre temperature (°C)	ΔE (NBS units)		Frying time (min)	Pan temperature (°C)	Liquid loss (%)
4 (1 + 1 + 1 + 1)	180	9·3	58	8·5		1·9	160	11·0
6 (1 + 1 + 2 + 2)	180	14·7	69	11·1		2·4	165	14·5

Even in comparison with deep-fat frying (Table 3) the temperature rise at the centre is faster during double-sided contact frying, primarily because of the contact pressure achieved. This comparison also shows that, besides the shorter frying time needed in double-sided pan frying, cooking losses and colour changes are much lower. In fact, a very high pan temperature ($\sim 250\,°C$) would be necessary to reach the same combination of centre temperature and surface colour as in deep-fat frying, depending on the very high surface temperature reached during the latter.

For industrial purposes a continuous double-sided Teflon belt oven for surface browning precedes the microwave oven in the INPRO-process as mentioned earlier.

Infrared Frying

Infrared (IR) heating has also been used for frying purposes in the food industry and provides the greatest heat transfer effect of all forms of frying. Other important properties of this form of heat transfer is the transmission effect in the short-wave IR region, giving superior upstart and process characteristics, as reported by Ginsburg[16] and confirmed by our own experiments. From a consideration of the energy distribution of different IR sources shown in Fig. 17, it is evident that the high temperature short-wave quartz tubes are the most effective IR radiators for in-depth heating of foods. Figure 18 shows a comparison of ours by computer simulation between short-wave IR frying and double-sided contact frying at the same specific energy input. Results from experimental pilot plant work, particularly with regard to frying of meats in the catering field show very promising results.

Oven Frying

Oven heating or convection heating with air is primarily used for frying foods of larger dimensions, as the heat transfer medium has a considerably lower heat transfer capacity. Besides the oven temperature and degree of turbulence, the relative humidity will play an important part in this process, as illustrated in Fig. 19. For a moist meat surface, both surface temperature, thermal driving force and rate of evaporation will be controlled by the wet-bulb temperature. As shown in Fig. 20, Bengtsson *et al.*[17] determined temperature profiles for two different start temperatures and two different oven temperatures during oven frying of 55-mm thick, rectangular beef samples. Temperature profiles were steeper during the early stage of frying. The profiles also become steeper when using a higher oven temperature or when frying samples direct from the frozen state. The

TABLE 3
Comparison of double-sided and deep-fat frying

	Deep-fat frying					Double-sided frying			
Frying time (min)	Fat temperature (°C)	Liquid loss (%)	ΔE (NBS units)	Centre temperature (°C)	Frying time (min)	Pan temperature (°C)	Liquid loss (%)	ΔE (NBS units)
1	180	15·7	14·9	26	0·8	220	9·0	8·0
2	180	27·5	19·0	48	1·5	220	15·0	14·5
3	180	34·8	20·6	68	2·3	220	24·0	18·0

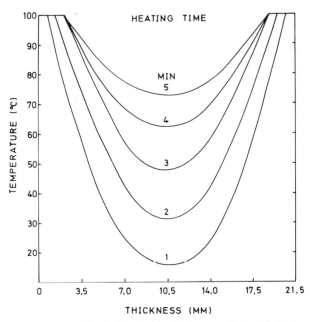

FIG. 16.   Temperature profiles during double-sided contact frying of a 21·5-mm thick slab of minced meat.

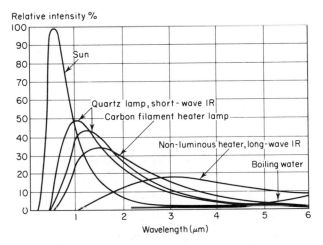

FIG. 17.   Relative intensity of some infrared sources. (Philips, *The IR Handbook.*)

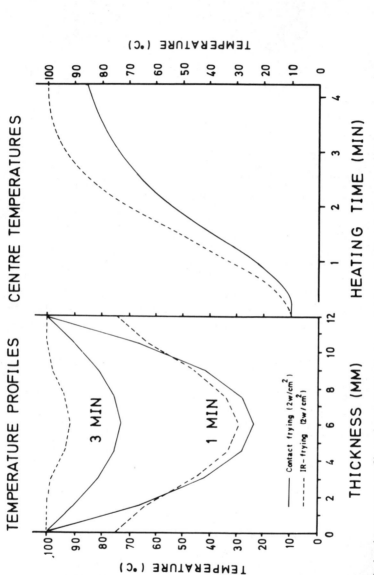

FIG. 18.   Calculated temperature profiles and temperature development at the centre of a slab of meat heated by short-wave IR and double-sided contact frying at equal power density.

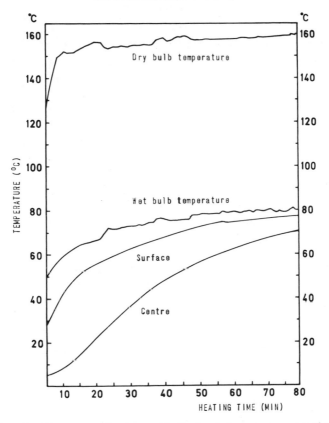

FIG. 19.   Temperature development in beef and air during oven roasting.[17]

maximum surface temperature was higher when frying at 225 °C than at
175 °C, while the maximum surface temperature differed but little when
frying from initial sample temperatures of − 20 °C or + 5 °C. Temperature
and moisture profiles varied inversely with each other, a high temperature
corresponding to a low moisture content, while the fat profiles seemed to be
affected more by raw material variation than by heating (Fig. 21). The
strong inverse correlation between temperature and water content indicates
the importance of avoiding unnecessary overheating. For industrial
purposes some equipment is in use, particularly for the frying of chicken
parts.

In addition to these 'pure' frying methods, combinations of methods may
be used and may be superior to the single methods. To be able to evaluate

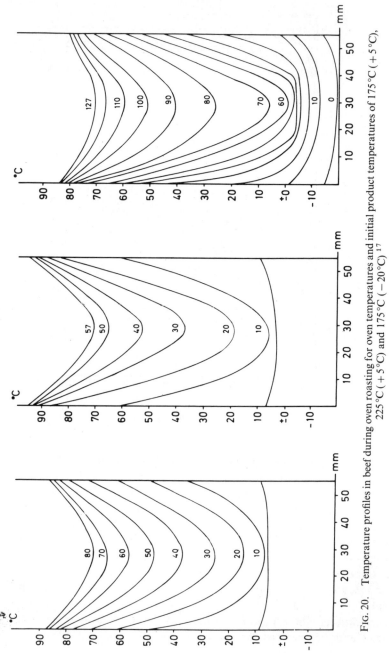

FIG. 20. Temperature profiles in beef during oven roasting for oven temperatures and initial product temperatures of 175°C (+5°C), 225°C (+5°C) and 175°C (−20°C) [17]

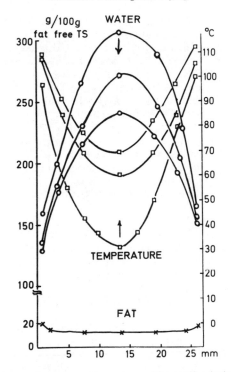

FIG. 21.   Temperature and corresponding water and fat profiles during oven roasting for an
oven temperature of 175°C and initial product temperature of 5°C.[17]

such combinations properly it is probably necessary to use computer
simulation screening to study the combined effect of all important process
and product variables and thereby reduce to a minimum expensive and
elaborate experimental studies in large-scale equipment.

REFERENCES

1. Leonard, S., Luh, B. S. and Simone, M. (1964). Aseptic canning of foods: 1.
   Preparation and processing procedures. *Food Technol.*, **18**, 81.
2. Mansfield, T. (1967). Engineering processing equipment to produce quality
   demands. SIK-Rapport No. 220.
3. Reichert, J. E. (1974). Optimal sterilisation temperature for ready to eat foods
   (German). *Fleischwirtschaft*, **54**, 1305.
4. Bengtsson, N. E. and Ohlsson, T. (1974). Microwave heating in the food
   industry. *Proc. of the IEEE* 62, **1**, 44.

5. May, K. N. (1969). Applications of microwave energy in preparation of poultry convenience foods. *J. Micr. Power*, **4**, 54.
6. Smith, D. P. (1969). Industrial microwave cooking comes of age: Chicken cooking operation $2\frac{1}{2}$ years. *Micr. Energy Appl. Newslett.*, **2**, 9.
7. Watanabe, W. and Tape, N. W. (1969). Microwave processing of wieners: 1. Composition and method of preparation. *Canad. Inst. of Food Technol.*, **2**, 64.
8. Bengtsson, N. E. and Jakobsson, B. (1974). Cooking of meat patties for freezing—A comparison of conventional methods and their combination with microwave heating. *Micr. Energy Appl. Newslett.*, **7**, No. 6, 3.
9. Ohlsson, T. (1975). Thawing and heating with microwaves: A process study using computer simulation. SIK:s Service-Serie No. 505.
10. Ohlsson, T. (1975). Microwave heating of foods. Applications and limitations (Swedish). SIK-Rapport No. 377.
11. Bengtsson, N. E. and Green, W. (1970). Radio-frequency pasteurisation of cured hams. *J. Food Sci.*, **35**, 683.
12. Svensson, B., Bodin, B. and Hermansson, S. (1975). Osco, a new boiling method for potatoes. 6th Triennial Conference of the European Ass. for Potato Research, Wageningen: Abstract of Conference papers, p. 54.
13. Jakobsson, B. and Bengtsson, N. E. (1972). Freezing of raw and cooked beef—studies of heat and mass transport during cooking from frozen and thawed state (Swedish). SIK-Rapport No. 303.
14. Walderhaug, R. (1974). Frying of foods in the Fodema contact fryer (Swedish). SIK-Rapport No. 368.
15. Dagerskog, M. and Bengtsson, N. E. (1974). Pan frying of meat patties—relationship among crust formation, yield, composition and processing conditions. *Lebensm. Wiss. u. Technol.*, **7**, 202.
16. Ginsburg, A. S. (1969). *Application of Infrared Radiation in Food Processing*, Leonard Hill, London.
17. Bengtsson, N. E., Jakobsson, B. and Dagerskog, M. (1976). Cooking of beef by oven roasting: A study of heat and mass transfer. *J. Food Sci.*, in press.
18. Bengtsson, N. E. (1976). Industrial food preparation. Preparation and preservation (Swedish). *Livsmedelsteknik*, **18**, 69.
19. Hallström, B. and Sörenfors, P. (1975). Heat transfer during thermal preparation of meat (German). *Fleisch*, **29**, 185.

7

Changes of Muscle Proteins During the Heating of Meat

Reiner Hamm

Bundesanstalt für Fleischforschung, D-8650 Kulmbach, Blaich 4, Federal Republic of Germany

INTRODUCTION

The heating of meat is accompanied by changes in appearance, flavour, texture and nutritive value. The most drastic changes in meat during heating such as shrinkage and hardening of tissue, release of juice and discoloration are caused by changes in the muscle proteins. The muscle cell consists of the myofibrils, the sarcoplasmic reticulum and some cell organella such as nuclei, mitochondria, lysosomes, ribosomes etc. This structural material is embedded in the fluid matrix of the sarcoplasma. The cell is surrounded by

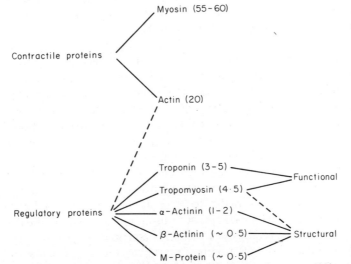

Fig. 1. Myofibrillar proteins of the skeletal muscle.[144] The protein content (% by weight) in rabbit myofibril is shown in parentheses.

the cell wall, the sarcolemma. The protein system of the myofibril which amounts to about two-thirds of the total muscle proteins, consists of the contractile proteins myosin and actin, the regulatory proteins tropomyosin and troponin and some other minor proteins (Fig. 1). Muscle also contains varying amounts of connective tissue proteins, particularly collagen, the thermostability of which is quite different from that of the myofibrillar and sarcoplasmic proteins. Of course, it is very difficult to investigate heat-induced changes of such a complicated and highly organised protein system so that we can understand these changes at a molecular level. The most recent research work in this field is reviewed in this article.

THE HEATING OF THE ISOLATED MYOFIBRILLAR PROTEINS

As we should expect, some of this research was carried out by studying the behaviour of isolated myofibrillar proteins, particularly myosin, actin and the actin–myosin complex (actomyosin), but also the subunits of the myosin molecule, namely 'light meromyosin' (LMM) and 'heavy meromyosin' (HMM). Some of the criteria used for studying the process of heat denaturation are protein solubility and viscosity, fluorescence studies, adenosine triphosphatase (ATPase) activity of myosin or HMM, number of sulfhydryl groups available for SH reagents such as N-ethylmaleimide (NEM) or p-chloromercuribenzoate (PCMB).

Biochemists define 'denaturation' as a change in the specific steric conformation of a protein, i.e. a change in the secondary and tertiary structure without a chemical modification of the amino acids. Thus denaturation is a physical process, not a chemical one. According to this definition, the oxidation of sulfhydryl (SH) to disulfide (SS) groups or the reduction of SS to SH groups should not be called 'denaturation'.[1]

As to myosin, which represents the major part of the myofibrillar proteins (Fig. 1), the heat-induced changes of the subunit LMM-Fr 1 were particularly studied because this highly α-helical tail portion of the myosin molecule is soluble at low ionic strength and is believed to govern the solubility of myosin. A drastic decrease of the solubility of LMM-Fr 1 occurs at 60 °C (pH 5·4).[2a] The unfolding of LMM during heat denaturation, measured by optical rotary dispersion, begins at about 30 °C and is finished at about 50 °C (Fig. 2).[3a] There is a decrease in stability to heat at alkaline pH values and the molecule is more stable in the acidic range of the isoelectric point (IP). With LMM-Fr 1 preparations obtained in the usual way by a short tryptic treatment of myosin, a depolymerisation of the protein to smaller molecular weight proteins and peptides as a result of heating was observed,[2b] but this effect was apparently due to the influence of contaminating trypsin[2a] because LMM prepared by a non-enzymatic procedure (LMM-C) did not depolymerise by thermal treatment.[3b]

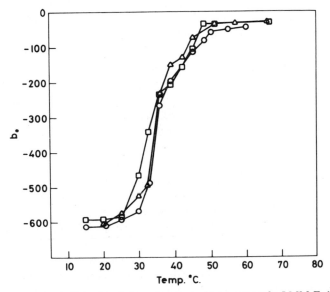

FIG. 2.   Optical rotary dispersion (b_0) as a function of temperature for LMM-Fr 1 prepared from tryptic digest 75 sec (Δ), 5 min ($\bigcirc$), and 30 min ($\square$) of rabbit myosin; 0·02M sodium tetraborate, 0·2M NaCl; pH 9·0.[3]

Actin, another major component of the myofibrillar proteins, exists in two forms: the monomolecular globular G-actin and the polymerised fibrillar F-actin, only the latter form being present in muscle tissue. Studies of the temperature dependence of fluorescence data revealed a heat denaturation transition which occurred with G-actin between 30° and 50°C, with F-actin between 40° and 60°C.[4]

In muscle, actin and myosin associate within a few hours after death of the animal during development of rigor mortis forming the actomyosin complex. Therefore, with regard to meat properties the thermal stability of actomyosin is more interesting than that of the components actin and myosin. The heat denaturation of actomyosin follows first-order kinetics as it was reported by several authors.[5-7] Heating to temperatures above 40°C removed the Ca^{2+} requirement for superprecipitation[8a] and the Ca^{2+} sensitivity (increase of ATPase activity by addition of Ca^{2+}) of actomyosin.[9] The possibility of a conformational change in actomyosin on heat denaturation has been suggested.[6]

As with other enzymes, the thermal effects on myosin and actomyosin ATPase are twofold. First, with increasing temperature the rate of ATP breakdown by these enzymes increases, but with higher temperatures inactivation by irreversible denaturation occurs.[8b, 10, 11] The maximum ATPase activity of 'natural actomyosin' (complex of actomyosin, tropomyosin and troponin)[9] and of pure actomyosin[12] was reported to be within the range of 43–47 °C.

In most of the previous studies of heat denaturation of myosin or actomyosin the possibility existed that some of the changes reported were due at least in part to the presence of oxygen or trace metal ion impurities present in reagent grade salts since both of these factors have been shown to affect the stability of myosin[13] and actomyosin.[14] In addition it is possible that inclusion of troponin and tropomyosin as contaminants in the actomyosin preparation also affected the experimental results. In the studies of Jacobson and Henderson[12] these factors have been eliminated. As these authors found, the temperature of maximum changes in the overall conformation remained unaffected by the binding of F-actin to myosin. For both myosin and actomyosin the 'melting temperature' T_M (temperature of maximum slope in transition region of kinematic viscosity) was 43 ± 2 °C. However, the range of temperature, over which large conformational changes were observed, was affected by the binding of F-actin to myosin. For myosin changes were observed between 37–50 °C while for actomyosin large changes occurred between 20–50 °C. The temperature of maximum myosin ATPase activity of actomyosin (45 ± 2 °C) corresponded to the T_M. However, for myosin the temperature of maximum enzyme activity was 33 °C, considerably below T_M (Fig. 3). So, actin offers considerable protection to temperature inactivation of the actin site of myosin though the F-actin–myosin complex is highly dissociated in the presence of ATP. Also other authors reported that binding of actin to myosin provides some degree of stabilisation of myosin.[7, 15] Yasui et al.[16] concluded from their ATPase studies that the protective action of F-actin against denaturation of myosin is exclusively achieved through the interaction of myosin and actin. However, there is no significant stabilisation of myosin by F-actin in terms of the temperature sensitivity of the overall conformation.[12] Apparently, the Mg^{2+}–actomyosin complex is. more stable to heat than the uncomplexed protein or the Ca^{2+} protein complex.[17]

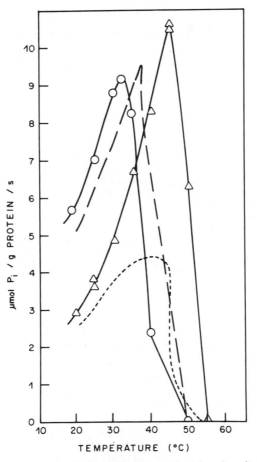

FIG. 3. Temperature sensitivity of myosin and actomyosin adenosine triphosphatase. Specific enzymatic rate as a function of temperature. All solutions contained 0·2 mg/ml protein in 0·6M KCl, 10mM $CaCl_2$. 0·05M Tris at pH 7·4 and 1·0mM ATP. (◯) Myosin solution; (△) initial rate of the natural actomyosin solution; (-----) rate for the same solution 500–800 sec after ATP addition; (– – – –), a myosin solution which had been shown by superprecipitation to be contaminated with a small amount of F-actin.[12]

With actomyosin, after heating up to 60–70 °C a marked irreversible increase in the SH groups titratable with PCMB[12] or reacting with NEM[18–20] was found (Fig. 4). This SH increase was explained by the fact that NEM or PCMB react only part of the SH groups of the native protein and that heating causes an unfolding of the protein molecules during heat denaturation which makes more SH groups available for the reagent.[18]

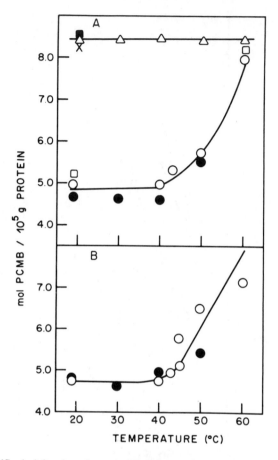

Fig. 4.   A. Sulfhydryl titration of myosin (Δ) and natural actomyosin ($\bigcirc$, $\bullet$) in 0·6M KCl, with 0·05M Tris at pH 7·4. The myosin concentration was 0·2 mg/ml, while the actomyosin concentration was 0·4 mg/ml. ($\times$) Natural actomyosin; ($\square$, $\blacksquare$) synthetic actomyosin in 8M urea (0.2 mg/ml myosin and 0·1 mg/ml actin). B. effect of heating, then cooling, on actomyosin (experimental conditions similar to A).[12]

There obviously does not exist a specific temperature but a wide temperature range in which SH groups, previously masked in the native protein, are exposed.[1] The finding of Dubé *et al.*[21] that by heating extracts of myofibrillar proteins in the presence of urea at temperatures between 60 and 90 °C the number of SH groups available for Ellman's reagent decreases, disagrees with the results just mentioned. This disagreement is perhaps due to the effect of a strong denaturing reagent (urea) during heating, so that heating could not cause any further structural changes of the protein molecules that would render more SH groups available to the reagent.

The unfolding of the actomyosin molecule, demonstrated by the release of SH groups, begins at the same temperature at which the maximum ATPase activity and the maximum change in overall conformation are reached ($\sim 45\,°C$).[12] The total number of SH groups is not changed during heating of actomyosin up to about 70 °C. Between 70 and 120 °C, however, the total number of SH groups decreases with increasing temperature which is mainly due to the oxidation of SH to SS by oxygen.[18, 19]

Hamm and Hofmann[18] reported that oxidation of SH to SS was not observable up to 70 °C although coagulation of actomyosin had already occurred from 45 °C. They concluded that the heat coagulation of myofibrillar proteins cannot be due to the oxidation of SH groups but is due to intermolecular association of other side-groups on the molecules. This conclusion was supported by the results of Samejima *et al.*[22] who showed that neither a reducing reagent such as thioglycol or Na_2SO_3 nor a SH-blocking reagent such as PCMB prevents a myosin solution from coagulation.

The extent of protein–protein interaction in beef actomyosin solution between 15 and 70 °C, as indicated by a change in light scattering absorbance, was studied by Deng *et al.*[23] The Arrhenius plot showed a strong temperature dependent protein–protein interaction at temperatures below 40 °C, less changes between 40 and 60° and a further increase of interaction with rising temperatures above 60 °C (Fig. 5).

As was demonstrated by optical rotary dispersion measurements, the unfolding of the highly α-helical molecules of the myofibrillar protein tropomyosin begins at about 35 °C and is finished at about 60 °C.[3a] Like LMM this protein is also less stable to heat at alkaline pH values than in the acidic pH range or at the IP. Intrinsic fluorescence studies revealed that when the temperature was raised, a thermal melting of the tropomyosin helix occurred at 54 °C.[24]

Summarising these results obtained by heating the isolated proteins, it can be stated that the most drastic changes of the myofibrillar proteins occur between 30 °C and 50 °C, reaching almost completion at 60 °C. These changes are characterised by an unfolding of the protein molecules, accompanied by an association of molecules and resulting in coagulation and loss of enzyme activity. At temperatures above 70 °C more severe changes such as oxidation of SH groups occur.

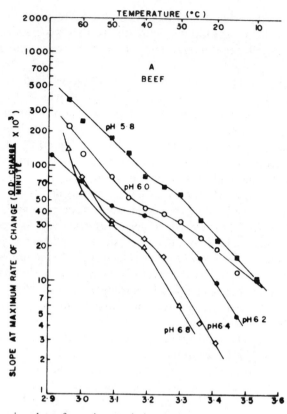

FIG. 5.   Arrhenius plots of protein–protein interaction rates (measured by light scattering absorbance) in beef actomyosin solutions at various pH.[23]

CHANGES IN MYOFIBRILLAR PROTEINS DURING THE HEATING OF MEAT

The heat stability of myofibrillar proteins might not be the same in the intact fibre and in the isolated state. In numerous papers the changes of these proteins during the heating of muscle tissue or muscle fibres has been studied. Some of this work has already been reviewed.[19, 25-28]

The greatest decrease in the solubility of myofibrillar proteins during heating of meat occurs at temperatures between 40 and 60 °C; above 60° these proteins become almost insoluble (Fig. 6),[19, 29-32] but at 55 °C even 7 h heating did not cause a complete desolubilisation.[33] The decrease of protein solubility as well as of Ca^{2+}-activated ATPase and Mg^{2+}-activated ATPase in myofibrils at different pH values follows first-order kinetics. All three reactions seem to occur simultaneously.[7]

Between 40 and 60 °C breakdown of myosin into smaller compounds was indicated by DEAE cellulose chromatography in the presence of 7M urea. LMM seemed to be more stable to heating than HMM. Helical changes in

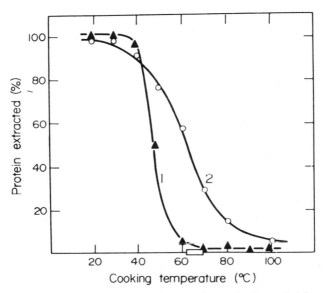

FIG. 6.   Myosin and sarcoplasmic protein denaturation and collagen shrinkage. Curve 1, percentage denaturation of myosin in the myofibril. Curve 2, percentage denaturation of the sarcoplasmic proteins in the muscle. Oblong horizontal area, the temperature zone of collagen shrinkage.[32]

the actin molecule were found. Tropomyosin was the most stable protein.[34] During heating of muscle tissue the thermostability of actin is higher than that of myosin as it was demonstrated by polyacrylamide SDS electrophoresis,[35] starch gel electrophoresis[36] and electron micrography.[34] When bound to actin, myosin is more heat resistant in the myofibril.[7, 36]

According to these results, the temperature range of maximum changes in the solubility of myofibrillar proteins seems to be about the same in the intact muscle fibre as in the isolated state. This, however, seems not to be valid for ATPase activity because during storage at 35 °C the ATPase activity of bovine myosin decreased much faster in solution than in the muscle fibre[15] (Fig. 7).

Besides losses in solubility and ATPase activity some other interesting reactions accompany the heat-induced changes of myofibrillar proteins during the heating of meat from different species. During the heating of muscle from 30 to 70 °C the SH groups available to NEM and other SH reagents increase, indicating an unfolding of protein molecules as was observed with isolated myosin and actomyosin.[18, 37-42] Similar to the

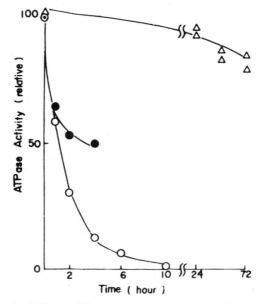

FIG. 7.   Changes in ATPase activity during storage in 0·1M KCl at pH 7·0 and 35 °C of isolated myosin A(○), myosin B(●) and myosin A extracted from glycerol-treated fibre bundles.[15]

isolated proteins, the total number of SH groups in proteins does not change by heating the muscle tissue up to 70 °C,[18, 37, 41-3] but decreases with temperatures above 70 °C.[18, 44, 45] It should be mentioned that these changes in the SH/SS system of meat concern the proteins because 97 % of the SH and SS content of muscle tissue is bound to proteins.[1]

At temperatures above 80 °C a loss of total SH plus SS groups occurs by oxidation to cysteic acid[43] or by the splitting off of H_2S.[18, 46-9] This release of H_2S, which increases exponentially with rising temperature[18, 47] does not originate from SS groups or methionine but from the protein SH groups.[18, 48] The amount of H_2S does not increase essentially,[50] when the

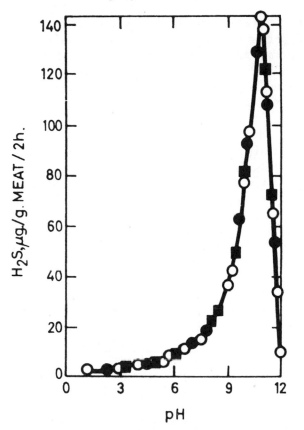

FIG. 8. Effect of variation of pH on H_2S content of volatiles produced by heating meat: ○ mutton, ● beef, ■ pork.[51]

temperature exceeds 120 °C but the pH value has a strong effect on the formation of H_2S during the heating of meat[51, 52] (Fig. 8).

It was found that the heating of meat with higher fat content produced significantly more H_2S than the heating of lean meat.[53] In most of these studies the H_2S formed by the heating of meat was transferred into the solution of the H_2S reagent. Marchenko and Kasenjasheva[54] developed a procedure for separate determination of the levels of H_2S and mercaptans being volatilised during meat cooking and those left in cooked meat. They found that after heating meat up to 80 °C the main part of the volatile sulfur compounds had been left in the cooked meat.

The reaction mechanism of H_2S formation is not yet fully understood. In general the release of H_2S during sterilisation of meat is a serious problem because of the corrosion of tins, discoloration of cans (marbling) and their contents and an unfavourable or even offensive smell when the can is opened. H_2S development means also a decrease in the nutritive value of meat products because it indicates the destruction of essential sulfur-containing amino acids.

The heating of muscle tissue as well as of myofibrils results in an increase of pH[19, 55, 56] which depends on the initial pH and starts at about 30 °C; the maximum pH increase was observed to be between 40 and 60 °C. Simultaneously the IP of the myofibrillar proteins—measured as the pH of the minimum of water-holding capacity—is shifted to higher values (e.g. from 5·0 to 6·0 after heating at 80 °C). As could be demonstrated by dye-binding experiments, the increase in pH as well as the shift of IP are due to an increase in available basic protein groups. Apparently some imidazolium groups of histidine, which are masked in the native myofibrils, are released by the unfolding of actomyosin during heating.[19]

The denaturation of myofibrillar proteins during heating of tissue is connected with a release of protein-bound magnesium and calcium which occurs mainly between 40 and 60 °C. Apparently unfolding the peptide chains lowers the sequestering power of the muscle proteins.[19]

From all these results it can be concluded that the decrease in solubility of myofibrillar proteins between 30 and 60 °C is accompanied by an unfolding of the protein chain. Probably an association of the unfolded peptide chains causes protein coagulation. The question arises whether this association is the result of the formation of new cross-linkages between adjacent molecules. As already mentioned, the formation of SS bonds by oxidation of SH groups cannot be the reason for coagulation because coagulation occurs at temperatures much below the temperature at which the formation of SS begins.[18] As Hamm[19, 57] concluded from studies on the influence of pH on the water-holding capacity of raw and heated meat, between 30 and 50 °C relatively unstable cross-linkages between the unfolded protein chains are formed leading to a tighter network of protein structure within the isoelectric range of pH. At temperatures above 50 °C new cross-linkages are formed which are quite stable and cannot be split by addition of weak base or acid. The chemical nature of these cross-linkages, however, is still unknown; perhaps hydrophobic groups of protein side-chains play an important role.[58, 59] Karmas and Dimarco[60] suggested in the discussion of thermal profile studies on beef muscle tissue that the first peak of the thermal profile at 60 °C may be strongly related to the protein portion of the water–protein system but that the second peak at 82 °C is due to changes in water structure. They hypothesise that protective, semicrystalline water structures surrounding the non-polar amino acid radicals of proteins were destroyed by heat, followed by formation of hydrophobic bonding, yielding an aggregated denatured state.

The antigenic activity of muscle proteins decreases with increasing degree of denaturation. But in pork heated at 90 °C the antigenicity had not yet been completely destroyed.[61]

When a given temperature is attained by means of microwave energy less detrimental protein changes occur than by conventional heating, but the type of changes in both cases remains the same.[55] Heating of meat for 10 h at 45–60 °C causes some solubilisation of myofibrillar proteins resulting from the effect of proteolytic enzymes present in muscle tissue.[62, 63]

After the heating of muscle at 120 °C for longer time (7–45 h) some of the coagulated muscle protein are broken down to free amino acids and peptides of low molecular weight.[64]

CHANGES IN CONNECTIVE TISSUE PROTEINS DURING HEATING

It has to be realised that, besides myofibrillar proteins, meat contains several connective tissue proteins. Changes in these proteins by heating can influence the quality of the product. The most important thermolabile protein of this type is collagen. There are several reports in the literature indicating that collagen fibres shrink from a third to a quarter of their initial length at a temperature around 60 °C; at higher temperatures collagen is transformed into water-soluble gelatin.[19, 27, 28, 65] The transition of native collagen to an almost irreversibly denatured state occurs in water within a wider temperature range and more slowly than the quick shrinking process: first, a product is formed, the molecules of which can easily renature; prolonged heating leads to a practically irreversible denaturation[66] in which a dissociation of the collagen fibre into its α, β and γ components takes place.[67–9] The maximum changes in solubility and degree of denaturation during heating of collagen fibres occur between 55 and 65 °C.[66]

As mentioned above, collagen fibres shrink at a temperature around 60 °C. Laakkonen,[28] however, concluded in his review that this does not seem the case for collagen in meat; on the contrary, at 60 °C the collagen fibres seem to shrink less than at higher temperatures. Most of the studies on heat-induced changes of collagen have been carried out with tendon collagen. However, according to Mohr and Bendall[70] the internal connective tissue (endomysium) of muscle, which accounts for about 74 % of the muscle collagen, differs remarkably from tendon collagen in the following ways: it develops higher tension at thermal denaturation; the thermal contraction begins about 2–5 °C higher than tendon collagen (64·5 °C instead of 62 °C); it has a higher total number of cross-links with most of these being thermostable, whereas a considerable proportion of those in tendon are labile.

The shrinkage and partial solubilisation of the endomysial collagen of bovine muscle is initiated at about 60 °C and completed at approximately 70 °C; on the other hand, perimysial connective tissue shrinkage requires an initial temperature of 70 °C and higher before any significant fibre changes can be observed.[32, 65, 71–3] A microscopic observation of muscle tissue showed that a complete disruption of connective tissue occurred at 90 °C.[34]

Heat treatment affects not only the collageneous fibres but also the reticular tissue. As microscopic studies revealed, at 66 °C only the former tissue but not the latter one was changed. At 85 °C a decrease in the visible amount of reticular fibres was evident. The elastin fibres, however, were not altered.[74]

At 70 °C microwave energy appeared to be more effective in solubilising collagen than any of the other conventional cooking procedures (broiling, braising, roasting).[75] During prolonged heating at temperatures not above 60 °C the proteolytic activity of the tissue (collagenases) might contribute to the solubilisation of collagen.[63]

For some meat products such as canned ham the gel formation of the solubilised collagen after cooking is important. The gelation process occurs in two steps. In the first step more or less collagen is transformed to soluble gelatin; then, during cooking, a new helicoidal structure is formed.[65]

In general, it can be concluded that the denaturation of muscle collagen, resulting in shrinkage and solubilisation, occurs at higher temperatures than the denaturation of myofibrillar proteins.

RELATION OF HEAT-INDUCED PROTEIN CHANGES TO TENDERNESS AND WATER-HOLDING CAPACITY OF MEAT

The most drastic changes during heating of meat such as shrinkage and hardening of tissue and the release of juice, are caused by changes in the meat proteins. The question arises in which way that heat denaturation of the muscle proteins and of the connective tissue proteins, which was discussed in the preceding pages, is related to the heat-induced alteration of meat. This question is of practical interest in connection with the tenderness and water-holding capacity (WHC) of meat and for optimising the procedures used in meat cooking.

During heating, the muscle fibres shrink and the meat becomes tougher and harder.[28, 55, 76, 77] Heating up to 40 °C does not remarkably affect the

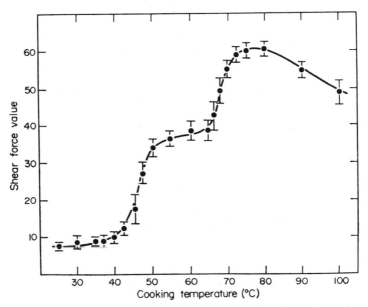

FIG. 9. The two-phase effect of cooking temperature on shear-force values. Standard deviations are given by vertical lines. Each point is the mean of 8–16 determinations from the muscle of four bulls.[32]

mechanical properties of meat.[27] Two distinctly separate phases of toughening of beef develop with increasing cooking temperature. A three- to four-fold toughening occurs in the first cooking phase between 40 and 50 °C followed by a further doubling in the second phase between 65 and 75 °C (Fig. 9). The first phase is associated with denaturation in the contractile system, indicated by the loss of actomyosin solubility (Fig. 6). The second phase is closely associated with collagen shrinkage (Fig. 10); this apparently induces shortening along the meat fibre, forcing out meat juice.[27, 32] In the temperature range between 50 and 60 °C collagen shrinkage may occur without appreciable hardening reaction and collagen solubilisation.[27] In other experiments in the temperature range between 50

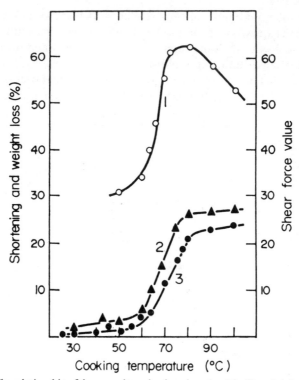

FIG. 10. The relationship of the second toughening phase (see Fig. 8) to shrinkage and weight loss. Curve 1, shear force value (mean of 6 determinations); 2, shrinkage along the fibres; 3, weight loss.[32]

and 60 °C no step but a minimum of shear force, i.e. a maximum of tenderness was observed.[27, 78, 79] The increase in tenderness found in these studies was attributed to shrinkage and fragmentation of connective tissue.[72] However, with regard to the heat-induced changes in connective tissue mentioned elsewhere, it is hard to understand that changes in the mechanical properties of meat at temperatures below 60 °C should be due to a weakening of the intramuscular connective tissue. Paul *et al.*[73] assume that in the temperature range between 60 and 82 °C the tenderness change is governed by the increasing coagulation of contractile proteins rather than by the breakdown of collageneous tissue. But as it was shown above, the coagulation of the myofibrillar proteins is almost finished at 60 °C. It is probably not coagulation but an increasing formation of stable cross-linkages in the coagulated myofibrillar proteins which contributes to such changes during heating at temperatures above 55 °C.[19]

The stepwise changes in meat tenderness during heating as demonstrated in Fig. 9 are closely related to changes in fibre diameter.[80, 81] Decrease in width of muscle fibre occurred soon after heating began and was essentially complete by 62 °C; this may be due to coagulation of myofibrillar and globular proteins. Shortening of muscle fibres proceeded slowly until about 54 °C. Between 60 and 70 °C muscle fibres shortened rather suddenly which might be due to changes in connective tissue proteins. The greater amount of shortening on heating is associated with higher shear values.[82] Above 70 °C the fibres continue to shorten, but gradually, perhaps because of the formation of such stable cross-linkages as disulfide bonds within the actomyosin system.[18, 21]

Laakkonen *et al.*[83] found that during very slow heating of bovine muscle the major decrease in tenderness does not occur if the temperature is below that at which collagen shrinks (60 °C); if the temperature is higher, then severe coagulation will cause a high weight loss, and more tightly packed, less tender tissue will be formed. The increase in tenderness during prolonged heating at 60 °C might be due to the effect of tissue collagenases.[83]

The sarcomere length of the muscle fibres of raw meat, which indicates the contractile state of muscle, is closely related to the tenderness of cooked meat,[28, 84-7] but this relationship should not be discussed here because it concerns rather the existence of a correlation of the shear force values between raw and cooked meat than the nature of meat protein changes on heating.

The considerable decrease of water-holding capacity (WHC) during heating of meat, which results in the release of juice, is due to a tightening of the myofibrillar network by heat-denaturation of the proteins.[19, 57, 88] Changes in WHC during heating are closely connected with alterations in tenderness[79, 89, 90] and rigidity of tissue.[19, 55] Most decrease in WHC occurs between 30 and 50 °C (Fig. 11),[19, 79] i.e. in a temperature range in which the myofibrillar proteins coagulate. This range corresponds to the first phase of the toughness curve presented in Fig. 8. From this correspondence it can be concluded that up to 50 °C the toughening effect of heating is caused by a dehydration of the myofibrillar system. It is interesting that a decrease in width but not in length of the muscle fibres accompany this maximum loss of WHC.[81] A smaller proportion of juice is released between 55 and 90 °C.[19, 55, 79] This might be partially due to a shortening of muscle fibres by changes in the connective tissue (Fig. 10), and also to an increased formation of new cross-linkages in the coagulated myofibrillar system.[19]

The mechanism of the fluid discharge from meat by cooking is determined by the degree of pre-rigor shortening (e.g. 'cold shortening').[86] The changes in the mechanical properties of meat and in WHC do not proceed continuously. Increase in toughness (Fig. 9) as well as decrease in WHC (Fig. 11) show a characteristic step between 50 and 55 °C[19, 32] which indicates a stop or retardation of changes in this temperature range. It is

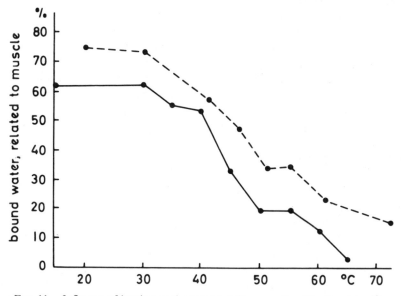

FIG. 11.   Influence of heating on the water-holding capacity of beef muscles.[19]

interesting that a number of chemical changes occur during heating such as an increase in pH, decrease of available acidic protein groups and release of Ca^{2+} and Mg^{2+} from proteins, indicating also a step between 50 and 55 °C.[19, 57] The time–temperature curve measured during the heating of meat also reveals a step between 50 and 55 °C.[91] In this temperature range the coagulation of myofibrillar proteins is almost completed but remarkable changes in connective tissue proteins have not yet started. Therefore, the step decreases in tenderness and in WHC seem to indicate the transition from actomyosin denaturation to collagen denaturation. It should also be mentioned, however, that with actomyosin preparations a stepwise course of heat-induced changes such as a decrease in available acidic groups[19] or protein–protein interaction[23] (Fig. 5) were observed. Therefore, it seems possible that the step of heat-induced changes between

50 and 55 °C marks the transition from the coagulation process of actomyosin to a closer protein–protein interaction under the formation of new stable cross-linkages within the actomyosin.

In conclusion it can be stated that changes of tenderness, rigidity and WHC of meat on heating occur in two phases: the first phase being between 30 and 50 °C, and the second between 60 and 90 °C. In the temperature range between 50 and 55 °C negligible changes occur. Changes in the first phase are due to heat coagulation of the actomyosin system. The second phase seems to be due to denaturation of the collageneous system (shrinking and solubilisation of collagen) and/or to the formation of new stable cross-linkages within the coagulated actomyosin system. Further work will be necessary to decide to what extent the actomyosin changes and collagen denaturation contribute to tenderness changes and decrease in WHC in the second phase (temperatures above 55 °C).

DENATURATION OF SARCOPLASMIC PROTEINS DURING HEATING OF MEAT: CHANGES IN MEAT COLOUR

The remarkable changes of meat quality factors such as tenderness, and WHC during heating are mainly due to protein alteration in the myofibrillar and collageneous system of tissue, as was discussed in the preceding pages. Heat denaturation of the sarcoplasmic proteins, i.e. of muscle proteins soluble in pure water or at low ionic strength, is certainly of minor importance in view of these changes. But it seems possible that the denaturation of the sarcoplasmic proteins may contribute to some extent to the consistency of cooked meat. Heat coagulation of these proteins results in the formation of gels which can link structural muscle elements.[32]

Most of the sarcoplasmic proteins coagulate between 40 and 60 °C.[19, 55, 90] According to Bolshakov et al.[30] about 85 % of the protein extractable at low ionic strength from pork is precipitated at 60 °C. Davey and Gilbert[32] found with bull neck muscle, however, that in contrast to the distinct heat denaturation curve of myosin, that heat denaturation of the sarcoplasmic proteins extended from approximately 40 to 90 °C (Fig. 6). Thus denaturation of those proteins spans both the toughness phases described above (Fig. 9), and it is not possible to associate it with one or the other. The role of denaturation of sarcoplasmic proteins, if any, in the changes of tenderness and WHC is not yet clarified.

Lee and Grau[92] showed that the bovine sarcoplasmic proteins which migrated with the highest velocity in an electric field were more easily denatured at 50 and 66 °C and that the cathodic proteins were more stable. Also Laakkonen et al.[93] observed that anodic proteins coagulate first, but the slow-moving fraction coagulates faster than the fast-moving one. Apparently the type and number of protein ionic charges influence protein stability against heating. This explains also the strong influence of the pH of meat on the extent of WHC loss by heating.[19, 57]

The heat denaturation of sarcoplasmic haem proteins (mainly myoglobin) is of particular importance because it determines the change of meat colour during cooking from red to greyish brown. Myoglobin in muscle coagulates at about 65 °C, i.e. at a temperature lower than the temperature at which the pigment in pure solution is denatured.[19, 56, 94] Myoglobin apparently co-precipitates with other muscle proteins and part of the thermal denaturation behaviour of myoglobin has to be attributed to the denaturation of the other meat proteins.[94, 95] Heat brings about the transfer of haematin from metmyoglobin to any of several proteins found in meat. In cooking meat the haem proteins are mainly di-imidazole complexes, the imidazole residues being supplied by the histidine group of the bound proteins.[96] Draudt[94] found that the precipitation of myoglobin in meat depends on the nature of the ligand of the sixth position of the porphyrine and on the oxidative state of the iron.

The three myoglobins evident in the electrophoretic pattern are not affected by heating bovine muscle to 55 °C for 5 min.[93] After heating, more haem-positive bonds appeared in the electropherogram of the sarcoplasmic extract,[56] an observation which is not yet fully understood.

Roberts and Lawrie[55] suggested that the observed initial resistance of myoglobin to heat denaturation and its subsequent loss of electrophoretic mobility between 75 and 88 °C could be developed as an index of the temperature attained by meat products.

RELATIONSHIP OF HEATING MEAT TO THE ACTIVITY OF MUSCLE ENZYMES: DETERMINATION OF THE DEGREE OF HEATING

The heating of meat causes a denaturation of muscle enzymes which results in a loss of enzymatic activities.[19] In meat canning it is important not only to destroy microorganisms but also to eliminate the activity of such muscle enzymes as proteinases which may exert a detrimental influence on meat quality. The stability of the numerous muscle enzymes to heating is quite different. The decrease in ATPase activity of myosin and actomyosin during thermal treatment has already been discussed above.

Considering the importance of the detrimental effect of tissue proteinases in insufficiently sterilised meat, it is surprising that very little information is published on the heat inactivation of this group of enzymes. The porcine muscle cathepsins were found to be stable at temperatures under 37 °C. The optimum temperature of these enzymes was about 48 °C,[97] but these experiments were carried out not with meat but with the enzyme fraction isolated from muscle. As Penfield and Meyer[63] found, during thermal treatment of beef the proteolytic activity of tissue measured by the effect on cowhide-azo dye (collagenase activity) remained unchanged until 60 °C and decreased remarkably between 60 and 70 °C. Apparently, tissue proteinases are inactivated at 70–73 °C.[19]

Lipases and carboxyl esterhydrolases do not survive the usual harsh conditions of sterilisation (110 °C) of pork even in very fatty tissues.[98] The inactivation of enzymes responsible for oxygen reception in porcine muscle takes place at 65 °C within 10 min.[99] Inactivation of the diphosphatases and triphosphatases of bovine muscle by heat occurs at about 40 °C. Between 55 and 60 °C these enzymes, which dephosphorylise added tri- and di-phosphate to monophosphate, are almost completely inactivated.[100]

The activities of several enzymes have been suggested for monitoring the pasteurisation and sterilisation of meat and meat products. The acidic phosphatase was thoroughly studied and found to be of some use particularly for quality control of canned ham (Fig. 12).[29, 101–4] This method, however, is not very reliable.[102, 103] Therefore, Suvakov *et al.*[105] concluded from their results that the determination of acid phosphatase activity should not be adopted as a method for proof of maximum attained temperature in the centre of canned ham.

Another group of enzymes used as an indicator of sufficient thermal treatment of meat products are the acetyl and butyryl esterases, the activity of which is measured with indoxylbutyrate, α-naphthylacetate or indoxylacetate as substrates. These enzymes are recommended for testing canned ham, frankfurters and liver sausages.[106–111] However this method is problematical; so, during storage of meat products some reactivation of carbesterase can occur.[111] Another enzyme suitable for this purpose is peroxidase which allows a simpler and faster procedure than the determination of acidic phosphatase activity.[112, 113] The relatively high heat-stability of aspartate aminotransferase of muscle[19] has not yet been proved as an indicator of sterilisation temperatures reached in meat.

All these enzyme tests, however, bring about difficulties because of the

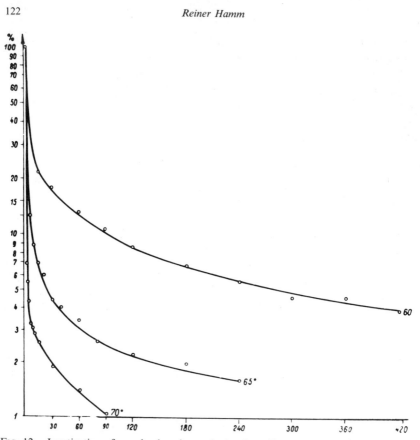

FIG. 12.   Inactivation of muscle phosphatase by heating of ham, wrapped in thin cellothen foil. (*M. semimembranosus* in the post rigor state, pH 6.2). 3 % NaCl was added to the minced meat. The activity of the unheated meat was 1151 phosphatase units.[101] Ordinate: percent of the original activity; abscissa: time of heat treatment (min).

influence of different factors such as pH, storage of meat before processing, storage of can etc., and the effects of NaCl, $NaNO_2$ and other additives on the activity of the enzyme and its heat stability.

BROWNING REACTIONS

Besides heat-induced changes of haem pigments, non-enzymatic browning reactions of the Maillard type (interaction of carbonyl compounds with amino groups)[58, 59] may also contribute to the colour changes during the heating of meat and meat products. Meat contains small amounts of carbohydrate originating mainly from glycogen (glucose, hexose phosphates and other intermediates of the glycogen metabolism) and nucleotides (ribose). As Hamm[19] found, the heating of tissue and myofibrils at temperatures above 80 °C results in a loss of dye-binding basic groups. It is possible that this decrease is the result of Maillard reactions. Browning of muscle proteins begins at about 90 °C and increases with temperature and time of heating, causing a loss of amino acids and of sugars from the tissue.[19]

The browning of meat depends on the pH. An increase or decrease in pH from its normal level increases the intensity of brown colour formation on heating.[114] The browning of pork with or without added sugars is strongest when roasting in fat, lower when heating in air and minimum when boiling. The addition of sucrose or glucose yielded the same browning intensity in thermal processing.[115] During the cooking of pork up to 100 °C under different conditions Savic and Kepcija[116] observed that the emergence of browning was greatest at 60 °C. This result is surprising; probably in these experiments the true browning reaction was not observed but the colour change by haem pigment alteration.

As Ziemba[114] mentioned in his review, the browning of meat and fish is characterised by the participation of metabolites from oxidative phosphorylation. Browning of meat might be due to a combination of Maillard reactions and caramelisation of sugars.[114, 117] Browning due to heating of pork results from an amino–sugar reaction. This was proved by blocking amino groups with acetylation and carbonyl group of sugars by bisulfite or hydroxylamine addition; by using yeast fermentation to remove free sugars naturally present in meat; or by adding glucose oxidase.[117] However, a small but significant amount of browning seemed to be due to pyrolysis of the natural meat sugars. The predominant role of reducing sugars in the browning of meat was also confirmed by other authors.[118–20] But an absolute classification of sugars and proteins with regard to the Maillard reaction is not possible.[119] Of course, the temperature and duration of heating are very important for the extent of Maillard reactions in meat.[121] Sugar–protein reactions of the Maillard type mean a loss of amino acids. So, during the manufacturing of meat meal by batch dry rendering (maximum temperature about 130 °C) the browning reaction takes place and the availability of the essential amino acids lysine, histidine, threonine, methionine and cystine is reduced.[122] Removal of the reducing sugars in the raw material used for meat meal manufacturing by yeast or sulfite, prevents the Maillard reactions. The resultant meat meals have a similar available amino acid content and nutritive value to chicks as the raw material.[120] The loss of amino acids by the heating of meat will be discussed in the next section.

Maillard reactions to some extent give meat a desirable colour but they also contribute to the development of the flavour and aroma of cooked meats. Extensive work has been carried out in order to elucidate the compounds responsible for the formation of meat flavour by heating; in numerous heating experiments with meat, meat extracts and model systems a great number of volatile compounds have been isolated which could potentially contribute to flavour.[123] It is not clear, however, whether the free amino acids and also the proteinaceous amino acids—and if so, to what extent—participate in the formation of volatile meat flavour compounds. Certainly the sulfur-containing amino acids of meat play a predominant role as precursors for meat flavour components.[123] In this connection H_2S, split off from proteins by heating, may be of importance.

INFLUENCE OF HEATING ON THE AMINO ACID CONTENT OF MEAT

Maillard reactions and also the release of H_2S from proteins may cause loss of essential amino acids on heating meat and, therefore, a decrease in its nutritive value. The question arises whether such detrimental effects occur during normal cooking and canning of meat and meat products.

As Hofmann[124, 125] reported in his reviews on the effect of heating on the amino acid content of meat, normal cooking does not have much influence on the essential amino acids. This conclusion was confirmed by additional research work.[39, 121, 126–8] Also ultra-high frequency treatment of meats did not influence the amino acid composition.[128]

Such a small effect of cooking is observed if the amino acid content is related to protein (N × 6·25).[129] But if the amino acid content is related to 100 g raw meat, the cooking of beef in a fluid (NaCl solution) at 100 or 120 °C causes a significant decrease of all amino acids (2–28 %), particularly of methionine and cystine. Therefore, the normal cooking of meat causes some loss in nutritive value,[129] but except for the sulfur-containing amino acids the loss of the other essential amino acids was less than 10 %. The loss of the non-essential amino acids was of the same magnitude. Similar results were obtained with broiler meat.[130] The decrease of methionine, cystine, lysine and available lysine during moist heating of meat at 120 °C was also noted by Gaudy and Landis;[131] the loss of the other essential amino acids was rather slight.

The baking of meat resulted in a higher loss of amino acids—particularly cysteine, tryptophan, tyrosine and glutamic acid—than boiling, whereas arginine, threonine and aspartic acid were most stable.[126, 132]

In cooking meat dishes the destruction of amino acids can be higher than in cooking the meat itself because of the detrimental influence of added carbohydrates present in sauces, breadcrumbs etc. Frozen meat dishes rewarmed at 90 °C showed lower contents of essential amino acids, particularly of phenylalanine, tryosine and isoleucine, than those rewarmed at 70 °C; keeping the dishes for one hour at 70 °C caused an additional loss.[133] Beef treated with egg and breadcrumbs and roasted, contained 35 % less lysine than the same meat after normal roasting.[134]

Heating at higher temperatures for a longer time, as is often used for full sterilisation of canned meat, damages essential amino acids, particularly cystine, cysteine, methionine and lysine.[130, 135-40] Sterilisation of beef caused a 10–15 % decrease in lysine content; in the presence of glucose, 90 % was lost.[134]

As Hofmann[124, 125] stated, the digestibility of meat proteins can be lowered by heating even if the amino acid content is not changed. Changes in the spatial arrangement of the molecular protein structure without changes in the primary structure are probably sufficient to lower the digestibility of proteins.[59] As to the normal cooking procedures, in a number of experiments a small loss in digestibility was found[39, 58, 124, 125, 141, 142] whereas in other experiments none was found.[124, 125, 143] At higher temperature, however, as used in meat sterilisation, a decrease in digestibility and in the biological value was regularly observed.[124, 125, 144]

Severe heating also causes damage to proteins in the absence of carbohydrates. It is believed that heat causes the formation of new enzyme-resistant cross-linkages within the protein molecule, so reducing the digestibility and the biological availability of some constituent amino acids. The most likely cross-linking reaction seems to be the formation of new 'isopeptide' bonds by reaction of the α-amino group of lysine with either the carboxyl group of aspartic or glutamic acids or more probably with the amide groups of glutamine and asparagine. Other cross-linkages may result from the degradation of cystine. Heat can cause fission of the SS bond, yielding dehydroalanine which may condense with cysteine to form lanthionine.[142] In heated chicken muscle ε-N-(β-L-aspartyl)-L-lysine and ε-N-(γ-L-glutamyl)-L-lysine were determined; their content was found to increase as the material was subjected to more treatment (autoclaving at 121 °C for up to 27 h). Lanthionine, however, was not found.[142]

In conclusion, normal cooking of meat results only in minor changes of amino acids and digestibility, but under the conditions of heat sterilisation a proportion of the essential amino acids is damaged and the digestibility is reduced. Sulfur-containing amino acids seem to be the most sensitive to thermal processing at higher temperatures.

REFERENCES

1. Hofmann, K. and Hamm, R. Sulfhydryl and disulfide groups in meats. *Advances in Food Research*, in press.
2a. Samejima, K. and Yasui, T. (1975). Some additional studies on the thermal denaturation of light meromyosin fraction 1. *J. Food Sci.*, **40**, 451.
2b. Yasui, T., Yazawa, Y., Takahashi, K. and Samejika, K. (1972). Thermostability of light meromyosin. *Japan. Bull. Meat and Meat Products*, **7**, 34.
3a. Woods, E. F. (1969). Studies on the denaturation of tropomyosin and light meromyosin. *Protein Res.*, **1**, 29.
3b. Samejima, K., Morita, J., Takahashi, K. and Yasui, T. (1972). Denaturation of myosin and its subfragments: II. Effect of thermal treatment on light meromyosin. *J. Biochem.* (Tokyo), **71**, 661.
4. Lehrer, C. S. and Kerwar, G. (1972). Intrinsic fluorescence of actin. *Biochem.*, **1**, 1211.
5. Yasui, T., Fukazawa, T., Hashimoto, Y., Kitagawa, S. and Sasaki, A. (1958). Denaturation of myosin B. *J. Biochem.* (Tokyo), **45**, 717.
6. Blum, J. J. (1960). Interaction between myosin and its substrates. *Arch. Biochem. Biophys.*, **87**, 104.
7. Penny, J. F. (1967). The effect of post-mortem conditions on the extractability and ATPase activity of myofibrillar proteins of rabbit muscle. *J. Food Technol.*, **2**, 325.
8a. Levy, H. M. and Ryan, E. M. (1967). Effect of temperature on the rate of hydrolysis of adenosine triphosphate and inosine triphosphate by myosin with or without modifying additions. *Science*, **156**, 73.
8b. Levy, H. M., Sharon, N., Ryan, E. M. and Koshland, D. E. (1962). Inhibition and reversal of heat inactivation of the relaxation site of actomyosin by dithiothreitol. *Biochim. Biophys. Acta*, **56**, 118.
9. Hartshorne, D. J., Barns, E. M., Parker, L. and Fuchs, F. (1972). The effect of temperature on actomyosin. *Biochim. Biophys. Acta*, 267, 190.
10. Barany, M. (1967). ATPase activity of myosin correlated with speed of muscle shortening. *J. Gen. Physiol.*, **50** (No. 6, part 2), 197.
11. Kominz, D. R. (1970). Studies of adenosine triphosphatase activity and turbidity in myofibril and actomyosin suspensions. *Biochem.*, **9**, 1792.
12. Jacobson, A. L. and Henderson, J. (1973). Temperature sensitivity of myosin and actomyosin. *Canad. J. Biochem.*, **51**, 71.
13. Buttkus, H. (1971). The sulfhydryl content of rabbit and trout myosins in relation to protein stability. *Canad. J. Biochem.*, **49**, 97.

14. Jacobson, A. L. and Henderson, J. (1971). Autoxidation of actomyosin. *Canad. J. Biochem.*, **49**, 1264.
15. Yasui, T., Gotoh, T. and Morita, J. (1973). Influence of pH and temperature on properties of myosin A in glycerol-treated fibre bundles. *J. Agric. Food Chem.*, **21**, 241.
16. Yasui, T., Kawakami, H. and Morita, F. (1968). Thermal inactivation of myosin A-ATPase in the presence of F-actin. *Agric. Biol. Chem.*, **32**, 225.
17. Young, L. L. (1974). Heat stability of chicken actomyosin. *J. Food Sci.*, **39**, 389.
18. Hamm, R. and Hofmann, K. (1965). Changes in the sulfhydryl and disulfide groups in beef muscle protein during heating, *Nature*, **207**, 1269.
19. Hamm, R. (1966). Heating of muscle systems. In *The Physiology and Biochemistry of Muscle as a Food*, ed. by E. J. Briskey, R. G. Cassens and J. C. Trautman, University of Wisconsin Press, Madison, Wisconsin.
20. Schrott, F. (1974). The effect of salts on myofibrillar protein during thermal treatment (German). Thesis. University of Munich.
21. Dubé, G., Bramblett, V. D., Judge, M. D. and Harrington, R. B. (1972). Physical properties and sulfhydryl content of bovine muscles. *J. Food Sci.*, **37**, 23.
22. Samejima, K., Hashimoto, Y., Yasui, T. and Fukazawa, T. (1969). Heart gelling properties of myosin, actin, actomyosin and myosin subunits in a saline model system. *J. Food Sci.*, **34**, 242.
23. Deng, J. C., Toledo, R. T. and Lillard, D. A. (1976). Effect of temperature and pH on protein–protein interaction in actomyosin solution. *J. Food Sci.*, **41**, 273.
24. Sato, A. and Mihashi, K. (1972). Thermal modification of structure of tropomyosin: I. Changes in the intensity and polarisation of the intrinsic fluorescence (tyrosine). *J. Biochem.* (Tokyo), **71**, 597.
25. Hamm, R. (1965). Influence of heating on animal tissues (German). *Dechema Monograph*, **56**, 159.
26. Hamm, R. (1970). Properties of meat proteins. In *Proteins as Human Food*, ed. by R. A. Lawrie, Butterworths, London.
27. Draudt, K. N. (1972). Changes in meat during cooking. *Proc. 25th Ann. Reciprocal Meat Confer.*, Amer. Meat Sci. Assoc., Ames, Iowa, p. 243.
28. Laakkonen, E. (1973). Factors affecting tenderness during heating of meat. *Advances in Food Research*, **20**, 257.
29. Cohen, E. H. (1966). Protein changes related to ham processing temperatures: I. Effect of time–temperature on amount and composition of soluble proteins. *J. Food Sci.*, **31**, 746.
30. Bolshakov, A., Khlebnikov, V. and Mitrofemov, N. (1968). Alterations in muscle and connective tissue proteins and cured pork during cooking (Russian). *Myas. Ind. SSSR*, **39**(6), 34; *Chem. Abstr.*, **69**, 85519y (1968).
31. Acton, J. C. (1972). Effect of heat processing on extractability of salt-soluble protein, tissue binding strength and cooking loss in poultry meat. *J. Food Sci.*, **37**, 244.
32. Davey, C. L. and Gilbert, K. V. (1974). Temperature-dependent toughness in beef. *J. Sci. Food Agric.*, **25**, 931.

33. Mihalyi, V. (1967). Protein changes occurring at heat treatment of cured meat products. *Yearbook of the Hungarian Meat Res. Inst.*, **1967**, 95.
34. Chrystall, B. B. (1971). Macroscopic, microscopic and physico-chemical studies of the influence of heating on muscle tissues and proteins. *Dissertation Abstr.*, *Internat. Sect. B, Sci.* and *Engng.*, **31**, 6050.
35. Hofmann, K. (1973). Identification and determination of meat proteins and added proteins by the sodium dodecylsulfate polyacrylamide gel electrophoresis (German). *Z. analyt. Chem.*, **267**, 355.
36. Kako, Y. (1968). Studies on muscle proteins: II. Change of beef, pork and chicken protein during the meat products manufacturing processes. *Mem. Fac. Agric., Kagoshima University*, **6**, 175.
37. Kovaleva, T. A. (1967). Changes in the number of SH groups in muscles on heat alteration (Russian). *Tsiotologiya*, **9**, 1018; *Chem. Abstr.*, **68**, 46210a (1968).
38. Malyutin, A. F. (1969). Changes in the content of sulfhydryl groups in meat under super-high frequency heating. (Russian). *Primen. Sverkhvysokochastatn. Nagreva Obshch. Pit.*, **1969**, 29; *Chem. Abstr.* **75**, 18805z (1971).
39. Dworschak, E. (1970). Studies on the biological value of the protein of raw and roasted meat (German). *Z. Lebensmittel-Untersuch. Forsch.*, **143**, 166.
40. Randall, C. J. and Bratzler, L. J. (1970). Changes in various protein properties of pork muscle during the smoking process. *J. Food Sci.*, **35**, 248.
41. Tinbergen, B. J. (1970). Sulfhydryl groups in meat proteins. *Proc. 16th European Meat Research Workers' Meeting*, Varna (Bulgaria), Vol. 1, 576.
42. Bowers, J. A. (1972). Eating quality, sulfhydryl content and TBA values of turkey breast muscle. *J. Agric. Food Chem.*, **20**, 706.
43. Bognar, A. (1971). Contribution to the evaluation of the nutritive value of meat and its dependence on the thermal treatment (German). Thesis. University of Hohenheim.
44. Krylova, N. M. and Kusnezova, W. M. (1964). Changes of the sulfhydryl groups in meat as influenced by kind of treatment. *Proc. 10th European Meat Research Workers' Meeting*, Roskilde (Denmark), Report G 12.
45. Marchenko, A. P. (1968). Change of sulfur-containing compounds during sterilisation of beef and lamb (Russian). *Tr. Vses. Nauch.-Issled. Inst. Myas. Prom.* **1968**, 84; *Chem. Abstr.*, **74**, 12376j (1971).
46. Fraczak, R. and Pajdowski, Z. (1955). The decomposition of sulfhydryl groups in meat by thermal processes (Polish). *Przemysl Spozywczy*, **9**, 334.
47. Mecchi, E. P., Pippen, E. L. and Linewweaver, H. (1964). Origin of hydrogen sulfide in heated chicken muscle. *J. Food Sci.*, **29**, 393.
48. Parr, L. J. and Levett, G. (1969). Hydrogen sulphide in cooked chicken meat. *J. Food Technol.*, **4**, 283.
49. Lendvai, J., Mihalyi-Kengyel, V. and Zukal, E. (1973). Changes in the content of hydrogen sulfide of heat-treated meat protein preparations (Hungarian). *Elelmiszervizsgálati Közlemények*, **19**, 185.
50. Sowa, T. (1968). Studies on the decomposition of thiol groups in meat during thermal treatment at 90–150 °C (Polish). *Przemysl Spozywczy*, **22**, 27.
51. Johnson, A. R. and Vickery, J. R. (1964). Factors influencing the production of hydrogen sulphide from meat during heating. *J. Sci. Food Agric.*, **15**, 695.

52. Krylowa, N. N. and Marchenko, A. P. (1969). Effect of pH value on changes in sulphur-containing compounds in meat during sterilisation (Russian). *Tr. Vses. Nauch-Issled. Inst. Myas. Prom.*, **1969**, 97; *Food Sci. Technol. Abstr.*, **4**, 4 S 369 (1972).
53. Kunsman, J. E. and Riley, M. L. (1975). A comparison of hydrogen sulfide evolution from cooked lamb and other meats. *J. Food Sci.*, **40**, 506.
54. Marchenko, A. P. and Kosenjasheva, T. (1974). Changes in the amount of hydrogen sulfide and mercaptans in beef and pork (Russian). *Trudy* (Moskva), **29**, 87.
55. Roberts, P. C. B. and Lawrie, R. A. (1974). Effects on bovine l. dorsi muscle of conventional and microwave heating. *J. Food Technol.*, **9**, 345.
56. Fogg, N. E. and Harrison, D. L. (1975). Relationships of electrophoretic patterns and selected characteristics of bovine skeletal muscle and internal temperature. *J. Food Sci.*, **40**, 28.
57. Hamm, R. (1972). *Colloid Chemistry of Meat* (German), Verlag P. Parey, Hamburg and Berlin.
58. Baltes, W. (1971). Influence of heat-sterilisation on proteins (German). *Wissenschaftl. Veröffentlich. d. Dtsch. Gesellschaft für Ernährung*, **21**, 112.
59. Baltes, W. (1976). Changes of foods during canning (German). *Fleischwirtschaft*, **56**, 298.
60. Karmas, E. and Dimarco, G. R. (1970). Denaturation thermoprofiles of some proteins. *J. Food Sci.*, **35**, 725.
61. Ozawa, S., Yano, S., Abe, T., Mogi, K. (1972). Differentiation of animal species by immunological method: II. Experiments on cooked meats. *Nippon Chikusan Gakkai-ho*, **43**, 395 (1972); *Food Sci. Technol. Abstr.*, **5**, 1 S 9 (1973).
62. Paul, P. C., Buchter, L. and Wierenga, A. (1966). Solubility of rabbit muscle protein after various time–temperature treatments. *J. Agric. Food Chem.*, **14**, 490.
63. Penfield, M. P. and Meyer, B. H. (1975). Changes in tenderness and collagen of beef semitendinosus muscle heated at two rates. *J. Food Sci.*, **40**, 150.
64. Mihalyi-Kengyel, V., Zukal, E. and Körmendy, L. (1975). The breakdown of myofibrillar proteins by severe heating. *Acta Alimentaria*, **4**, 367.
65. Kopp, J. (1971). Influence of temperature, cooking-time and salt concentration on the solubility of collagen in porcine muscle (German). *Fleischwirtschaft*, **51**, 1647.
66. Hörmann, H. and Schlebusch, H. (1968). Denaturation of collagen within the fibre, investigated on a molecular base (German). *Hoppe Seyler's Z-Physiol. Chem.*, **349**, 179.
67. Tristram, G. D., Worrall, J. and Steer, D. C. (1965). Thermal denaturation of soluble cowhide collagen. *Biochem. J.*, **95**, 350.
68. Grassmann, W. (1965). Denaturation and renaturation of collagen (German). *Leder*, **16**, No. 2, 32.
69. Monaslidze, D. R. and Bakradze, N. G. (1970). Intermolecular collagen (procollagen) melting (Russian). *Soobshch. Akad. Nauk. Gruz. SSR*, **58**, No. 3, 58; *Chem. Abstr.*, **73**, 126963s (1970).
70. Mohr, V. and Bendall, J. R. (1969). Constitution and physical chemical properties of intramuscular connective tissue. *Nature*, **223**, 404.
71. Goll, D. E., Hoekstry, W. O. and Bray, R. W. (1964). Age-associated changes

in bovine muscle connective tissue: I. Exposure to increasing temperature. *J. Food Sci.*, **29**, 515.

72. Schmidt, J. G. and Parrish, F. C. (1971). Molecular properties of post-mortem muscle: 10. Effect of internal temperature and carcass maturity on structure of bovine longissimus. *J. Food Sci.*, **36**, 110.

73. Paul, P. C., McCrae, S. E. and Hofferber, L. M. (1973). Heat-induced changes in extractability of beef muscle collagen. *J. Food Sci.*, **38**, 66.

74. Deethardt, D. and Tuma, H. J. (1971). A histological evaluation of connective tissue in longissimus dorsi muscle from raw and cooked pork. *J. Food Sci.*, **36**, 563.

75. McCrae, S. E. and Paul, P. C. (1974). Rate of heating as it affects the solubilisation of beef muscle collagen. *J. Food Sci.*, **39**, 18.

76. Hegarty, P. V. J. and Allen, C. E. (1975). Thermal effects on the length of sarcomeres in muscle held at different tensions. *J. Food Sci.*, **40**, 24.

77. Hegarty, P. V. J. and Allen, C. E. (1976). Comparison of different post-mortem temperatures and dissection procedures on shear values of unaged and aged avian and ovine muscles. *J. Food Sci.*, **41**, 237.

78. Schmidt, J. G., Kline, E. A. and Parrish, F. C. (1970). Effect of carcass maturity and internal temperature on bovine longissimus attributes. *J. Animal Sci.*, **31**, 861.

79. Bouton, P. E. and Harris, P. V. (1972). The effects of cooking temperature and time on some mechanical properties of meat. *J. Food Sci.*, **37**, 140.

80. Hostetler, R. L. (1967). A method to study the effect of heat on muscle fibres. *Proc. 20th Ann. Reciprocal Meat Confer.*, Amer. Meat Sci. Assoc., Lincoln, Nebraska, p. 221.

81. Hostetler, R. L. and Landmann, W. A. (1968). Photomicrographic studies of dynamic changes in muscle fibre fragments: I. Effect of various heat treatments on length, width and birefringence. *J. Food Sci.*, **33**, 468.

82. Field, R. A., Pearson, A. M. and Schweigert, B. S. (1970). Labile collagen from epimysial and intramuscular connective tissue as related to Warner–Bratzler shear values. *J. Agric. Food Chem.*, **18**, 280.

83. Laakkonen, E., Sherbon, J. W. and Wellington, G. H. (1970). Low-temperature, long-time heating of bovine muscle: 3. Collagenolytic activity. *J. Food Sci.*, **35**, 181.

84. Hegarty, P. V. J. and Allen, C. E. (1972). Rigor-stretched turkey muscles: effect of heat on fibre diameter and shear value. *J. Food Sci.*, **37**, 652.

85. Bouton, P. E., Harris, P. V., Shorthose, W. R. and Ratcliff, D. (1974). Changes in the mechanical properties of veal muscles produced by myofibrillar contraction state, cooking temperature and cooking time. *J. Food Sci.*, **39**, 869.

86. Davey, C. L. and Gilbert, K. V. (1975). Cold shortening and cooking changes in beef. *J. Sci. Food Agric.*, **26**, 761.

87. Davey, C. L. and Gilbert, K. V. (1975). The tenderness of cooked and raw meat from young and old beef animals. *J. Sci. Food Agric.*, **26**, 953.

88. Hamm, R. (1960). Biochemistry of meat hydration. *Advances in Food Research*, **10**, 355.

89. Rogers, P. J., Goertz, G. E. and Harrison, D. L. (1967). Heat-induced changes of moisture in turkey muscles. *J. Food Sci.*, **32**, 298.

90. Laakkonen, E., Wellington, G. H. and Sherbon, J. W. (1970). Low-temperature, long-time heating of bovine muscle: 1. Changes in tenderness, water-holding capacity, pH and amount of water-soluble compounds. *J. Food Sci.*, **35**, 175.
91. Tyszkiewicz, St., Tyszkiewicz, I. and Dukalwska, M. (1966). Model experiments on the influence of the main parameters of the pasteurisation process upon the meat exudate and the denaturation of meat protein. *Roczniki Instytutu Przemyslu Miesnego*, **3**, No. 1, 24.
92. Lee, A. and Grau, R. (1966). Behaviour of bovine sarcoplasma during heating (German). *Fleischwirtschaft*, **46**, 1239.
93. Laakkonen, E., Sherbon, J. W. and Wellington, G. H. 1970. Low-temperature, long-time heating of bovine muscle. 2. Changes in electrophoretic patterns. *J. Food Sci.*, **35**, 180.
94. Draudt, K. N. (1969). Effect of heating on the behaviour of meat pigments. *Proc. 22nd Ann. Reciprocal Meat Confer.*, Amer. Meat Sci. Assoc., Chicago, Illinois, p. 180.
95. Ledward, D. A. (1971). On the nature of cooked meat haemoprotein. *J. Food Sci.*, **36**, 883.
96. Ledward, D. A. (1974). On the nature of the haematin–protein bonding in cooked meat. *J. Food Technol.*, **9**, 59.
97. Deng, J. C. and Lillard, D. A. (1973). The effect of curing agents, pH and temperature on the activity of porcine muscle cathepsins. *J. Food Sci.*, **38**, 299.
98. Hottenroth, B. (1974). On the heat inactivation of the lipase and other carboxyester hydrolases in canned meat products with high fat content (German). *Fleischwirtschaft*, **54**, 1071.
99. Chomiak, D. (1971). Investigations on thermal inactivation of oxygen reception process in porcine muscle. *Roczniki Instytutu Przemyslu Miesnego*, **8**, No. 2, 73.
100. Neraal, R. (1975). On the enzymatic breakdown of tripolyphosphate and diphosphate in minced meat (German). Thesis. University of Giessen.
101. Gantner, G. and Körmendy, L. (1968). Method for checking the sufficient heating of meat products by means of the phosphatase assay (German). *Fleischwirtschaft*, **48**, 188.
102. Körmendy, L. and Jacquet, B. (1968). Methods for the determination of the degree of heating of ham by means of the phosphatase activity (French). *Ann. Technol. Agric.*, **17**, 159.
103. Cohen, E. H. (1969). Determination of acid phosphatase activity in canned hams as indicator of temperatures attained during cooking. *Food Technol.*, **23**, 961.
104. Cantoni, C. and Frittoli, A. 1974. Use of enzyme methods for determining the degree of cooking of cooked hams (Italian). *Industrie Alimentari*, **13**, 80.
105. Suvakov, M., Visacki, V., Marinkov, M. and Korolija, S. (1967). Relation of acid phosphatase content to heat treatment of cured meat. *Technol. mesa*, **8**, 306.
106. Purr, A. (1965). Test paper for the assay of esterases in animal and plant tissues and micro-organisms (German). *Nahrung*, **9**, 445.
107. Pfeiffer, G., Kotter, L., Böhm, D. and Glatzel, P. (1969). On the

determination of preceding heating of meat and meat products using a new carbesterase test (German). *Fleischwirtschaft*, **49**, 209.

108. Glatzel, P. (1970). A simple test for the assay of indoxyl acetate cleaving enzymes (German). Thesis. University of Munich.

109. Singrün, B. (1970). A simple carbesterase test using naphthyl acetate as substrate for the determination of the degree of heating of meat. Thesis. University of Munich.

110. Brehmer, H. and Forschner, E. (1974). A modified carbesterase screening-test for controlling the degree of heating of frankfurters and liver sausages (German). *Fleischwirtschaft*, **54**, 85.

111. Pfeiffer, G., Wellhäuser, R. and Gehra, H. (1970). Further experiences in the carbesterase heating-test with frankfurter-type sausages (German). *Arch. Lebensmittelhyg.*, **21**, 250.

112. Finogenova, N. V., Markova, O. V., Mikhailova, L. and Trusevich, Z. G. (1973). Control methods in heat treatment of meat products. (Russian). *Gigiena i Sanitariya*, **1973**, Nr. 10, 106; *Food Sci. Technol.*, *Abstr.*, **6**, 1 S 82 (1974).

113. Livshits, O. R. (1968). Control over the quality of thermal processing of meat products by determining the peroxidase activity (Russian). *Vop. Pitan.*, **27**, No. 2, 80; *Chem. Abstr.*, **69**, 18073*k* (1968).

114. Ziemba, Z. (1966). Browning reactions in foods (Polish). *Przemysl Spozywczy*, **142**, 82; *Z. Lebensmittel-Untersuch.-Forsch.*, **142**, 82 (1970).

115. Cavlek, B. (1969). Study on the influence of thermal processing on meat. *Rim*, **1**, No. 2, 13.

116. Savic, J. and Kepcija, D. (1966). Influence of heating technique upon browning of pork. *Acta Veterinaria*, **16**, No. 1, 3.

117. Pearson, A. M., Tarladgis, B. G., Spooner, M. E. and Quinn, J. R. (1966). The browning produced on heating of fresh pork: II. The nature of the reaction. *J. Food Sci.*, **31**, 184.

118. Bowers, J. A., Harrison, D. L. and Kropf, D. H. (1968). Browning and associated properties of porcine muscle. *J. Food Sci.*, **33**, 147.

119. Frangne, R. and Addrian, J. (1972). The Maillard reaction: VI. Reactivity of various purified proteins (French). *Ann. Nutrit. Aliment.*, **26**, No. 4, 97.

120. Skurray, G. R. and Cumming, R. B. (1974). Prevention of browning during batch dry rendering. *J. Sci. Food Agric.*, **25**, 529.

121. Dvorak, Z. and Vognarova, J. (1965). Effect of the Maillard reaction and smoking on the biological qualities of meat and meat products. *Prumysl. Potravin*, **16**, No. 4, 172; *J. Sci. Food Agric.*, **16**(ii) 206 (1965).

122. Skurray, G. R. and Cumming, R. B. (1974). Physical and chemical changes during batch dry rendering of meat meals. *J. Sci. Food Agric.*, **25**, 521.

123. Wilson, R. A., Mussinan, C. J., Katz, I. and Sanderson, A. (1973). Isolation and identification of some sulfur chemicals present in pressure-cooked beef. *J. Agric. Food Chem.*, **21**, 873.

124. Hofmann, K. (1966). On the nutritive value of meat and its change during heating (German). *Fleischwirtschaft*, **46**, 1121.

125. Hofmann, K. (1970). Influence of heating on the nutritive value of meat protein (German). *Rim*, **2**, No. 1, 43.

126. Bontscheff, N. (1965). On the changes in the protein and amino-acid content

of some foods during thermal treatment and enzyme treatment (German). *Nahrung*, **9**, 161.

127. Blum, A. E., Lichtenstein, H. and Murphy, E. W. (1966). Composition of raw and roasted lambs and mutton: II. Amino acids. *J. Food Sci.*, **31**, 1001.

128. Kozmina, E. P. and Malyutin, A. F. (1969). Biological value of meat proteins, prepared in super-high frequency apparatus. (Russian). *Primen. Sverkhvysokochastatn. Nagreva Obshch. Pit.*, **1969**, 36; *Chem. Abstr.*, **75**, 18806*a* (1971).

129. Bognar, A. (1971). Influence of thermal treatment on the amino-acid content of beef (German). *Ernährungs-Umschau*, **1971**, 200.

130. Grabowski, T. and Sczaniecka, E. (1969). Contents of some available amino acids in poultry meat subjected to different heat treatments. (Polish). *Postepy Drobiarstwa*, **11**, No. 2, 45; *Food Sci. Technol. Abstr.* **4**, 4 S 452 (1972).

131. Gaudy, N. and Landis, J. (1973). Influence of different heat treatments of some carcass parts on the total amino-acid content and the amino-acid content of enzyme hydrolysates (German). *Mitt. Gebiete Lebensmittelunters. u.Hyg.*, **64**, 133.

132. Bontscheff, N. (1968). Changes in the amino-acid composition of some food products during thermal treatment and enzyme action (Bulgarian). *Izv. Inst. Khranene, Bulg. Akad. Nauk.*, **7**, 79; *Chem. Abstr.*, **70**, 36534*m* (1969).

133. Rogowski, B. (1975). Influence of storage and heating of frozen meals on the amino acids of meat (German). *Fleischwirtschaft*, **55**, 343.

134. Czeremski, K. and Jarzabek, K. (1964). Changes in the biological value during thermal processing of meat under different conditions, measured by means of the fluorodinitrobenzol determination of available lysine (Polish). *Przemysl Spozywczy*, **18**, 714.

135. Zoldowsga, A. (1967). The effect of thermal processes upon the flavour of canned meat with special reference to the changes of amino acids. *Roczniki Instytutu Przemyslu Miesnego*, **4**, 91.

136. Ziemba, Z. and Mälkki, V. (1969). Losses of available lysine in canned beef as influenced by the severity of heating. *Proc. 15th European Meat Research Workers' Meeting*, Helsinki, p. 461.

137. Markh, A., Flaumenbaum, B. and Chirkina, T. (1971). Change in the nitrogenous substances of meat under thermal action (Russian). *Myas. Ind. SSSR*, **42**, No. 5, 35; *Chem. Abstr.*, **75**, 108650*z* (1971).

138. Hofmann, K. (1972). Preservation of meat: Influence on nutritive value and possible effect of human health (German). *Berichte über Landwirtschaft*, **50**, 700.

139. Hofmann, K. (1972). The influence of preservation techniques on the quality of meat and meat products (German). *Fleischwirtschaft*, **52**, 1403.

140. Piskarev, A., Dibirasulaev, M. and Korzhenko. (1972). Change in the level of free amino acids and nucleotides in fresh killed and aged meat during sterilisation (Russian). *Myas. Ind. SSSR*, **1972**, No. 2, 34; *Chem. Abstr.*, **76**, 152146*j* (1972).

141. Morisio, G. L. and Losano, C. (1968). The effect of cooking on the digestibility of meat. *Minerva Dietol.*, **8**, 10; *Chem. Abstr.*, **69**, 6623*h* (1968).

142. Hurrel, R. R., Carpenter, K. J., Sinclair, W. J., Otterburn, M. S. and Asquith, R. S. (1976). Mechanism of heat damage in proteins: 7. The significance of

lysine-containing isopeptides and of lanthionine in heated proteins, *Brit. J. Nutr.*, **35,** 383.

143. Golovkin, N. A. and Meluzova, L. A. (1974). Effect of temperature conditions during meat storage on the enzymatic degradation of myofibrillar proteins *in vitro. Proc. 4th Intern. Congress of Food Sci. and Technol.*, **7a,** 36.

144. Ziemba, Z. (1970). Loss of digestibility of canned beef protein as influenced by the sterilisation value (Polish). *Przemysl Spozywczy*, **24,** No. 10, 429.

145. Ebashi, S. and Nonomura, Y. (1973). Proteins of the myofibril. In *The Structure and Function of Muscle*, Vol. III, 2nd edn., ed. by G. H. Bournes, Academic Press, New York and London, p. 285.

8

Physical, Chemical and Biological Changes Related to Different Time–Temperature Combinations

I. D. Morton

*Food Science Department, Queen Elizabeth College, University of London,
Campden Hill, London, W8 7AH, England*

CHANGES IN FAT

Fats are important ingredients of food. They provide energy, serve as a vehicle for essential fatty acids and for the fat-soluble vitamins; they also improve flavour, modify texture and add satiety to meals.

Edible fatty materials used for deep-fat frying include fats which are normally liquid at room temperature as well as some fats which are solid. Frying oils and fats are unavoidably exposed in varying degrees to adverse conditions during the frying process which produce certain undesirable changes. These include an increase in viscosity, a darkening in colour and the formation of polymeric materials. The methods available for investigation of changes in oils and fats are summarised in Table 1. The polyunsaturated fatty acid content in the triglycerides and possibly the nutritive value of such oils are decreased. There is some evidence in the literature to indicate that the undesirable changes which occur in heated fats may have a deleterious effect on human health.[1] Much of the research effort in the past decades has been expended in a study of the mechanism of the deterioration of fats and the formation of reaction products from heated fatty acids and triglycerides. There are, however, little reported data on the quantitative changes which occur in the composition of the unsaponifiable components of fats and oils.

If we look at the changes in the structures of the triglycerides present in fats and oils, we find that the greatest changes take place in those oils which are unsaturated. Figure 1 shows a simplified formula of an intact triglyceride fat molecule with glycerol connected to three fatty acids. Three different types of reactions can occur in heated fats:

(*a*)   Hydrolysis results in the formation of free fatty acids which at

135

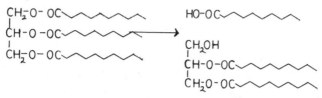

FIG. 1.   Hydrolysis of a triglyceride.

higher concentrations can noticeably affect the taste of an oil. However, as they are normal intermediates in fat metabolism, they are presumably no problem in so far as health is concerned.

(b)   Reactions with oxygen lead to the formation of hydroperoxides, epoxides, hydroxides and ketones, illustrated in Fig. 2. All these may subsequently undergo fission into smaller fragments. On the other hand, oxidised fatty acids can remain in the triglyceride molecule and cross-link with each other. This leads to dimeric and even higher polymeric triglycerides. The oxidation products of fat can be classified as being made up of volatile, monomeric and polymeric compounds.

(c)   Cross-linking does not necessarily take place only via oxygen bridges as illustrated in Fig. 3. Particularly in the absence of

TABLE 1
Analytical procedures for determining changes in oils and fats

Colour	Iodine colour scale
	Lovibond colour scale
	Transmission 360–500 nm
Conjugation	Extinction, 232 nm
	Extinction, 268 nm
	Refractive index
Polymerisation	Iodine value
	Density
	Viscosity
	Gel permeation chromatography
Oxidation	Peroxide value
	Epoxide value
	Aldehyde value
	Alkali colour value
	Light petroleum insoluble oxidised fatty acids
	Volatile carbonyl compounds
	Liquid chromatography

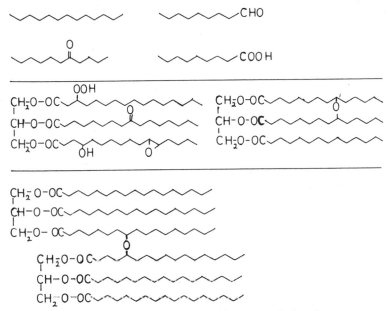

FIG. 2. Some oxidation products from triglycerides. Top: volatiles. Centre: monomeric. Bottom: dimeric and polymeric.

oxygen, heating can form new carbon–carbon bonds. If these bonds are formed within one fatty acid, cyclic fatty acids are obtained. New bonds between two different fatty acids leads to the formation of dimeric acids either within one triglyceride molecule or between two molecules. This latter reaction is the first step in the polymerisation of fats.

Figures 2 and 3 illustrate that heated fats may contain considerable amounts of polymerised material as well as oxidation products. If animal experiments are to be used to investigate the possible toxicity of heated fats, it is essential that the material used is well characterised with regard to the degree of oxidation and polymerisation. Unfortunately many of the experiments described in the older literature have not followed this basic rule and many of the feeding experiments reported must now be regarded as meaningless due to the lack of this background information.

In characterising the degree of deterioration of heated fats, many methods are available. Some of them are listed in Table 1 but they do not all necessarily give conclusive results. Most information is derived from a combination of at least two if not more types of analysis. Modern work has

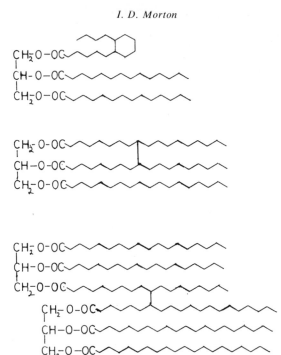

FIG. 3.   Dimerisation and polymerisation of triglycerides. Top: cyclic fatty acids. Centre: dimeric fatty acids. Bottom: dimeric triglycerides.

stressed the importance of gel permeation chromatography which gives quantitative data about the distribution of molecular sizes. Experimental work with gel permeation chromatography has been adequately described in the literature.[2] The equipment is expensive but the results can only be obtained using such equipment. It is certainly the preferred method to measure the amount of polymerised material in a heated oil. Liquid chromatography permits semi-quantitative determination of the oxidised material. Published work[3] indicates that whilst natural fats and oils do not contain higher polymerised material, the refining of an oil can lead to the presence of such compounds. In the refining of soya bean oil, heating to 190 °C during deodorisation can cause formation of 0·5% dimeric triglycerides. By the use of gel permeation chromatography, the history of an oil so far as preheating is concerned, can be discovered.[4] Extensive heating during the frying process leads to the presence of a large number of peaks in the gel permeation chromatogram. The published literature shows the results of frying with sunflower oil.[5] Approximately 11% of the original oil can be destroyed during the heating and frying process.

Liquid chromatography can be used, provided suitable solvents are available, to obtain fractions of different degrees of oxidation from heated oils. Increasing the heating time of an oil such as groundnut oil can increase the amount of oxidised material.[6]

For many years, research on an international scale has been directed at solving the question of the possible toxicity of heated fats. Four different approaches have been used for this problem:

Heating under Normal Conditions

Some research workers have looked at the fats heated under normal household or commercial conditions. Samples of fats are heated in the laboratory or in an industrial fryer with and without the presence of food. It can be assumed from previous work, that these fats would contain 10–20% of polymerised material. Heated fats of this type when fed to test animals, even in quite high amounts, do not produce any adverse effects.

Abused and Over-heated Fats

Initially the lack of adverse results led some research workers to study the effect of feeding over-heated fats. In these experiments, fats and oils, sometimes highly unsaturated oils, were heated for long periods of time up to extremely high temperatures. In some cases oxygen was blown into the oils to increase the oxidation reaction. It could be assumed that such preparations might contain up to 50% or even more of polymerised material. The feeding of these samples to animals can produce severe irritations in the gastro-intestinal tract and diarrhoea. Normally retardation of growth is observed. In extreme cases, of course, the animals die. Despite these spectacular effects it is not possible to draw absolute, final conclusions about the toxicity of over-heated fats. No generalisations are possible and the question is still open as to how long an oil or fat should be used for frying before it is spoiled.

Isolation of Possible Toxic Material

Other investigators have used the approach of looking for individual compounds which could then be formed during heating and have tried to evaluate their toxicity. This is a very difficult study but a very instructive and interesting paper was published recently.[7] Various fats were heated under the normal conditions and over 130 compounds were identified in the heated fats. Despite all their efforts, the authors did not find even one compound showing unexpectedly high toxicity. A summary sentence from their paper reads: 'We have found that actual used frying fats contain only very small quantities of substances which are toxic when administered in large doses to weaning rats, and that the fats themselves produce no appreciable ill-effects on animals consuming them'.

Long-term Feeding Experiments

Unilever in Germany, has undertaken long-term feeding experiments with heated fats in co-operation with Professor Lang from the University of Mainz. They selected two types of oils with quite different properties. One was 'Biskin', a partially hardened groundnut oil which is stable during heating, and the second oil used was soya bean oil. This latter is highly unsaturated and therefore oxidises readily if heated with access of air. A summary of their results[8] indicates that heated fats are as healthy as unheated ones. Survival rate was only one of the parameters studied; histological investigations were also made. In all these experiments no significant difference was observed between heated and unheated fats. The final conclusion can be drawn that heated fats have no detrimental effects upon the health of animals, even if they contain considerable amounts of polymerised and oxidised material. This conclusion is at variance with the conclusions drawn by earlier workers in publications in the period from about 1930 to 1950.[9]

The quality of used frying fats and oils was discussed at a round-table conference and in a series of lectures organised by the Deutsche Gesellschaft für Fett-Wissenschaft held at Münster in 1973. These discussions and lectures have been published in full.[10] The general conclusions reached at this conference, which dealt with used frying fats and oils and how to establish when they are in a state of deterioration, are as follows:

(a)   If the odour and taste are unacceptable.

(b)   If the assessment by organoleptic means is doubtful, then if the smoke point is at 170 °C or lower and the concentration in light petroleum of insoluble oxidised fatty acids is 0·7 % or greater.

(c)   If the organoleptic assessment is doubtful and the concentration of light petroleum insoluble oxidised fatty acids is 1·0 % or higher.

These three rules, which depend on data as well as organoleptic means, give a clear indication as to whether the oil is acceptable or not. It is interesting to note that the German Society for Fat Research selected other criteria than gel permeation chromatography or liquid chromatography. The answer, of course, is simple. These methods are expensive and need skilled technicians to operate them. Both the smoke point and the content of light petroleum insoluble oxidised fatty acids can be determined without expensive equipment.

The stability of frying oils has been investigated by the German research workers at the Unilever research laboratory in Hamburg.[10] They found that nearly half of the samples of frying oils from commercial fryers must be regarded as deteriorated, particularly if pure soya bean oil has been used. Hardened soya bean oil performed very much better. Their results can be summarised with the statement that 'oils with a high content of polyunsaturated fatty acids can be used for deep frying. However, they are easily oxidised and they deteriorate faster than those oil blends which contain lesser amounts of unsaturated fatty acids'.

In our own work, we were particularly interested in the unsaponifiable fraction of vegetable oils and the importance of this fraction on oil stability. The unsaponifiable fraction contains sterols, 4α-methyl sterols, triterpene alcohols, hydrocarbons and pigments. The commonly occurring phytosterols which include β-sitosterol, campesterol and stigmasterol have been claimed to be effective agents in reducing serum cholesterol in experimental animals and man.[11] There are a number of patents issued[12] which deal with the suitability of fats containing phytosterols in concentrations as high as 6% for use as hypocholesteremic materials. We were particularly interested in olive oil and the sterols which are contained in this oil. Olive oil is obtained from the fruit of varieties of *Olea Europaea*, the well-known olive tree. Virgin olive oil is obtained from the fruit of the olive tree by mechanical or other physical means under *mild* conditions which do not lead to any alteration of the oil. Generally, olive oil is considered to contain principally oleic acid, linoleic, palmitic, palmitoleic and stearic acids in the triglycerides. There are trace amounts of other fatty acids present.

In our early experiments, we heated samples of virgin olive oil at $180 \pm 5\,^{\circ}C$ for periods of eight hours with overnight cooling. After different periods of heating, samples were removed, stored under nitrogen in a refrigerator and then analysed. The unsaponifiable fraction of the different oils was examined, initially by thin-layer chromatography and then by gas–liquid chromatography (GLC) using a Pye Unicam, Model 104, instrument, fitted with a glass column packed with Diatomite CQ coated with 2% OV–17. It was relatively straightforward to determine the sterol content by GLC. The same technique could not be used for determination of the total 4α-methyl sterol content. This is because the concentration of these compounds in olive oil is very small and quantitative recovery from thin-layer plates is difficult. In addition, one of the 4α-methyl sterols had

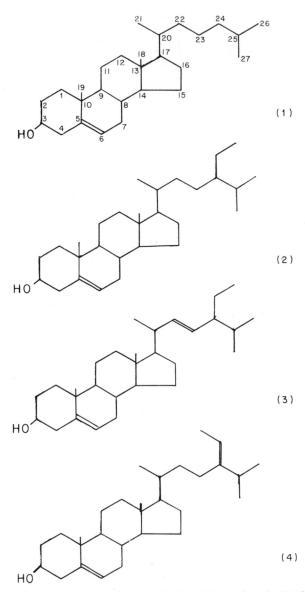

FIG. 4. Formulae of the common plant sterols, 4α-methyl sterols and related compounds: Ī, Cholesterol; 2, β-sitosterol; 3, stigmasterol; 4, Δ⁵-avenasterol; 5, fucosterol; 6, squalene; 7, campesterol; 8, cycloartanol; 9, 24-methylene cycloartanol; 10, 24-methyl iophenyl; 11, citrostadienol; 12, gramisterol; 13, obtusifoliol; 14, cycloeucalenol.

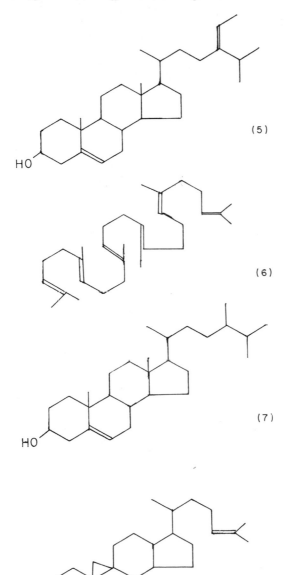

Fig.4—*continued*.

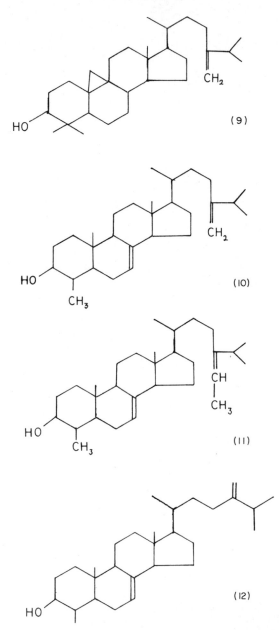

(9)

(10)

(11)

(12)

Fig. 4—*continued.*

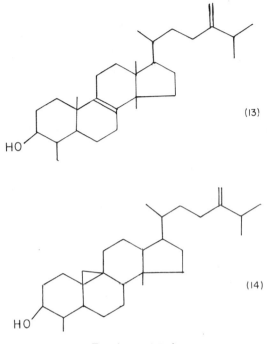

FIG. 4—*continued.*

a retention time very close to that of β-sitosterol. If we used stigmasterol or cholesterol as an internal standard in GLC, this difficulty was overcome. By investigating the percentage distribution of the individual sterols in both fresh and heated samples, we were able to show that citrostadienol deteriorates more rapidly on heating than the other components of the four α-methyl sterol fraction. Among sterols, Δ^5-avenasterol appeared more susceptible to air oxidation at high temperatures[13] (see Fig. 4).

The antioxidant property of the true sterol fraction is clearly due to the presence of Δ^5-avenasterol. β-Sitosterol is ineffective in the initial stages and becomes slightly pro-oxidant after prolonged heating. Stigmasterol (isomeric with Δ^5-avenasterol) is also ineffective. In further work we attempted to elucidate the relationship between the structure of the sterol and the antioxidant activity. An ethylidene side-chain is important for the antioxidant effect. This is not the complete answer as there is a difference in behaviour between Δ^5-avenosterol and Δ^7-avenasterol. There is also a difference between Δ^5-avenasterol and fucosterol. This latter relationship is purely a *cis-trans* isomeric one. To confirm these results, we added in

TABLE 2
Common plant sterols and 4α-methyl sterols

Some Common Plant Sterols
   Campesterol
   Stigmasterol
   β-Sitosterol
   Δ^5-Avenasterol
   Fucosterol

Some Common Plant 4α-methyl Sterols
   Obtusifoliol
   Gramisterol
   Citrostadienol
   Cycloeucalenol

realistic ratios, pure sterol fractions including both the sterols and the 4-methyl sterols to cottonseed oil, and subjected the oil with these added sterols to the same heating conditions as for olive oil. We obtained similar results for stability of the cottonseed oil with the added sterols as for olive oil.[14]

The stability of olive oil at high temperatures has been attributed to its low linoleic acid content,[15] but the present study shows that the concentration and composition of the unsaponifiable fraction does affect to a very great extent the decomposition rate of polyunsaturated fatty acids in triglycerides. In this connection it is important to note that olive oil is principally used as the virgin oil or as pure olive oil. Both these grades retain major portions of the unsaponifiable fraction.

There is need to protect polyunsaturated fatty acid containing oils during heating in air at high temperatures. Antioxidants such as BHA and propyl gallate have been employed to improve the oxidative stability of edible oils.[16] They have less effect on the darkening and polymerisation of the oils and the loss of nutritive value during prolonged heating. A number of patents relating to the use of antioxidants in frying oils have appeared recently. The complexity of the admixtures used indicates that none of the antioxidants used can efficiently protect the heated fats during prolonged heating and there is need for compounds capable of reducing oxidation during heating of fats.

Substituting olive oil for more saturated vegetable fatty materials could have a beneficial effect in the human diet. The presence of a high level of Δ^5-avenasterol provides an oil with a proved resistance at cooking temperatures and as the sterol is sensitive at high temperatures, its protection at different stages of refining (heating, decolorisation, deodorisation) can improve the quality of the refined oil.

We have also studied the activity of other Δ^7-sterols with a double bond also at C-24 including citrostadienol and gramisterol. Obtusifoliol, a Δ^8-sterol with a methylene side-chain and cycloeucalenol with no double bond in ring D but containing a cyclopropane ring, have been included.[17]

In other work we have been concerned with the kinetics of the autoxidation of oleic acid and linoleic acid mixtures and have measured the rate of oxygen uptake. Mixtures with different molar fractions were prepared, keeping the total number of moles constant. Oxidations were carried out at 30 °C and graphs were made of the percentage of oxidation against time. We plotted the overall first-order rate constant against molar fraction and the overall second-order rate constant against molar fraction. It was found that during the monomolecular decomposition of hydroperoxides the reaction rates decrease with decreasing linoleic acid molar fraction. No pro-oxidant or antioxidant effects were observed during this period. The predominant effect was that of oleic acid acting as a diluant for the oxidation of linoleic acid. When the reaction enters the period of bimolecular decomposition of hydroperoxides, a slight antioxidant effect was found in mixtures containing less than 25% oleic acid. With increasing oleic acid concentration, this effect levels off and gives way to a more marked pro-oxidant effect when the oleic acid is above 75%.[18] This work is being extended with the aim of trying to explain the rate of oxidation of triglycerides, but it is of course a very difficult field to study.

A recent paper shows that thermal treatment of palm oil does not materially influence the content of polycyclic aromatic hydrocarbons. Concern has been expressed[19] that during the conventional, heat-bleaching at 200–240 °C in vacuum, the carotenoid pigments in palm oil can be converted into polycyclic non-volatile compounds which remain in the refined oil. This would imply that there is an increase in the polycyclic aromatic hydrocarbons arising during processing. A large number of oils have been investigated from Malaysia, the Ivory Coast and Zaire. The content of polycyclic aromatic hydrocarbons remaining in the refined palm oil after deterioration is of the same order as that found in other edible oils such as soya, corn and peanut oil.

With the growth in the consumption of 'snack' foods the intake of frying oil is on the increase. Many of these snack foods have been deep-fat fried and little work has been recorded in the literature on the absorption of oil into the fried food. Our own work, admittedly with only a limited number of products, including fried fish, as such, fried fish with batter and snack products such as mushrooms has shown that the oil absorbed into the fried food is more oxidised than the frying oil medium itself.[20]

We fried snack products in cottonseed oil and in groundnut oil. Samples of the frying oil were taken and stored under nitrogen until analysed. The fried food was also carefully extracted with solvent under nitrogen to obtain the absorbed oil without any further oxidation. The absorbed oil and the frying media were examined for oxidised triglyceride content after the frying bath had been used for varying times. We found that the frying oil itself after 24 hours use, contained 10 % oxidised material but the oil from the fried food contained 15 % oxidised material. Examination of the fatty acid composition of these oxidised oils showed that, as we expected, the linoleic acid content was decreased. We also found a small decrease in the oleic acid content. We can hypothesise that the food appears to 'cleanse' the oil of its more oxidised material but we are left with the problem of finding out just how important is the absorption of oxidised fat into the fried foods. We carried out an experiment to study the uptake of oil into a fried food. We took as a model substance hydrated cotton wool in a glass test tube and placed on top of the cotton wool oil containing red Sudan dye. We then plunged the test tube into heated oil at 180 °C to allow it to come up to the same temperature and noticed that the oil remained on top of the hydrated cotton wool. We assume that this is due to the effect of the steam from the wool (as a food product) holding the oil at bay. When the test tube was removed from the heating source the oil was immediately absorbed into the cotton wool as it cooled down. We assume that in a food product, little absorption of the oil takes place during the frying process but as soon as the food is removed from the oil, exposed to cooling in the air, absorption begins immediately. We consider that the work should be repeated and expanded but we believe that the question of absorption of the frying oil into the food product affects the nutritional status of such snack foods.[21]

In the refining of fats and oils the minor constituents including phosphatides, carotenoids, sterols, gums and mucilages, are normally removed. The other important material that is often mainly removed in the refining process is vitamin E or the tocopherols. With the rise in the intake of the polyunsaturated fatty acids in the diet there is a need for increased intake of tocopherol at the same time.[22] Tocopherols in oils are not stable to heat. An interesting paper[23] indicates that the loss of α-tocopherol was slightly greater than that of γ-tocopherol during the refining procedure. γ-Tocopherol is the major part of vitamin E present in oils, generally accounting for about 70 % of the total. Refining and subsequent storage, however, does result in over 70 % loss in total tocopherols.[23] Tocopherols are not stable to heat in oils and we must be sure that there is, in our food, adequate intake of vitamin E.

Much has been written and spoken recently about self-sufficiency. In Great Britain we could be almost self-sufficient for both carbohydrate and protein foodstuffs. It is clear that we cannot be self-sufficient with regard to fat. Based on energy considerations we can produce sufficient carbohydrate and protein for our needs but we must import a certain amount of edible oil.[24] More work should be undertaken on the production of edible oils from plants which will grow in these latitudes and the latitudes of Northern Europe to make sure that as good a yield of satisfactory oil as possible is obtained. One could mention here the Canadian work on the production of a rape seed oil free or almost free of erucic acid.[25] It is also essential that further research is carried out to make sure that the changes take place in these oils due to temperature, time or biological action should be minimal. If we have these oils, we must also make sure that the minor constituents including the sterols and other unsaponifiable materials are such as to ensure that the polyunsaturated character of the oil is not destroyed by any unwanted changes nor are compounds produced which show unexpected high toxicity.

ACKNOWLEDGEMENT

I should like to thank my colleagues, Dr H. Alim, Dr D. C. Boskou, Dr A. R. Rosas Romero and Mr S. S. Hussain for many helpful discussions.

REFERENCES

1a. Crampton, E. W., *et al.* (1953). Studies to determine the nature of the damage to the nutritive value of some vegetable oils from heat treatment: III. The segregation of toxic and non-toxic material from the esters of heat-polymerised linseed oil by distillation and by urea adduct formation. *J. Nutr.*, **49**, 333–46.
1b. Crampton, E. W., Common, R. H., Pritchard, E. T. and Farmer, F. A. (1956). Studies to determine the damage to the nutritive value of some vegetable oils from heat treatment. IV. Ethyl esters of heat polymerised linseed, soybean and sunflower seed oils. *J. Nutr.*, **60**, 13–24.
1c. Kaunitz, H., *et al.* (1956). Biological effects of the polymeric residues isolated from autoxidised fats. *J. Amer. Oil Chem. Soc.*, **33**, 630–4.
1d. Kaunitz, H., *et al.* (1956). Nutritional properties of the molecularly distilled fractions of autoxidised fats. *J. Nutr.*, **60**, 237–44.

2. Maley, L. E., Richman, W. B. and Bombaugh, K. J. (1967). Aqueous and prep. scale gel permeation chromatography. *Amer. Chem. Soc. Div. Polym. Chem. Preprints*, 1967, **8**(2), 1250–8. *Chem. Abstr.*, 1969, **70**, 10,001*q*.
3. Loncin, M. (1975). Refining of palm oil. *J. Amer. Oil Chem. Soc.*, **52**, 144A–6A.
4. Billek, G. (1973). Alteration of fats under conditions of deep-frying and their analytical detection: Artefact formation in frying fats during deep-frying (German). *Fette Seifen Anstrichmittel*, **75**, 582–6.
5. Unbehend, M., Scharmann, H., Strauss, H. J. and Billek, G. (1973). The application of gel permeation chromatography to the investigation of heat-oxidised fats (German). *Fette Seifen Anstrichmittel*, **75**, 689–96.
6. Unbehend, M. and Scharmann, H. (1973). Nutritional–physiological properties of frying oils: III. Gel permeation chromatography of fish-frying oils (German). *Z. Ernährungswiss*, **12**, 134–43.
7. Artman, N. R. and Smith, D. E. (1972). Systematic isolation and identification of minor components in heated and unheated fat. *J. Amer. Oil Chem. Soc.*, **49**, 318–26.
8. Lang, K., Henschel, J., Waibel, J. and Billek, J. (1973). Nutritional–physiological properties of frying oils: IV. Influence on life-expectancy of experimental animals (German). *Z. Ernährungswiss.*, **12**, 241–7.
9. Peacock, P. R. (1948). Heated fats as a possible source of carcinogens. *Brit. J. Nutr.*, **2**, 201–4.
10. The following papers are collected together and published by the German Society for Fat Research 1974, under the title *Bratfette und Siedefette* (65 pp.):
10a. Klages, R. (1973). Significance and Technology of Frying: Scope, Economic Significance and Future Developments (German). *Fette Seifen Anstrichmittel*, **75**, 579–81.
10b. Dostmann, W. (1974). Significance and Technology of Frying: Frying equipment for commercial purposes and for the household (German). *Fette Seifen Anstrichmittel*, **76**, 58–60.
10c. Niemann, A. (1974). Significance and Technology of Frying: Commercial Production of Fried Foods (German). *Fette Seifen Anstrichmittel*, **76**, 60–2.
10d. Billek, G. (1973). Alteration of fats under conditions of deep-frying and their analytical detection: Artefact formation in frying fats during deep-frying (German). *Fette Seifen Anstrichmittel*, **75**, 582–6.
10e. Pardun, H., Blass, J. and Kroll, E. (1974). Alterations of fats under conditions of deep-frying and their analytical detection: Evaluation of the quality of frying fats and their analysis (German). *Fette Seifen Anstrichmittel*, **76**, 97–104 and 151–8.
10f. Guillaumin, R. (1973). Determination of new chemical species formed during the heating of oils (French). *Rev. franç. Corps gras*, **20**, 285–90.
10g. Guillaumin, R., Gente, M. and Desrieux, M. (1973). Comparative study of the influence of the frying procedure on the percentage of new chemical species in different oils (French). *Rev. franç. Corps gras*, **20**, 413–9.
10h. Mankel, A. (1974). Analysis and Assessment of Frying Fats: Observations from the practice of food surveillance (German). *Fette Seifen Anstrichmittel*, **76**, 20–5.
10i. Lang, K. (1974). Nutritional Effect of Heated Fats: Long-term animal experiments (German). *Fette Seifen Anstrichmittel*, **76**, 145–51.

10j. Wurziger, J. (1974). Evaluation of frying fats from the view-point of food regulations: Criteria for assessment by the Food Chemist (German). *Fette Seifen Anstrichmittel*, **76**, 52–7.

10k. Klein, G. (1974). Evaluation of frying fats from the view-point of food regulations: Legal interpretation of the regulations of the Food Law (German). *Fette Seifen Anstrichmittel*, **76**, 16–19.

10l. Anon. (1974). Investigation and utilisation of used frying fats and oils: Recommendations of the German Society of Fat Research (German).

11a. Anon. (1964). Hypocholesteremic Effect of Sitosterol. *Nutr. Rev.*, **22**, 326–8.

11b. Sims, R. J., Fioriti, J. A. and Kanuk, M. J. (1972). Sterol additives as polymerisation inhibitors for frying oils. *J. Amer. Oil Chem. Soc.*, **49**, 298–301.

12a. Erickson, B. A. Ger. Pat. 2,035,069. Hypocholesteremic Shortening. *Chem. Abs.*, **74**, 1971, 86594j.

12b. Jandacek, R. J. (1975). US Pat. 3,865,939. Hypocholesteremic oils. *Food Sci. Technol. Abstr.*, **7**, 1975, 9N, 353.

13. Boskou, D. and Morton, I. D. (1975). Changes in the Sterol Composition of Olive Oil on heating. *J. Sci. Food Agric.*, **26**, 1149–53.

14. Boskou, D. and Morton, I. D. (1976). Effect of Plant Sterols on the Rate of Determination of Heated Oils. *J. Sci. Food Agric.*, **27**, in press.

15. Montefredine, A. and Luise, M. (1969). Modification of oils and edible fats used in normal processes of food cooking (Italian). *Ind. Aliment.*, **8**, 69–76.

16. Peled, M., Gutfinger, T. and Letan, A. (1975). Effect of Water and BHT on stability of Cottonseed Oil during Frying. *J. Sci. Food Agric.*, **26**, 1655–66.

17. Boskou, D. and Morton, I. D. (1976). Unpublished work.

18. Rosas Romero, A. R. and Morton, I. D. (1975). A kinetic study of the competitive oxidation of oleic acid–linoleic acid mixtures. *J. Sci. Food Agric.*, **26**, 1353–6.

19. Rost, H. E. (1976). Influence of thermal treatments of palm oil on the content of polycyclic aromatic hydrocarbons. *Chem. and Ind.*, **1976**, 612–13.

20. Hussain, S. S. and Morton, I. D. (1976). Absorption by food of frying oil. *Proc. 4th Internat. Congress Food Sci. and Tech.*, Madrid 1974, Vol. I in press.

21. Alim, H. and Morton, I. D. (1976). Oxidation in foodstuffs fried in edible oils: *Proc. 4th Internat. Congress Food Sci. and Tech.*, Madrid 1974, Vol. I in press.

22. Aylward, F. and Morton, I. D. (1971). Vitamin Fortification of Foods. *Proceedings SOS/70, 3rd Intern. Congress of Food Sci. and Technol.*, pp. 192–9.

23. Walker, B. L. and Slinger, S. J. (1975). Effect of processing on the tocopherol content of Rapeseed Oils. *Canad. Inst. Food. Sci. Tech. J.*, **8**, 179–80.

24. Pereira, H. C. (1976). Future Plans for Britain's Food Supplies. *J. Sci. Food Agric.*, **27**, 700.

25. Krzymanski, J. and Downey, R. K. (1969). Inheritance of fatty acid composition in winter forms of rapeseed, *Brassica rapas. Canad. J. Plant Sci.*, **49**, 313–19.

9

Chemical, Physical and Biological Changes in Carbohydrates Induced by Thermal Processing

GORDON G. BIRCH

*National College of Food Technology, University of Reading,
St George's Avenue, Weybridge, Surrey, KT13 0DE, England*

INTRODUCTION

The cyclic, polyhydroxy and hemiacetal characteristics of sugar molecules are basically responsible for all their physical, chemical and biological properties, and for the changes in these induced by heat. Modern scientific techniques such as nuclear magnetic resonance have established that these cyclic molecules, whether as simple monosaccharides or as chains of monosaccharide residues in polymers, tend to exist in 'chair' conformations[1,2] usually as an energetically preferred type (Fig. 1). We are therefore at last in a position to interpret all the chemical, physical and biological properties of food carbohydrates in terms of a defined stereostructure.

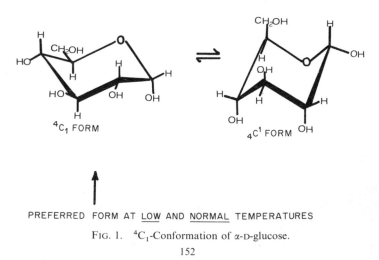

PREFERRED FORM AT <u>LOW</u> AND <u>NORMAL</u> TEMPERATURES

FIG. 1. $^4C_1$-Conformation of α-D-glucose.

152

Clearly the nutritional value of food carbohydrate, which is one of the major concerns of the food scientist, is an integral result of its chemical, physical and biological properties and we are now able to recognise that food carbohydrates differ extensively in their metabolic effects.

Heat affects the properties of different carbohydrates in foods to different extents, and not always in the same direction. Such differences should all eventually be predictable from basic stereostructures as depicted in Fig. 1. Many have already been profusely catalogued.

EFFECTS OF HEAT ON PHYSICAL PROPERTIES OF SUGARS

It is undoubtedly true that all the physical properties of carbohydrates are affected by the hydrogen bonding which results from their polyhydroxyl character. This is basically of two types, intra- and inter-molecular hydrogen bonding (Fig. 2). In addition, oligomers and polymers may exhibit inter-residue (still intramolecular) hydrogen bonding, and all types are affected by the application of heat to a greater or lesser extent, depending on the configuration of the different hydroxyl groups, which in

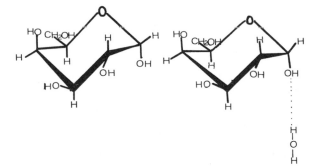

FIG. 2.   Intra- (left) or inter-molecular (right) hydrogen bonding with sugar and/or water molecules.

turn governs the strength of the hydrogen bond complex. Intermolecular hydrogen bonding capacity is high in all sugars which means that they form strong crystalline lattices which require considerable heat to disrupt. Thus sugars have high melting points (often 200 °C or higher) and many types are even known to decompose without melting at over 300 °C. Similarly they have high boiling points, even at very low pressures, and consequently exhibit no vapour pressure or contribute no odour to foods. This strong tendency to intermolecular hydrogen bonding gives rise also to the high solubility of crystalline sugars in water after the application of heat. Here water molecules replace sugar molecules and produce intimate associations of sugar and water at quite moderate temperatures. Pure crystalline sucrose, for example, dissolves in half its own weight of water at 20 °C.

Heat will also affect the proportions of different tautomeric forms of reducing sugar which are present, and hence the tendency of any one of these to crystallise. Nickerson[3] has studied the tendency of α-lactose monohydrate to crystallise at different temperatures. Such studies may facilitate the control of texture in milk products, e.g. by prevention of grittiness from α-lactose monohydrate.

In polymers intramolecular (inter-residue) hydrogen bonding can contribute to the rigidity of the polysaccharide chain, and thence indirectly to its association with neighbouring chains. Cellulose, for example, which differs from starch only in the configuration of its inter-residue linkage is stabilised by extensive intra- and inter-molecular hydrogen bonding and no amount of heat will effect the slightest dissolution in water.

Perhaps one of the most important modern applications of polysaccharide–water interaction in food technology is the process of gelatinisation. This occurs to some extent with most polysaccharides and implies a change in rheological properties as polysaccharide and water molecules interact to alter and re-form hydrogen-bonded networks. In the case of starch, which is the principal food polysaccharide, the term 'gelatinisation' has come to be synonymous with the process of swelling and eventual bursting of the starch granules, i.e. the minute organelles which are characterised by a particular shape in each particular source of dietary starch. It is noteworthy here that potato starch possesses a quite remarkable high swelling power, due to the manner in which the polymer is organised

TABLE 1

Resistance of starches to heat/acid treatment or amylase treatment in relation to phosphorus content[28,37]

Starch	P Content (mg P/g)	% Acid hydrolysis (pH 1·5, 25 min 15 psi)	% Amylase hydrolysis (24 h)
Potato	0·76	48·4	7·8
Rice	0·64	42·7	62·6
Wheat	0·59	49·2	35·4
Maize	0·18	54·3	24·3

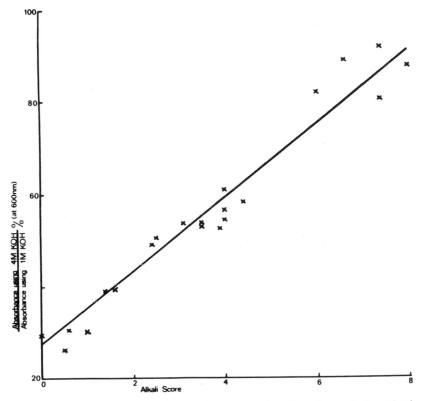

FIG. 3. Relationship between alkali-released amylose and quality index as alkali number in rice varieties.[24e]

within the granule,[4] and this may be responsible for its peculiar resistance to hydrolysis by protons or enzymes, and some of its biological properties which will be reported below (Table 1).

During the cooking of starchy foods the granules swell, then burst and release their polysaccharide components into the surrounding aqueous environment. This entire process governs the textural quality of the food and Priestley has developed an elegantly simple method for assessing degree of gelatinisation and rice quality by colorimetric determination of released amylose and the application of alkali (Fig. 3). The proportion of amylose and amylopectin released during cooking will affect the stickiness of rice grains and their adherence to one another.

Whilst simple sugars have remarkably high water-solubility, polysaccharides often have low solubilities and solubility rates. Malto-dextrins

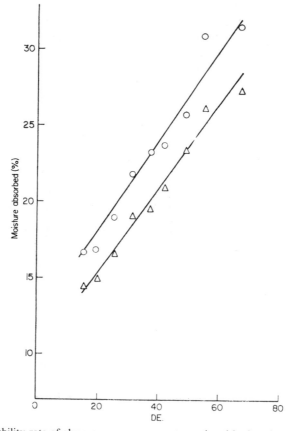

FIG. 4.   Solubility rate of glucose syrup components produced by low degree of heat/acid
treatment.[6]

naturally have solubility rates intermediate between those of D-glucose and
starch (Figs. 3 and 4).[5–7] Since solubility rate is a function of hydration of
sugar molecules it is related to hygroscopicity. Both physical properties are
important in regard to the processing of foods and their storage under
different conditions of temperature and humidity. Indeed the humectant
properties of sugars, polysaccharides and related polyols have become of
even greater significance in the field of intermediate moisture foods.[8]

An important physical property of carbohydrates in the context of their
analysis and specification for food use is optical rotation. This depends on
the type of carbohydrate and is radically altered by chelation of the
carbohydrate with inorganic salts[6] as well as temperature.

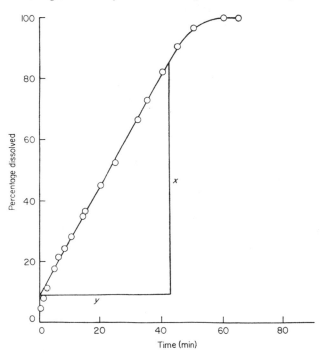

Fɪɢ. 5.   Hygroscopicity of glucose syrup components produced by high or low degrees of heat/acid treatment.[7]

There is no doubt that the conformation of carbohydrates and the configuration of individual hydroxyl substituents in each sugar residue will govern the hydrogen bonding which underlies all their important properties in foods. Solubility, gelatinisation, crystallisation, melting point and boiling point all result from this, and at the present state of knowledge it seems likely that the best methods for understanding these phenomena are the modern techniques of nuclear magnetic resonance, electron spin resonance, and mixed phase studies.[9–12]

EFFECTS OF HEAT ON THE CHEMICAL PROPERTIES OF CARBOHYDRATES

The effect of heat on the chemical properties of carbohydrates may be considered under four headings, taste, caramelisation, pyrolysis and interaction with other food components. These naturally are not independent but are related to a greater or lesser extent, depending on chemical structure.

Taste

Although this is a biological property it is an attractive example of the application of stereochemistry to foods and the relationship between taste and chemical structure of sugars has received considerable attention during the past 15 years.[13] The best working hypothesis relates the sweetness to hydrogen bonding which occurs between suitably disposed α-glycol groupings of sugars and taste receptor sites.[14] Changes of taste which occur during heating can be explained on the assumption that (1) hydrogen bonds rupture as the temperature is raised and (2) differences of the proportions of isomers of reducing sugars exist at different temperatures.[2]

It is important when considering the effect of temperature on taste, to distinguish between *relative* and *absolute* sweetness. The former is usually determined with reference to sucrose, and the *relative* sweetness of most sugars decreases with increasing temperature. However, the *absolute* sweetness of sugars usually increases with increasing temperature. Therefore the absolute sweetness of sucrose must increase faster than the absolute sweetness of the other sugars.

Fructose differs from other food sugars in that its absolute sweetness decreases with increasing temperature[15] and this is related to the fact that the fructopyranose molecule (Fig. 6) exists in the alternate (inverted) $_4C^1$-conformation.[16] Thus, as the temperature is raised the intramolecular hydrogen bond between the axial hydroxyl group at C-5 and the ring oxygen atom is ruptured, and the relative advantage conferred on the anomeric proton for hydrogen bonding to the taste receptor site is lost. These considerations may be important in assessing the value of the new high fructose corn syrups in food formulations.

Caramelisation

As sugar molecules are heated, conformational inversion is facilitated and many sugars in the alternate $_4C^1$-conformation are able to form an intramolecular 1,6- or 3,6-anhydro bridge by condensation of one molecule of water (Fig. 7).

Such reactions are involved in the process of 'caramelisation' or scorching or carbohydrates, and caramel, which is the brown product resulting from this process, is used as a colouring material in modern food technology. Caramel may be defined as a brown substance with a characteristic smell and uncertain composition produced by the partial thermal breakdown

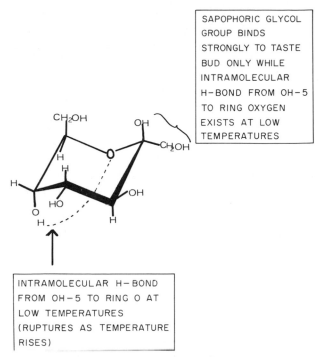

SAPOPHORIC GLYCOL
GROUP BINDS
STRONGLY TO TASTE
BUD ONLY WHILE
INTRAMOLECULAR
H-BOND FROM OH-5
TO RING OXYGEN
EXISTS AT LOW
TEMPERATURES

INTRAMOLECULAR H-BOND
FROM OH-5 TO RING O AT
LOW TEMPERATURES
(RUPTURES AS TEMPERATURE
RISES)

FIG. 6.   β-D-Fructopyranose. Absolute sweetness decreases as temperature rises.

of carbohydrates. It consists of dehydration products, pyrolysis products, polymers and coloured substances, the proportions of which depend on the extent of heating, the presence of catalysts such as acids, alkalies and salts, and the type of carbohydrate which is used. Obviously those sugars (e.g. fructose), which exist normally in the less stable $_4C^1$-conformation, are likely to caramelise more quickly and easily than those which do not.[17-19]

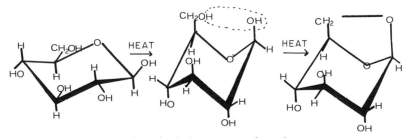

FIG. 7.   Formation of anhydro sugars at elevated temperatures.

Pyrolysis

After polymerisation, depolymerisation and dehydration reactions secondary thermal reactions cause carbon–carbon bond cleavage and many pyrolysis products of low molecular weight are produced in a similar way from both mono- and poly-saccharides.[20] It has therefore been suggested that all carbohydrates undergo similar degradative reactions, and by the combined use of GLC and MS, over 100 products from glucose-containing carbohydrates have been identified and in some cases quantitatively separated.[20] These are generally carboxylic acids, ketones, diketones, aldehydes, unsaturated ketones and aldehydes, furan derivatives, alcohols, hydrocarbons and aromatic compounds, as well as carbon monoxide and dioxide. The formation of carboxylic acids and carbon dioxide has been cited by Houminer[20] as evidence that oxidation proceeds even in the absence of air. Work on the pyrolysis products of low molecular weight from carbohydrates has progressed so far that even the percentage of each individual carbon atom in the D-glucose molecule leading to the formation of the pyrolysis products is known[20] (Table 2).

TABLE 2
Percentage contribution of individual carbon atoms of
glucose in pyrolysis products[20]

Product	C-1	C-2	C-6
Carbon monoxide	67	16	9
Carbon dioxide	57	12	7
Formaldehyde	15	5	65
Formic acid	Almost 100	—	—
Acetic acid	Mainly C-1 and C-2		
Furfural	80% without C-6		
	20% without C-1		
	100% with C-2		

Interaction with Other Food Components

The most important reaction of carbohydrates with other food components is the so-called non-enzymic browning or Maillard reaction[21] which occurs between reducing sugars and nitrogenous compounds, in particular amino acids and proteins. Although the reaction may be inhibited to some extent by food additives such as sulphur dioxide[22] it is highly temperature dependent[23] and may increase 2–3 times for each 10 °C rise in temperature in model systems, and even more than this in natural systems. Of course as temperature rises caramelisation reactions may occur along with Maillard reactions and hence fructose decomposition may increase 5–10 times with each 10 °C rise in temperature.[2] The effect of temperature on Maillard reactions must, however, be considered in relation to other variables such as acidity and water activity. The activation energy for Maillard reactions decreases with increasing water activity[23] so that the browning rate increases. It also increases with increasing pH.

Carbohydrates are not normally considered to react significantly with lipids but small amounts of lipid material may in fact play an important

TABLE 3
Changes in calorific value caused by drying sugar/citric acid mixtures in a vacuum at 65–70 °C for 32 h[25,36]

Mixture	% Increase in calorific value (i.e. observed − calculated)
20% Sucrose/10% citric acid	+5·4***
20% Sucrose/5% citric acid	+13·1***
Commercial frozen orange juice concentrate	+7·3
Special low-calorie orange juice concentrate (34° Br)	+6·0

part in stabilising polysaccharides. Priestley[24a] believes, on the basis of X-ray analysis, that an amylose/lipid complex may control many of the physical properties, such as gelatinisation, retrogradation and solubility in cooking water, which are involved in the cooking quality of rice.

Of more recent interest is some research at the National College of Food Technology[25] which has revealed a change in calorific value of sucrose/citric acid mixtures when dried prior to combustion in a bomb calorimeter (Table 3). It is quite feasible that the increase which occurs represents some esterification because GLC peaks in the region expected for sucrose esters have been identified in these mixtures and the drying process is accompanied by an increase in saponification value. Such effects can lead to confusing calorific values in products (such as fruit juices) consisting largely of sugar and acid.[26]

The interaction of carbohydrates with enzymes at different temperatures is too vast a subject to be dealt with in this paper but it is clear that the stability of amylases at different temperatures during cooking will govern the course of dextrinisation and saccharification that occurs in baking.[27]

Similar basic considerations, but also involving the stabilising and activating effects of minerals, may apply to the digestion or indigestion of raw starches in the gut. In this case the situation is complicated by the dual role of any minerals which are involved. That is to say that some minerals may activate or stabilise the enzyme, while the same or other minerals may chelate with the carbohydrate. Furthermore, these chelated complexes may well possess enhanced stability.

EFFECTS OF HEAT ON THE BIOLOGICAL PROPERTIES OF FOOD CARBOHYDRATE

The consequences of heat treatment on the biological properties of food carbohydrate may be considered at four distinct levels in the case of starch and the common food sugars:

- (*a*)   The intact starch granule.
- (*b*)   The sugars and polymers.
- (*c*)   The caramel and pyrolysis products.
- (*d*)   The products of interaction with other food components.

The biological activity caused by the carbohydrate may then be considered either at the pre-absorption stage, i.e. the direct action on the alimentary system, or at the stage of post-absorption metabolism. Although carbohydrates have traditionally been regarded as all metabolically similar, and as 'fillers' in the diet, there is now abundant evidence that they differ markedly in their biological effects, and that these differences are dependent on chemical structure.

Biological Changes Induced by Heat in the Raw Starch Granule

It seems logical to assume that the process of cooking softens and disrupts cellulosic, hemicellulosic and pectic tissues so that mastication and digestion are facilitated. However, some very recent research[28] has shown that a more profound effect of the raw starch granule itself may be manifested. Raw potato starch, when incorporated into the diet of albino rats causes enlargement of the caecum by as much as 1500–1700% over the course of a few days. The effect is lethal, because the grossly enlarged caecum causes pressure on the diaphragm and the rats die of respiratory failure in about 10 days unless the diet is altered. The enlarged caecum effect does not occur with other raw starches which have been tested, neither does it occur with *heated* (gelatinised) potato starch. At the current stage of investigation it appears likely that the raw potato starch is insufficiently digested by amylases in the ileal mucosa. It therefore passes down the gut to the caecum where it causes a disturbance in the microflora population which leads to the enlargement.

Sugars and Polymers

Although mild heat treatment does not affect sugars or polymers alone, in the presence of acids heat is used to convert starches and dextrins into glucose syrups. These substances are now widely used in foods and the relative proportions of monomers and oligomers which they contain (which in turn depend on the degree of *heat/acid* treatment they have undergone) do certainly affect their metabolic fate. Although all carbohydrates may be ultimately metabolised to lipid they differ in their rate of conversion to serum triglycerides and cholesterol over short time periods. In male subjects, for example, the consumption of glucose syrups of higher molecular weight (i.e. which have received *lower* heat/acid treatments) may in this sense be advantageous[29] (Table 4).

TABLE 4

Serum lipid elevation by glucose syrup components produced by high or low degrees of heat/acid treatment[29]

Glucose syrup component	% Change after 6 days feeding		
	Cholesterol	Phospholipid	Triglyceride
D-Glucose (high acid/heat treatment)	− 35	− 27	+ 1·6
Moderate heat/acid treatment fraction	− 26	− 28	+ 29
Low heat/acid treatment fraction	− 31	− 22	− 22

An even newer discovery is that carbohydrate ingestion is able to elevate serum tryptophan levels and thence to increase the concentration of the neurotransmitter serotonin in the brain. Related to this at least some carbohydrates (usually again those with higher molecular weight, i.e. *lower* heat/acid treatment) cause higher IQ scores in intelligence tests[30] and recently it has been shown[31] that a mid-morning carbohydrate drink, administered over several months to industrial workers in a foundry, was able to lower the accident incidence rate.

Caramelisation and Pyrolysis Products

As well as imparting characteristic flavour to food products caramelisation and pyrolysis products may produce toxic effects in experimental animals. Unfortunately the situation is complicated in the former case by the ammoniation process which is employed in commercial caramel manufacture. This causes pyrazine and imidazole formation as well as other nitrogenous products, and the Food Protection Committee of the National Research Council of the US National Academy of Sciences

TABLE 5

Some biological properties and recommendations for caramelised, pyrolysed and Maillard products

Substance	Taste	Other reported biological properties or recommendations	Ref.
Laevoglucosan	Astringent	—	39
Hydroxymethyl furfural	Pungent	LD_{50} 1·9 g/kg mice LD_{50} 3·1 g/kg rats (1·6 mg/kg drinking water had no effect)	33
Caramel	Bitter	Generally 3000 ppm recommended in foods and beverages Under investigation (BIBRA, UK)	32
Maillard products	Various	40% drop in growth rate of rats fed 0·2% of a glucose/glycine reaction mixture LD_{50} drops from 20 g/kg to 4·1 g/kg when glucose/lysine mixtures are heated Organ enlargement	23

has recommended upper limits for caramel in foods and beverages (Table 5).[32] Dehydration products of carbohydrates, such as laevoglucosan and hydroxymethyl furfural (HMF) possess astringent and pungent sensory properties, and some Russian work[33] has shown that the LD_{50} value of HMF is 1·9 and 3·1 g/kg in mice and rats, respectively.

Maillard and Other Products

Despite improving flavour and appearance of prepared foods, there is no doubt that Maillard products depress nutritional value and in some high dose treatments they may actually produce toxic effects.[23] Bender[34] has pointed out that as little as 3% glucose in dried egg containing 83% protein will cause deterioration both in flavour and nutritional value, and if the glucose is removed by glucose oxidase the deterioration is prevented.

The interaction of sugars with minerals is as yet one of the least well understood aspects of food science. However, the selective chelation of particular sugars with particular metals (e.g. glucose with sodium and fructose with calcium) gives rise to products which may well reach optimum formation at certain temperatures, and have interesting biological properties. It is already reported, for example, that the complexing of fructose with iron gives stable complexes[35] and this may improve the absorption of the sugar and the metal.

CONCLUSIONS

Carbohydrates outstrip other major food components in the degree of isomeric complexity which they exhibit. This in turn leads to an enormous diversity of chemical reactions under the influence of heat, and an interesting variety of chemical, physical and biological effects. Although modern carbohydrate chemistry has laid the foundations for understanding these effects at the molecular level, there is little doubt that this area of research will entertain food scientists for some considerable time.

REFERENCES

1. Birch, G. G. and Shallenberger, R. S. (1973). In *Molecular Structure and Function of Food Carbohydrate*, ed. by G. G. Birch and L. F. Green, Applied Science, London.
2. Shallenberger, R. S. and Birch, G. G. (1975). *Sugar Chemistry*, Avi, Westport, Conn.
3. Nickerson, T. A. and Moore, E. E. (1974). Alpha lactose crystallisation rate. *J. Dairy Sci.*, **57**, 160–4.
4. Banks, W. and Greenwood, C. T. (1975). *Starch and its Components*, Edinburgh University Press, Edinburgh.
5. Kearsley, M. W. (1976). Physico-chemical and Physiological Properties of Glucose Syrup Fractions produced by Reverse Osmosis. Thesis. Reading University, UK.
6. Kearsley, M. W. and Birch, G. G. (1975). Selected physical properties of glucose syrup fractions obtained by reverse osmosis: I. Specific rotation, average molecular weight, solubility rate. *J. Food Technol.*, **10**, 613–23.
7. Kearsley, M. W. and Birch, G. G. (1975). Selected physical properties of glucose syrup fractions obtained by reverse osmosis. II. Hygroscopicity. *J. Food Technol.*, **10**, 625–35.
8. Davies, R., Birch, G. G. and Parker, K. (1976). *Intermediate Moisture Foods*, Applied Science, London.

9. Suggett, A. (1976). Molecular motion and interactions in aqueous carbohydrate solutions: III. A combined nuclear magnetic and dielectric relaxation strategy. *J. Solution Chem.*, **5**, 33–46.

10. Suggett, A., Ablett, S. and Lilliford, P. J. (1976). Molecular motion and interactions in aqueous carbohydrate solutions: II. Nuclear magnetic relaxation studies. *J. Solution Chem.*, **5**, 17–31.

11. Suggett, A. and Clark, A. H. (1976). Molecular motion and interactions in aqueous carbohydrate solutions: I. Dielectric relaxation studies. *J. Solution Chem.*, **5**, 1–15.

12. Mynott, A. R., Higgs, S. J. and Jowitt, R. (1975). Relationship between composition and physical properties of a three-component sugar system. *J. Sci. Food Agric.*, **26**, 135–40.

13. Birch, G. G. (1976). Structural relationships of sugars to taste. *Critical Reviews in Food Science and Nutrition*, in press.

14. Shallenberger, R. S. (1964). Sweetening agents in foods. *Agric. Sci. Rev.*, **2**, 11–20.

15. Lee, C. K., Mattai, S. E. and Birch, G. G. (1975). Structural functions of taste in the sugar series: IV. Sugar–amino acid interaction as an index of sugar sweetness. *J. Food Sci.*, **40**, 390–3.

16. Lindley, M. G., Birch, G. G. and Khan, R. (1976). Sweetness of sucrose and xylitol: Structural considerations. *J. Sci. Food Agric.*, **27**, 140–4.

17. Feather, M. S. and Harris, J. F. (1973). Dehydration reactions of carbohydrates. In *Advances in Carbohydrate Chem.*, **28**, 161–224, ed. by R. S. Tipson and D. Horton, Academic Press, New York.

18. Sugisawa, H. and Sudo, K. (1969). The thermal degradation of sugars. The initial products of browning reaction in glucose caramel. *Canad. Inst. Food Technol. J.*, **2**, 94–7.

19. Örsi, F. (1969). Fractionierung von Caramel-Farbstoff durch Gelfiltration. *Die Nahrung*, **13**, 53–6.

20. Houminer, Y. (1973). Thermal degradation of carbohydrates. In *Molecular Structure and Function of Food Carbohydrates*, ed. by G. G. Birch and L. F. Green, Applied Science, London.

21. Hurrell, R. F. and Carpenter, K. J. Maillard Reactions in Food. This volume, pp. 168–84.

22. McWeeny, D. J. (1973). The role of carbohydrate in non-enzymic browning. In *Molecular Structure and Function of Food Carbohydrates*, ed. by G. G. Birch and L. F. Green, Applied Science, London.

23. Williams, J. (1976). Chemical and non-enzymic changes in intermediate moisture foods. In *Intermediate Moisture Foods*, ed. by R. Davies, G. G. Birch and K. J. Parker, Applied Science, London.

24a. Priestley, R. J. (1973). Physico-chemical properties of rice starch. Thesis. Reading University, UK.

24b. Priestley, R. J. (1976). Studies of parboiled rice: Part I. *Food Chemistry*, **1**(1), 5–14.

24c. Priestley, R. J. (1976). Studies of parboiled rice: Part II. *Food Chemistry*, **1**(2), 139–48.

24d. Priestley, R. J. (1977). Studies of parboiled rice: Part III. *Food Chemistry*, **2**(1), 43–50.

24e. Priestley, R. J. and Birch, G. G. (1973). An alternative to the alkali test for assessing the quality of milled rice. *Lebensmit. Wiss. u. Technol.*, **6**, 224–6.

25. Birch, G. G. and Renner, J. (1976). Calorific values of heated sucrose–citric acid mixtures, unpublished results.

26. Birch, G. G. and Priestley, R. J. (1973). Degree of gelatinisation of cooked rice. *Die Stärke*, **25**, 98–100.

27. Pyler, R. J. (1973). *Baking and Science Technology*, Vol. I, Siebel Publ. Co., Chicago, Ill.

28. El-Harith, E. A., Walker, R., Birch, G. G. and Sukan, G. (1976). Unpublished results.

29. Birch, G. G. and Etheridge, I. J. (1976). Short-term effects of feeding male subjects with glucose and syrup fractions and D-glucose. *Nutrition and Metabolism*, in press.

30. Birch, G. G. (1974). Dietary carbohydrate and intellectual performance. *Biochem. Soc. Trans.*, **2**, 300–1.

31. Brooke, J. D., Toogood, S., Green, L. F. and Bagley, R. (1973). Factory accidents and carbohydrate supplements. *Proc. Nutr. Soc.*, **32**, 94A.

32. Thompson, J. I. and Co. (1972). GRAS (generally recognised as safe) food ingredients—caramel. NTIS Rept. prepared for FDA.

33. Simonyan, T. A. (1969). Toxico-hygienic characteristics of oxymethylfurfural. *Voprosy Pitomiya*, **28**, 54–8.

34. Bender, A. E. (1966). Nutritional effects of food processing. *J. Food Technol.*, **1**, 261–90.

35. Barker, S. A., Somers, P. J. and Stevenson, J. (1974). Redissolvable ferric-D-fructose and ferric-D-glucose-D-fructose complexes. *Carbohydrate Res.*, **36**, 331–7.

36. Birch, G. G., Coulson, C. B., Paipradist, V. A. and Connett, D. A. (1973). Low-calorie natural orange juice. *Confructa*, **18**, 45–7.

37. Bhotiyakornkiat, V. A. and Birch, G. G. (1972). Phosphate content in glucose syrup conversion. *Process Biochem.*, **7**, 25–6.

38. Shallenberger, R. S. and Acree, T. E. (1967). Molecular theory of sweet taste. *Nature*, **216**, 480–2.

39. Birch, G. G. and Lee, C. K. (1971). The chemical basis of sweetness in model sugars. In *Sweetness and Sweeteners*, ed. by G. G. Birch, L. F. Green and C. B. Coulson, Applied Science, London.

10

Maillard Reactions in Foods

R. F. HURRELL AND K. J. CARPENTER

Department of Applied Biology, University of Cambridge, Pembroke Street, Cambridge CB2 3DX, England

Most people in the food-processing industry will at some time or other have encountered Maillard reactions, which occur between proteins, amino acids and amines with sugar, aldehydes and ketones and which appear to be the major cause of browning during the heating or prolonged storage of protein foods. Although they are necessary in certain foods for the development of flavours and odours, they can cause a severe reduction in the nutritive value of the protein component.

MECHANISM OF REACTION

Maillard reactions are complex and as yet not fully understood, although they do appear to follow common pathways.[1] The preliminary steps are

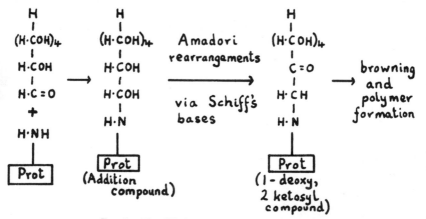

FIG. 1.   Simplified scheme of Maillard reactions.

shown in a simplified form in Fig. 1. The first step involves a condensation reaction between the carbonyl group of a reducing sugar with the free amino group from an amino acid or protein. The condensation product formed immediately loses a molecule of water and is converted into a Schiff's base. This is converted into the *N*-substituted aldosylamine, which is immediately converted to the 1-amino-1-deoxy-2-ketose by the Amadori rearrangement.

Following the formation of this deoxy-ketosyl derivative, the reactions leading to the formation of brown pigments are not too well defined. In these later stages however, amino acids can be degraded directly by reaction with other carbonyl compounds which are formed during the reactions.

FACTORS CONTROLLING REACTIONS

It is now well-known that Maillard reactions are greatly influenced by moisture content, pH, temperature and type of sugar used. Lea and Hannan[2] made the classic study of the reactions between proteins and reducing sugars under mild conditions. They measured the loss of free amino groups from a casein–glucose mixture and showed that the reaction rate was greatest at about 15 % moisture. Very low moisture contents and very high moisture contents almost stop the reactions completely. Lea and Hannan[2] also found that the reaction rate was increased by a rise in pH; it was 10 times faster at pH 7 than at pH 3. Temperature however, had the most dramatic effect, the reaction was 9000 times faster at 70 than at 10 °C and 40 000 times faster at 80 than at 0 °C.

Amongst carbohydrates, only reducing sugars can take part in Maillard reactions as they alone have the necessary carbonyl groups. Lewis and Lea[3] found that at 37 °C and 15 % moisture, the order of reactivity was xylose, arabinose, glucose, lactose, maltose, fructose. Fructose was about one-tenth as reactive as glucose.

The extent of physical content between the reacting substances also influences reaction rates under these 'dry' conditions. We have observed very different losses of lysine in materials heated under identical conditions of temperature, pH and moisture content. Materials mixed as a wet paste and then freeze-dried lost much more reactive lysine when heated under controlled conditions than if simply mixed in a 'powder' form. Presumably mixing to a paste and freeze-drying brings the sugar and protein molecules into more intimate contact.

NUTRITIONAL AND PHYSIOLOGICAL CONSEQUENCES

In food proteins most primary amino groups are represented by the ε-amino groups of lysine. Lysine reacts with reducing sugars under mild conditions to form biologically unavailable lysine–sugar complexes[4] and then with further heat there is total destruction of the lysine molecule. As it is an essential amino acid, destruction or loss of availability reduces the nutritive value of the corresponding protein.

Under mild conditions of heat such as during storage, the dominant effect on protein quality is a fall in the level of available lysine with little change in the overall digestibility of nitrogen.[5,6] After only 5 days at 37 °C and 15 % moisture, Lea and Hannan[7] reported that 70 % of the lysine groups in their casein–glucose system no longer reacted with fluorodinitrobenzene (FDNB), with only minimal losses of other amino acids.

Materials of this type show little actual browning and the sugar-bound lysine is thought to be mainly in the form of the deoxy-ketosyl derivative.[4,8] Finot and Mauron[9] synthesised ε-N-deoxy-fructosyl lysine from lysine and glucose and showed it to be unavailable as a source of lysine to rats,[4] 60–70 % of the orally administered compound appearing in the urine. With the peptide-bound moiety of lactulosyl lysine from heat damaged milk powder, he recovered only 6–9 % in the urine and 6–25 % in the faeces. It seems that, due to hindered trypsin action, only 10 % of the lysine–sugar units are released from the protein chain.[8] Once released, fructosyl lysine can be absorbed by passive diffusion;[10] however, any units which are not liberated from the peptide chain or not absorbed from the intestine can be destroyed by the intestinal flora of the hind gut.[11] This can give misleading results in digestibility experiments since unavailable compounds cannot be detected by faecal analysis.

With increasing severity of heat treatment Maillard reactions go past the deoxy-ketosyl stage. This affects not only the availability of lysine but the availability of all other amino acids and the digestibility of the protein as a whole. Advanced Maillard damage can destroy large proportions of lysine and arginine, and to a lesser extent tryptophan, cystine and histidine and can reduce the release of all others by *in vitro* enzymic hydrolysis.[8,12,13] Erbesdobler[8] measured the availability of all amino acids in casein–glucose water mixtures dried at different temperatures. The results are shown in Fig. 2. The mixture dried at 90 ° showed serious damage to lysine only, as judged by amino acid analysis of *in vitro* enzymic hydrolysates. The mixture dried at 105 °C however, showed much more general damage. Enzymic hydrolysis recovered 16, 26 and 31 % of the original lysine, arginine and histidine,

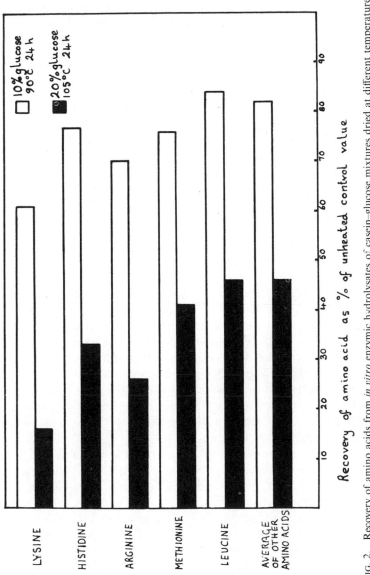

FIG. 2. Recovery of amino acids from *in vitro* enzymic hydrolysates of casein–glucose mixtures dried at different temperatures.[8]

respectively, compared to recoveries of between 34–55% for the other amino acids. This demonstration that the enzymic release of all amino acids, even those such as leucine with chemically inert side-chains, is retarded, indicates that the main cause of heat damage is a reduction in protein digestibility. However, this reduction in digestibility does not always completely explain the decreased nutritive value of the protein.

TABLE 1

*Protein quality values[14] on freeze-dried cod muscle heated with 10%
glucose at 14% moisture for 27 h at 85°C*

	A Unheated	B Heated with Glucose	B/A
Rat			
N digestibility (%)	96	41	0·43
Net protein utilisation (%)	91	18	0·20
Chick			
Available lysine (g/16g N)	11·2	0·2	0·02
Available methionine (g/16g N)	3·6	0·4	0·11

Table 1 from Miller, Carpenter and Milner[14] shows some protein quality values on freeze-dried cod muscle heated with 10% glucose. The N digestibility fell to 43% of its original value but the net protein utilisation and the availabilities of lysine and methionine to the chick fell much more. There was no destruction of methionine and the 35% destruction of lysine does not explain its low availability. It is possible that amino acids were absorbed from the gut in non-metabolisable forms and excreted in the urine. The formation of toxic compounds[15] has also been implicated as contributing to the reduced nutritive value of materials containing products of the Maillard reaction.

It has been suggested that profuse enzyme-resistant cross-linkages are introduced through reactions involving protein side-chains and the products of advanced Maillard reactions.[16] These cross-linkages appear to reduce the rate of protein digestion, possibly by preventing enzyme penetration or by masking the sites of enzyme attack.[17] The process of digestion is time limited and it seems that if more time could be spent in contact with digestive enzymes, such as *in vitro* studies, then more heat-damaged protein would be digested and less would appear in the faeces.[18]

As the S amino acids, methionine and cystine, are first limiting in the protein fraction of many human diets, the effect of Maillard reactions on these amino acids is obviously important. As reported above, methionine availability can be considerably reduced in severely heated protein–sugar mixes. However, methionine can also be damaged after relatively short periods of heating. Rao and McLaughlan[19] autoclaved a casein–glucose mixture at 121 °C for different times up to 40 min. Their results for FDNB-reactive lysine and *Streptococcus faecalis* available methionine are shown in

TABLE 2

Levels[a] of lysine and methionine in casein–glucose autoclaved[19] at 121 °C,

Time of autoclaving (min)	FDNB-reactive lysine	S. faecalis available methionine
0	100	100
5	92	93
10	57	83
20	35	61
40	32	53

[a] Percentage of unheated control value.

Table 2. It can be seen that, although lysine is damaged at a faster rate, available methionine was reduced to 61 % of its original value after only 20 min and to 53 % after 40 min. Cystine is heat labile and appears to be sensitive to the presence of sugars.[13] Although losses of available methionine will result as an indirect effect of reduced digestibility, it has been suggested by some workers that the oxidation of methionine to methionine sulphoxide or to methionine sulphone may be a further important factor in reducing methionine availability.[20–22]

QUALITY CONTROL PROCEDURES

It is important of course that quality control procedures should detect damage in both the early and the later stages of Maillard reactions. Early Maillard damage may occur before browning develops. Since lysine is normally the most severely damaged amino acid, many chemical procedures for the determination of reactive lysine in heated foodstuffs have been developed. It had been implicitly assumed that these methods would give similar results. However, it now appears that this is not always true when early Maillard compounds such as fructosyl lysine are present in the test materials.[23,24]

Fructosyl lysine, or rather its breakdown products on acid hydrolysis, are relatively easy to detect and can be used as an indicator of early Maillard damage. When it is hydrolysed in 6M HCl (Fig. 3) only 50% of its theoretical lysine content is recovered as lysine itself with a further 30% as

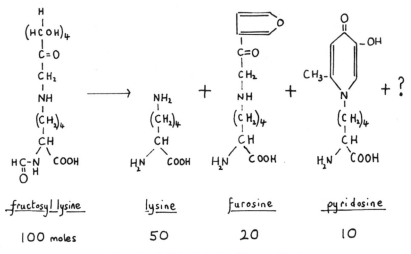

FIG. 3.   Acid hydrolysis of fructosyl lysine.

furosine and 10% as pyridosine.[4] Furosine is easy to detect by ion-exchange chromatography, where it appears just after arginine.

We have stored albumin with glucose (1:1) for 30 days at 37°C and 15% moisture so as to produce early Maillard damage. The presence of furosine in acid hydrolysates of this material was taken as confirmation that fructosyl lysine was present. The same material heated for 15 min at 121°C was used for studying advanced Maillard damage.

The heated and unheated materials were analysed by several methods for reactive lysine, but we shall consider only the fluorodinitrobenzene (FDNB) method [25,26] and the trinitrobenzene sulphonic acid (TNBS) method.[27] In the FDNB method, FDNB reacts with the free ε-amino groups of the protein and after acid hydrolysis, the yellow dinitrophenyl (DNP) lysine produced is measured colorimetrically. TNBS was suggested as an alternative reagent as, unlike FDNB, it is water soluble and less vesicant to the skin. It reacts like FDNB to produce TNP lysine.

Our results (Table 3) show that after 30 days at 37 °C, 24 % of the FDNB-reactive lysine remained compared to 22–34 % by rat and chick assays. The TNBS method however still predicted 67 % availability, more than was predicted by total lysine. With more severe heating the reactions appear to have gone past the deoxyfructosyl stage as less furosine was detected. TNBS and FDNB now gave similar predictions as to the extent of heat damage. It

TABLE 3
Lysine values[a] for heated albumin–glucose mixtures[24]

	Early Maillard 30 days at 37 °C	Advanced Maillard 15 min at 121 °C
Total	59	34
TNBS	67	21
FDNB	24	15
Rat and chick	22–34	—

[a] Percentage unheated control values.

has been proposed that TNBS can react with the ε-N portion of the Maillard compound and that on acid hydrolysis there is a high yield of TNP lysine.[23] Azo-dyes also appear to bind with fructosyl lysine, which makes them unusable as rapid predictors of early Maillard damage.[28]

It should be pointed out that in very indigestible materials, lysine residues with free ε-amino groups can pass through the digestive tract in indigestible peptides and so be unavailable. As they will still be measured as chemically reactive lysine, all reactive lysine procedures can overestimate biological availability in such materials.

HEAT TREATMENT OF FOODSTUFFS

Home cooking and commercial canning procedures have little effect on the nutritive value of food proteins. However, other industrial processes which involve the use of excessive heat at low moisture levels, such as during drying, and the subsequent storage of these products may result in severe nutritional damage.

Milk

Milk is the only important naturally occurring protein food that has a high content of reducing sugar. Liquid milk appears virtually unaffected by industrial pasteurisation[29] and only slightly affected by autoclaving and canning,[30] but the protein can deteriorate in dried milk during manufacture and storage.[5,31] Rolls and Porter[32] have reported that in efficiently spray-dried milk the availability of lysine is high (90–97 %), whereas in roller-dried milk it can vary from 60–95 % according to the manufacturing conditions. Moisture content greatly influences lysine damage during storage of dried milk and at ordinary temperatures it appears that moisture content greater than 5 % is necessary to produce a significant rate of damage.[5]

Damage is mainly of the early Maillard type; lysine reacts with lactose and is present mainly in the deoxyketosyl form, lactulosyl lysine.[8] Furosine and pyridosine were first detected in acid hydrolysates of dried milk damaged by heat during manufacture.[33−36]

Foods Containing Sucrose

Lea[37] reported that the non-reducing sugar sucrose did not react with casein when kept in contact with it for 6 months at 37 °C and 55 % relative humidity. Other workers however, have found that at higher temperatures foods containing sucrose can undergo Maillard reactions,[12,38] presumably as a result of the splitting of glycosidic bonds of sucrose to yield glucose and fructose.

We have investigated the loss of FDNB-reactive lysine from a 1:1 albumin–sucrose mix heated under controlled conditions in sealed glass ampoules.[39] The loss of reactive lysine (Fig. 4) from the albumin–sucrose mix increased as the heat process was increased. After 1 h at 100 °C, 94 % of the reactive lysine remained but this was reduced to 53 % after 1 h at 121 °C and to 13 % after 2 h at 121 °C. FDNB-reactive lysine in an albumin–glucose mix heated under the same conditions was reduced to 14 % of its original value after only 1 h at 100 °C. When albumin was heated with partially inverted sucrose (processed sucrose), lysine damage after the two milder heat treatments was between that which had occurred with sucrose and glucose. When albumin was heated alone there was virtually no reduction in reactive lysine even after 2 h at 121 °C.

As the moisture level of the albumin–sucrose mix was increased from 1–20 %, so the heat treatment (1 h at 121 °C) rendered progressively more lysine units inaccessible to FDNB (Fig. 5). At 1 % moisture there was little damage and 93 % of the reactive lysine remained after heating compared to 97 % in the albumin mix heated alone. However at 20 % moisture the reactive lysine remaining after heating was reduced to 42 % of its original value. As the pH was increased from 5–9 the loss of lysine by heating for 1 h at 121 °C and 15 % moisture was greatly reduced (Table 4). This is directly opposite to the effect of pH on the lysine–glucose reaction.

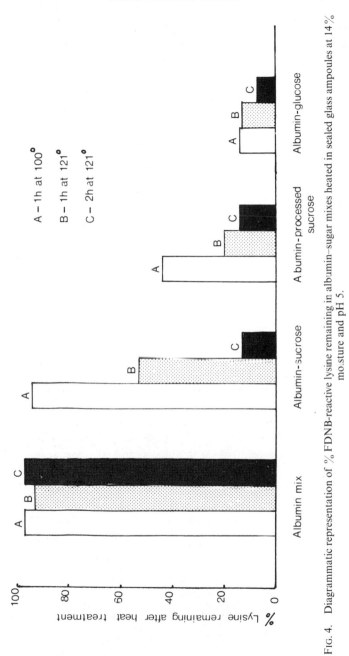

FIG. 4. Diagrammatic representation of % FDNB-reactive lysine remaining in albumin–sugar mixes heated in sealed glass ampoules at 14% moisture and pH 5.

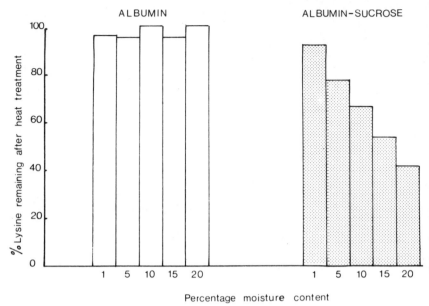

Fig. 5. Diagrammatic representation of % FDNB-reactive lysine remaining in albumin and in albumin–sucrose after heating for 1 h at 121 °C, pH 5 and different moisture levels.

The effect of time and temperature on the extent of lysine damage in protein–sucrose systems can be explained by an increase in heat treatments increasing sucrose inversion. Similarly moisture content and pH appeared to influence sucrose inversion, which seemed to be the rate-determining step at the relatively high temperatures used, rather than influencing the subsequent Maillard reactions with glucose and fructose. It is well known

TABLE 4

FDNB-reactive lysine values[a] of albumin–sucrose mix heated for 1 h at 121 °C, 15% moisture and different levels of pH

pH	Reactive lysine
5	30
7	71
9	87

[a] Percentage unheated control value.

that dilute acids increase sucrose hydrolysis and that dilute alkalis suppress it.

Maillard reactions involving protein and sucrose can occur in groundnuts either during processing to separate the oil from the protein concentrate or during roasting.[38] They can also occur during the production of soya bean meal.[12]

Bread and Cereals

Maillard reactions occur during the baking of bread and biscuits and during the production of breakfast cereals. Rosenberg and Rohdenburg[39a] reported a loss of 10–15 % lysine during bread baking. This loss is mainly in the crust.[40] Starch itself does not appear to react with the protein, but reducing sugars formed during the fermentation stage are not completely utilised by the yeast. Biscuits are an attractive vehicle for protein enrichment but are subjected to relatively high temperatures and are baked to a low moisture level. Carpenter and March[41] reported a 50 % loss of reactive lysine during the production of biscuits containing 15 % skim milk powder, which were intended for the treatment of kwashiorkor.

Sucrose does not appear to be inverted to any great extent by normal baking but can be completely inverted by prior fermentation with yeast. This became apparent when we were investigating the reasons for the very low nutritive value of a high protein cake mix prepared by Block *et al.*[42] The cake mix contained 50 % flour, 20 % albumin and 20 % sucrose as its main ingredients. When repeating this experiment (Table 5) we found that after

TABLE 5
Effect of baking and toasting on PER and lysine values of a cake mix and an albumin–sucrose mix

	Cake mix[a]	Cake mix[b]	Albumin–sucrose mix[b]
PER (rat growth)			
Uncooked	3·5	3·9	4·1
Baked and toasted	0·7	0·8	3·7
Baked and toasted + lysine	3·2	2·6	4·1
Lysine (g/16g N)			
Uncooked	6·3	7·2 (5·9)[c]	8·2 (8·0)[c]
Baked and toasted	5·3	3·9 (2·2)[c]	7·9 (6·4)[c]

[a] Ref. 42.
[b] Hurrell and Carpenter, unpublished results.
[c] FDNB-reactive lysine.

baking with yeast and toasting there was, as had been reported by Block *et al.*[42] an 80% fall in protein efficiency ratio (PER), which was mostly restored by the addition of lysine. However, when the albumin and sucrose fractions were baked and toasted in a similar way but without prior fermentation there was only a very small reduction in PER and the fall in lysine was also much less. Analysis of the fermented cake mix dough showed that sucrose had been completely inverted to glucose and fructose before baking. Maillard reactions during baking and toasting would explain the low nutritive value.

Animal Protein Concentrate

Low nutritive value of meat meals is now thought to be mainly due to the low quality of raw materials used for their production rather than heat damage.[43] Meat does contain small amounts of glucose and Skurray and Cummings[44] have reported that Maillard reactions do occur during the production of meat meals. They may also occur to a minor extent during the production of fish meals, which contain small amounts of reducing sugars. However, as reducing sugars and lysine react in a 1:1 ratio the effect on lysine is small. [45]

Blood too contains small amounts of glucose, but again Maillard damage is not thought to be responsible for the low nutritive value of some commercial blood meals. At the high temperatures during vat-drying (175 °C for 10–12 h) damage appears to be explained by reactions within the proteins themselves involving cross-linking between amino acid side-chains which greatly reduces protein digestibility.[46]

Damage to Free Amino Acids

The use of supplementation of foods with amino acids and feed mixtures, makes it necessary to consider the reactions of free amino acids with reducing sugars. Since free amino acids have free α-amino groups they can take part in Maillard reactions to be either bound or destroyed. Evans and Butts,[12] showed that free lysine and free arginine were damaged in a similar way to their peptide-bound forms but that methionine, valine, phenylalanine and aspartic acid were much more damaged in the free form. The losses of free lysine and methionine after enzymic hydrolysis of the heated protein–sucrose mixture were reported as 75% and 74%, respectively, compared with losses of 84% and 41% for the corresponding peptide-bound amino acids. When synthetic lysine was added to biscuits it appeared to be lost at the same rate as peptide-bound lysine.[47] With regard to methionine, Horn, Lichtenstein and Womack[48] have shown that the sugar-bound form, fructose–methionine is unavailable to the rat as a source of methionine and likewise, fructose–tryptophan and fructose–leucine have also been shown to be highly unavailable to rats.[49]

CONCLUSIONS

Finally, to put the Maillard reactions and their consequences into perspective, it would appear that there is little need in the well-fed Western World for food manufacturers to pay much attention to the protein quality of their products, for it is of little importance if one or two foods out of the many eaten lose a small part of their protein quality. Loss of available lysine in breakfast cereals, for instance, is more than made up for by the milk they are eaten with. Many of the flavours that we appreciate are produced at the expense of protein quality.

With baby foods and dried baby milks the need to control protein quality is greater since the baby is often dependent on a single food item. In underfed countries, any loss in protein quality represents an effective decrease in protein supplies. The controlling and monitoring of protein quality is also very important in the animal feedstuffs industry, where the compounder needs to know the biologically available content of those amino acids likely to be limiting.

Although lysine is the most sensitive amino acid to Maillard damage, especially under mild conditions, all amino acids can be reduced in availability when the heat process is increased. Non-reducing sugars such as sucrose can also take part in Maillard reactions by first being inverted to glucose and fructose by heat treatment or fermentation. The food manufacturer therefore needs to know under what conditions Maillard reactions occur, what analytical tests adequately monitor nutritional damage and, if necessary, what preventative measures can be taken.

REFERENCES

1. Hodge, J. E. (1953). Chemistry of browning reactions in model systems. *J. Agric. Food Chem.*, **1**, 928–43.
2. Lea, C. H. and Hannan, R. S. (1949). The effect of activity of water, of pH and of temperature on the primary reaction between casein and glucose. *Biochim. Biophys. Acta*, **3**, 313–25.
3. Lewis, V. M. and Lea, C. H. (1959). A note on the relative rates of reaction of

several reducing sugars and sugar derivatives with casein. *Biochim. Biophys. Acta*, **4**, 532–4.

4. Finot, P. A. (1973). Non-enzymic browning. In, *Proteins in Human Nutrition*, ed. by J. W. G. Porter and B. A. Rolls, Academic Press, London.
5. Henry, K. M., Kon, S. K., Lea, C. H. and White, J. D. C. (1948). Deterioration on storage of dried skim milk. *J. Dairy Res.*, **15**, 292–363.
6. Henry, K. M. and Kon, S. K. (1950). Effect of reaction with glucose on the nutritive value of casein. *Biochim. Biophys. Acta*, **5**, 455–6.
7. Lea, C. H. and Hannan, R. S. (1950). Studies of the reaction between proteins and reducing sugars in the 'dry' state: II. Further observations on the formation of the casein–glucose complex. *Biochim. Biophys. Acta*, **4**, 518–31.
8. Erbesdobler, H. (1976). Amino acid availability. In, *Protein Metabolism and Nutrition*, ed. by D. J. A. Cole, K. N. Boorman, P. J. Buttery, D. Lewis, R. J. Neale and H. Swan, Butterworths, London.
9. Finot, P. A. and Mauron, J. (1969). The blockage of lysine by the Maillard reaction: I. Synthesis of *N*-(1-deoxy-D-fructosyl)- and *N*-(1-deoxy-D-lactulosyl)-L-lysine (French). *Helv. Chim. Acta*, **52**, 1488–95.
10. Erbesdobler, H. *et al.* (1974). The absorption and distribution of fructose–lysine. *Zeitschrift für Tierphysiologie, Tierernährung und Futtermittelkunde*, **33**, 202–3.
11. Erbesdobler, H., Gunsser, I. and Weber, G. (1970). Degradation of fructose–lysine by the intestinal flora (German). *Zentralblatt für Veterinärmedizin*, **17**, 573–5.
12. Evans, R. J. and Butts, H. A. (1949). Inactivation of amino acids by autoclaving. *Science*, **109**, 569–71.
13. Miller, E. L., Hartley, A. W. and Thomas, D. C. (1965). Availability of sulphur amino acids in protein foods: 4. Effect of heat treatment upon the total amino acid content of cod muscle. *Brit. J. Nutr.*, **19**, 565–73.
14. Miller, E. L., Carpenter, K. J. and Milner, C. K. (1965). Availability of sulphur amino acids in protein foods: 3. Chemical and nutritional changes in heated cod muscle. *Brit. J. Nutr.*, **19**, 547–64.
15. Adrian, J. and Frange, R. (1973). Maillard reaction: 8. Effect of pre-melanoidins on digestibility of nitrogen *in vivo* and on proteolysis *in vitro* (French). *Annales de là Nutrition et de l'Alimentation*, **27**, 111–23.
16. Valley-Riestra, J. and Barnes, R. H. (1970). Digestion of heat-damaged egg albumen by the rat. *J. Nutr.*, **100**, 873–82.
17. Hurrell, R. F. *et al.* (1976). Mechanisms of heat damage in proteins: 7. The significance of lysine-containing isopeptides and of lanthionine in heated proteins. *Brit. J. Nutr.*, **35**, 383–95.
18. Buraczewski, S., Buraczewska, L. and Ford, J. E. (1967). The influence of heating of fish proteins on the course of their digestion. *Acta Biochimica Polon.*, **14**, 121–31.
19. Rao, M. N. and McLaughlan, J. M. (1967). Lysine and methionine availability in heated casein–glucose mixtures. *J. Assoc. Offic. Analyt. Chem.*, **50**, 704–7.
20. Ellinger, G. M. and Palmer, R. (1969). The biological availability of methionine sulphoxide. *Proc. Nutr. Soc.*, **28**, 42A.
21. Cuq, J. L., Provansal, M., Guilleux, F. and Cheftel, C. (1973). Oxidation of

methionine residues of casein by hydrogen peroxide: Effects on *in vitro* digestibility. *J. Food Sci.*, **38**, 11–13.

22. Slump, P. and Scheuder, H. A. W. (1973). Oxidation of methionine and cystine in foods treated with hydrogen peroxide. *J. Sci. Food Agric.*, **24**, 657–61.
23. Finot, P. A. and Mauron, J. (1972). The blockage of lysine by the Maillard reaction: II. Chemical properties of the *N*-(1-deoxy-D-fructosyl)- and *N*-(1-deoxy-D-lactulosyl)- derivatives of L-lysine (French). *Helv. Chim. Acta*, **55**, 1153–64.
24. Hurrell, R. F. and Carpenter, K. J. (1974). The reactive lysine content of heat-damaged material as measured in different ways. *Brit. J. Nutr.*, **32**, 589–604.
25. Carpenter, K. J. (1960). The estimation of available lysine in animal-protein foods. *Biochem. J.*, **77**, 604–10.
26. Booth, V. H. (1971). Problems in the determination of FDNB-available lysine. *J. Sci. Food Agric.*, **22**, 658–64.
27. Kakade, M. L. and Liener, I. E. (1969). Determination of available lysine in proteins: *Analyt. Biochem.*, **27**, 273–80.
28. Hurrell, R. F. and Carpenter, K. J. (1975). The use of three dye-binding procedures for the assessment of heat damage to food proteins. *Brit. J. Nutr.*, **33**, 101–15.
29. Budney, J., Chodkowska, B. and Rutkowski, A. (1964). Influence of heat on the available lysine content in milk (Polish). *Przemysl Spozywczy*, **18**, 153–7.
30. Porter, J. W. G. (1964). Reviews of the progress of dairy science. D. Nutritive value of milk and milk products. *J. Dairy Res.*, **31**, 201–20.
31. Erbesdobler, H. (1970). Loss of lysine during manufacture and storage of dried milk (German). *Milchwissenschaft*, **25**, 280–4.
32. Rolls, B. A. and Porter, J. W. G. (1973). Some effects of processing and storage on the nutritive value of milk and milk products. *Proc. Nutr. Soc.*, **32**, 9–15.
33. Erbesdobler, H. and Zucker, H. (1966). Lysine and available lysine in dried skim milk (German). *Milchwissenschaft*, **21**, 564–8.
34. Finot, P. A., Bricout, J., Viani, R. and Mauron, J. (1968). Identification of a new lysine derivative obtained upon acid hydrolysis of heated milk. *Experientia*, **24**, 1097–9.
35. Heyns, K., Heukeshoven, J. and Brose, K. H. (1968). Degradation of fructose amino acids to *N*-(2-furoylmethyl) amino acids: Intermediates in browning reactions. *Angew. Chemie, Intern. Ed.*, **7**, 628.
36. Finot, P. A., Viani, R., Bricout, J. and Mauron, J. (1969). Detection and identification of pyridosine, a second lysine derivative obtained upon acid hydrolysis of heated milk. *Experientia*, **25**, 134–5.
37. Lea, C. H. (1948). The reaction between milk protein and reducing sugar in the 'dry state'. *J. Dairy Res.*, **15**, 369–76.
38. Anantharaman, K. and Carpenter, K. J. (1971). Effects of heat processing on the nutritional value of groundnut products: II. Individual amino acids. *J. Sci. Food Agric.*, **22**, 412–18.
39. Hurrell, R. F. (1974). Studies of heat damage to foods containing protein and carbohydrate. Thesis, Cambridge, UK.
39a. Rosenburg, H. R. and Rohdenburg, E. L. (1951). The fortification of bread with lysine: The loss of lysine during baking. *J. Nutr.*, **45**, 593–8.

40 Gorbach, G. and Regula, E. (1964). The loss of essential amino acids in the baking process. *Fette: Seifen: Anstrichmittel*, **66**, 920–5.
41. Carpenter, K. J. and March, B. E. (1961). The availability of lysine in groundnut biscuits used in the treatment of kwashiorkor. *Brit. J. Nutr.* **15**, 403–10.
42. Block, R. J. *et al.* (1946). The effects of baking and toasting on the nutritional value of proteins. *Arch. Biochem.*, **10**, 295–301.
43. Atkinson, J. and Carpenter, K. J. (1970). Nutritive value of meat meals: 1. Possible growth depressant factors. *J. Sci. Food Agric.*, **21**, 360–5.
44. Skurray, G. R. and Cumming, R. B. (1974). Physical and chemical changes during batch dry rendering of meat meals. *J. Sci. Food Agric.*, **25**, 521–7.
45. Carpenter, K. J., Morgan, C. B., Lea, C. H. and Parr, L. J. (1962). Chemical and nutritional changes in stored herring meal: 3. Effect of heating at controlled moisture contents on the binding of amino acids in freeze-dried press cake and in related model systems. *Brit. J. Nutr.*, **16**, 451–65.
46. Waibel, P. E., Cuperlovic, M., Hurrell, R. F. and Carpenter, K. J. (1976). Processing damage to lysine and other amino acids in the manufacture of blood meal. *J. Agric. Food Chem.*, submitted for publication.
47. Clark, H. E., Howe, J. M., Mertz, E. T. and Reitz, L. L. (1959). Lysine in baking powder biscuits. *J. Amer. Dietetic Assoc.*, **35**, 469–71.
48. Horn, M. J., Lichtenstein, H. and Womack, M. (1968). Availability of amino acids: A methionine–fructose compound and its availability to micro-organisms and rats. *J. Agric. Food Chem.*, **16**, 741–5.
49. Sgarbieri, V. C., Amaya, J., Tanaka, M. and Chichester, C. O. (1973). Nutritional consequences of the Maillard reaction: Amino acid availability from fructose–leucine and fructose–tryptophan in the rat. *J. Nutr.*, **103**, 657–63.

11

Vitamin Losses during Thermal Processing

A. Benterud

Collett-Marwell-Hauge a/s, PO Box 204, N-1371, Asker, Norway

The current interest in nutritional labelling of food products has drawn increased attention to the question of what happens to the vitamins during thermal processing. A multitude of data has been published in the literature, covering studies on various conditions of thermal treatment. Excellent reviews from 1960 have been presented by Harris and von Loesecke in their book[1] and from 1971 by Bender.[2] In the following paper a brief summary of present knowledge will be given for the individual vitamins, with particular emphasis on more recent publications.

First, let us have a look at some thermal stability data of synthetic vitamins taken from a model experiment in the laboratory of Collett–Marwell Hauge. As a manufacturer of vitamin preparations our company was interested in studying the fate of vitamins when they are incorporated in a hot, practically water-free melt of carbohydrates in which they are protected from atmospheric oxygen. Figure 1 shows the retention of individual vitamins when they had been kept for exactly 15 minutes in that melt at a constant temperature, varying from 100 to 130 °C, after which the mixture was quickly cooled and analysed by chemical and physical methods.

It is clear from the graph that three vitamins are extremely heat resistant, namely vitamin E in the ester form, riboflavin and niacinamide, whereas thiamine is definitely the most labile of the vitamins. Some vitamins show a gradual degradation as the temperature is raised from 100 to 130 °C (vitamins A, D, B_{12} and C). With other vitamins there is a sudden drop above a certain temperature (with folic acid, above 110 °C; with pyridoxine and pantothenic acid, above 120 °C).

In food processing the vitamin stability situation is more complicated than this because other factors are involved in addition to temperature and

A. Benterud

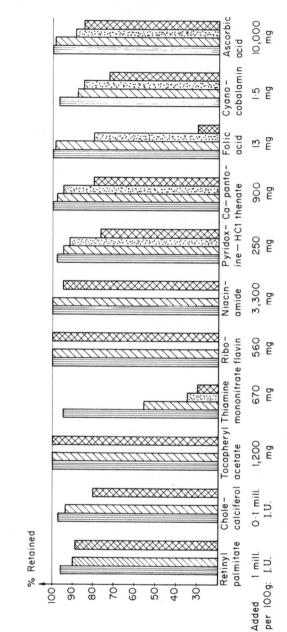

FIG. 1. Thermal destruction of synthetic vitamins.

time. Rate of decomposition is influenced by pH, presence of metal ions, oxidising or reducing agents, exposure to air, etc. Naturally-occurring antioxidants and protein complexes may have a protective effect, so that losses in foods may be less than what is found in solutions of vitamins.

VITAMIN A AND CAROTENE

Vitamin A and carotene lose activity when food is heated in the presence of oxygen, but they are fairly stable to ordinary cooking. Insignificant losses are found in milk during pasteurisation, sterilisation and spray-drying. By studies in our own laboratory we have found in the commercial baking of Norwegian vitaminised bread a vitamin A retention of 80–90% and a retention of 70–80% in biscuits and sweet rusks.

Carotene is well retained during the dehydration of vegetables if enzymes have been inactivated by blanching. According to earlier reports, the canning of vegetables causes little or no degradation of the naturally occurring carotenoid–protein complexes.[3] Sweeney and Marsh, however, found by separating carotene stereoisomers in a high-pressure liquid chromatograph that considerable isomerisation of all *trans*-to-*cis* forms may occur during canning, with a reduced biological potency as a result.[4] The vitamin A activity of green vegetables was lowered by 15–20%, the activity of yellow vegetables by 30–35%. On the other hand, an increased digestibility of the carotenoids after canning may improve the nutritional vitamin A value.

Reports from Hoffmann la Roche laboratories have demonstrated a good stability of synthetic β-carotene used as a colouring agent in commercially canned soups.[3] Retention in cakes baked at 175–200 °C was 90–95%, whereas in white bread it was only 50–80%. Degradation of added carotenoids in foods frequently causes a development of hay-like odours and flavours.

Frying of oils and fats may lead to considerable losses of carotene and vitamin A. Figure 2 illustrates the results of a study in our laboratory in which Norwegian vitaminised margarine was heated in a 2–3 mm layer in a frying-pan with occasional stirring.[5] The temperature of the melt was 130–160 °C in one experiment, 175–200 °C in another. Vitamin A was well retained as long as water was present, but after 24 and 12 min, respectively, 50% had been destroyed; after 40 and 20 min, respectively, about 90%. In the literature a 40% loss of carotene has been reported during frying of margarine for 5 min, and a 60% loss after 10 min.[6]

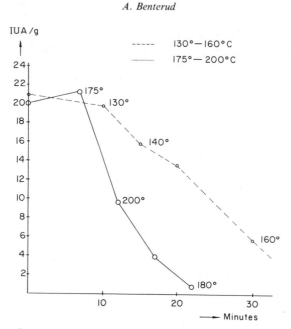

FIG. 2.   Losses of vitamin A in margarine heated in a frying pan.[5]

VITAMINS D AND E

Very little has been published about thermal stability of vitamin D in foods, obviously due to the lack of a simple and accurate analytical method. The vitamin is sensitive to the same factors as vitamin A, and the heat resistance of vitamin D is generally considered to be similar to that of vitamin A. This has been confirmed by authors who have reported a satisfactory vitamin D retention during the smoking of fish, spray-drying of egg and pasteurisation and sterilisation of milk.

Vitamin E in food products occurs in the form of free tocopherols, which are sensitive to oxidation by atmospheric oxygen. Since all eight tocopherols in food have different biological activities, reliable data on the stability of vitamin E can only be obtained by using modern analytical techniques. Bunnell et al.[7] studied a wide variety of foods and concluded that whereas ordinary home cooking processes do not involve large losses of tocopherol, considerable losses may result from commercial processing. Table 1 shows examples of canned vegetables, where losses have been up to 95%.

Vegetables supply very little of the human intake of vitamin E. Vegetable oils and cereals are more important dietary sources. The same authors found an 11 % loss of total tocopherol in vegetable oil during deep-fat frying, and potato chips and french fried potatoes prepared in the oil lost their tocopherol content very rapidly during storage. Heating vegetable oils for 3 h at 200 °C destroyed almost 100 % of the free tocopherols. When the

TABLE 1
Vitamin E content of fresh and canned vegetables[7]

	Total tocopherol (mg %)	α-Tocopherol (mg %)
Fresh green peas	1·73	0·55
Canned green peas	0·04	0·02
Frozen green beans	0·24	0·09
Canned green beans	0·05	0·03
Frozen kernel corn	0·49	0·19
Canned kernel corn	0·09	0·05

same oils were fortified with α-tocopheryl acetate, less than 20 % of the ester form was destroyed in the heating test.

Little information is available on the effect of baking on vitamin E. Moore *et al.*[8] observed that 47 % of the tocopherols in ordinary unbleached flour were destroyed in the baking of bread. If this should be a general trend, a fortification with tocopheryl acetate is worth consideration.

WATER-SOLUBLE VITAMINS

Data found in the literature on the losses of water-soluble vitamins cover a wide range. The conditions of processing are not always well defined, and even minor modifications may profoundly influence the vitamin content. Furthermore, calculations may have been based on data compiled from different sources, where much scatter is involved due to varying raw material characteristics and different analytical methods.

Some general rules may be given for minimising processing losses: slow heating, long cooking and slow cooling should be avoided, and so should the use of copper, copper alloy or black iron equipment. Vacuum deaeration or inert-gas treatment is recommended during processing wherever feasible. A major part of the loss is due to a leaching of soluble matter into the cooking water. Small pieces with a relatively large surface area therefore lose more vitamins than larger pieces. Precautions should be taken to minimise losses in vegetables during the scalding or blanching operations before canning or freezing. Steam blanching is the preferred method, and cooling with cold air is more satisfactory than cooling with water. Data published on electronic blanching indicate that dielectric heating does not provide significant advantages over steam blanching.

VITAMIN C

Blanching of fruits and vegetables inactivates the ascorbic acid oxidase and stabilises vitamin C. Losses of ascorbic acid due to blanching may vary from less than 10 % under the most favourable conditions and up to 50 % or more when the conditions are more severe. Similar fluctuations of vitamin C content may result in the domestic preparation of foods. Olliver[9] summarised the findings of many workers who studied different methods of cooking vegetables. As shown in Table 2, a higher loss of vitamin C is to be expected as the ratio of water to vegetable is increased. Steaming or pressure cooking causes a greater loss of vitamin than boiling in water, due to oxidation, but the amount lost by extraction is almost negligible. It has been reported by many authors that the resulting loss during pressure cooking is less than during boiling at atmospheric pressure, although in practice losses are highly dependent on the type of vegetable and the cooking time. To give an example, Fig. 3, adapted from Zacharias,[10] illustrates the distribution of ascorbic acid between vegetable and water versus time during cooking of

TABLE 2
Typical values of ascorbic acid losses in vegetables during household cooking[9]

Method	% Vitamin C		
	Destroyed	Extracted	Retained
Green vegetables			
Boiling (long time, much water)	10–15	45–60	25–45
Boiling (short time, little water)	10–15	15–30	55–75
Steaming	30–40	< 10	60–70
Pressure-cooking	20–40	< 10	60–80
Root vegetables (unsliced)			
Boiling	10–20	15–25	55–75
Steaming	30–50	< 10	50–70
Pressure-cooking	45–55	< 10	45–55

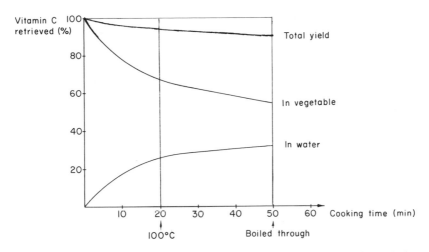

FIG. 3. Vitamin C distribution between vegetable and water during the cooking of sliced kohlrabi.[10]

sliced kohlrabi, a vegetable which is a significant dictary source of vitamin C in Norway. When the juice is eaten with the tissue, as in a mash or a stew, practically no loss takes place.

Considerable information has been published on the effect of commercial canning on the vitamin C content of fruits and vegetables. A summary of typical data is given in Fig. 4,[10] illustrating the range of 90 % of the findings. More than 90 % is retained in canned, high-acid citrus juices. In certain fruits and vegetables a protective effect may be expected from naturally-occurring flavonoid substances which are known to act as antioxidants. Small losses of ascorbic acid have also been reported during the extraction of juices and concentration of juices when proper precautions are taken.

The wide ranges indicated on the graph for vitamin C retention in canned vegetables are supposed to reflect the varying conditions of the blanching operation. High losses might also be due to the use of copper salts as colour stabilisers. It seems justified to conclude that with most types of canned foods it would be possible to keep vitamin C losses within 10–20 % by careful processing. This is corroborated by the experience of our company in the manufacture of strained baby food in glass jars. In fruit purées (pH about 3·5) where ascorbic acid has been added, we find a vitamin C loss of about 10 % after autoclave sterilisation (30 min at 95 °C).

In milk, vitamin C losses of about 20 % have been reported as the result of pasteurisation or ultra-high temperature sterilisation, and up to 50 % by

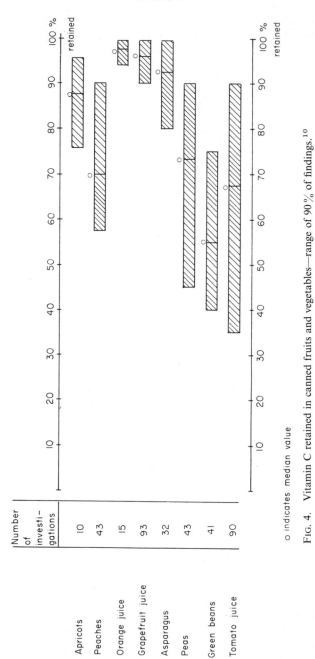

o indicates median value

FIG. 4.   Vitamin C retained in canned fruits and vegetables—range of 90% of findings.[10]

sterilisation in the bottle.[11] The stability is improved when oxygen is removed from the milk by incorporating a deaeration stage in the process.

Ascorbic acid is the most difficult of the vitamins to preserve during the dehydration of foods. Losses of 20 % have been reported in spray-dried and 30 % in roller-dried milk. The addition of sulphur dioxide to fruits and vegetables has a protective effect, and very satisfactory vitamin C retention

TABLE 3
Vitamin C losses during production of dehydrated mashed potatoes[12]

Add-back process		Freeze–thaw process	
Processing step	Loss (%)	Processing step	Loss (%)
Raw potatoes	0	Raw potatoes	0
Slicing and washing	9·3	Slicing	8·2
Blanching	16·9	Washing	14·5
Steam cooking	20·3	Steam cooking	17·4
Mashing (20 min, 60 °C)	57·4	Mashing (2 min, 80 °C)	17·6
Conditioning (cold air)	78·3	Freezing–thawing	16·2
Air-lift drying	63·6	Pre-drying	29·7
Fluid-bed drying	58·4	Granulation (fluid bed)	19·7
Fluid-bed cooling	57·0	Final drying (10 min, 60 °C)	18·9

has been found during vacuum drying. Vitamin C losses during the production of dehydrated mashed potatoes have recently been reported by Jadhav *et al.*[12] Table 3 gives the losses at each step of two different processing methods. The authors found before the final drying step a maximum vitamin C loss of 78·3 % in the add-back process, which is characterised by longer air exposure and heat treatment than the straight-through, freeze–thaw process, in which the maximum loss was 29·7 % (all losses calculated on dry weight). Strangely enough an apparent increase of the vitamin C content was observed in the final product (approximately 21 % for the add-back process and 11 % for the freeze–thaw process), which was attributed to a thermal formation of sugar–amino acid reaction products, which interfere with the analysis. Titration with 2,4-dinitrophenylhydrazine and the determination of vitamin C polarographically both showed this apparent increase. The authors' findings imply that literature data on the vitamin C content of instant potato granules may be open to question.

A greater destruction of vitamins usually occurs when vegetable foods are fried than when they are boiled. Losses of 30–50 % of ascorbic acid have been found in fried and baked potatoes. Newmark *et al.*[13] recently published an interesting report on vitamin C stability in bacon. The addition of ascorbic acid in the curing of meat products has been recommended as a means of blocking the formation of *N*-nitrosamines from the reaction of sodium nitrite with amines. Since nitrosamine formation, with the alleged carcinogenic effect, can possibly occur *in vivo* after eating bacon cured with nitrite, it is obviously important that ascorbic acid in the bacon survives the preparation for eating. In a frying test carried out at 170 °C for 6 min the authors found a retention of sodium ascorbate ranging from 67–81 %. Losses during storage of the bacon were about 1 % per week in the freezer, about 8 % per week in the refrigerator.

VITAMIN B₁

The chemistry and kinetics of thermal degradation of thiamine have been outlined in a series of interesting papers. Dwivedi and Arnold[14] in a comprehensive review showed that numerous reaction products may arise from a cleavage of the vitamin molecule to its pyrimidine and thiazole moieties. Volatile compounds formed by thermal breakdown as well as Maillard-type reaction products of thiamine are supposed to be important contributors to food flavour. Not a few patents have been issued on methods of producing meat-like flavours by heating thiamine together with carbohydrates, lecithin and other organic substances.

Farrer[15] in his review on the thermal destruction of vitamin B₁ in foods showed that the loss of thiamine could be generally described by a first-order rate equation. Teixeira *et al.*[16] applied a computer technique to optimise the influence of a canning process on thiamine retention while maintaining the required bacterial lethality. Figure 5 is taken from their study on canned green beans and shows the percentage of retained thiamine versus processing time and temperature. The curve indicates an optimum point after 90 min at a temperature of 120 °C, which is close to the conditions of normal practice.

Most reports on vitamin B₁ losses in canned vegetables show a better retention than this example. Figures ranging from 60 to 90 % retention are usual. In the blanching process losses due to leaching may amount to 20 %. In autoclaved fruits and tomato juice, where the pH is lower and processing time shorter, up to 100 % of thiamine may be retrieved. Losses reported in meat and fish products during canning vary between 20 and 70 %.

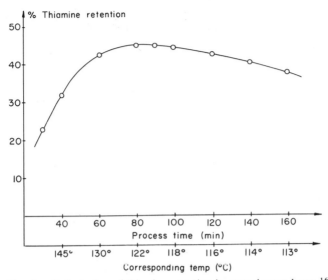

Fig. 5. Optimisation of thiamine retention in canned green beans.[16]

Everson *et al.*[17] compared aseptic canning with conventional canning and found a significantly improved thiamine retention in low-acid foods. Table 4 gives their results on strained beans and strained beef. Manufacturers of aseptic canning equipment even claim that thiamine losses can be kept below a 5 % level. In milk about 10 % is destroyed during pasteurisation or spray-drying and less than 10 % by ultra-high temperature treatment.[11]

Many workers have studied thiamine losses in meat prepared by different

TABLE 4

Vitamin B_1 losses during aseptic canning as compared with conventional canning[22]

Product	Container	Processing method	Thiamine loss (%)
Strained lima beans	Tin	HTST	15·8
	Glass	HTST	12·9
	Tin	Conventional canning	40·3
	Glass	Conventional canning	44·1
Strained beef	Tin	HTST	9·2
	Glass	HTST	5·0
	Tin	Conventional canning	21·6
	Glass	Conventional canning	18·2

TABLE 5
Vitamin B$_1$ retention in chicken prepared by different domestic methods[18]

Method	% Vitamin B$_1$ retained
Boiling	58
Pan-frying	65
Broiling on rack	78
Broiling on revolving skewer	76
Broiling in barbecue machine	65

kitchen operations. Such comparisons ought to be done with pieces of meat having identical size. Table 5 gives the results of a recent Norwegian study on chicken prepared by five commonly used methods.[18] The highest loss was found in boiled chicken, but the difference was retrieved in the cooking water, so the total yield was 100%. Broiling in conventional way gave significantly better retention than other methods of preparation.

In the baking of bread vitamin B$_1$ losses amount to 15–25%,[15] most of the losses occurring in the crust. Less destruction is found with white flour than with high-extraction flour. The effect is related to the content of minerals. In cakes where alkaline baking powder has been used, more than 50% may be lost.

Beetner *et al.*[19] studied the degradation of thiamine during extrusion cooking of cereals. As seen from Table 6, significantly higher losses were found as a result of temperature increase as well as increased velocity of rotation of the the screw.

The addition of sulphur dioxide in the dehydration of fruits and vegetables has a destructive effect on vitamin B$_1$. This is nutritionally less

TABLE 6
Retention of vitamins B$_1$ and B$_2$ during extrusion processing of corn grits[19]

Temperature (°C)	% Water before extrusion	Screw speed (rpm)	% Retained Vitamin B$_1$	% Retained Vitamin B$_2$
149	16	75	90·1	87·0
149	13	125	60·9	100·0
193	13	75	47·8	125·6
193	16	125	19·0	53·6

important since these types of dried foods are not major sources of dietary thiamine.

VITAMIN B_2 AND NICOTINIC ACID

Vitamin B_2 is considerably more resistant to high temperatures than vitamin B_1, as seen from the example in Table 4. Normally, therefore, negligible losses of riboflavin occur in food during thermal processing. Since the vitamin is highly sensitive to light, losses may be expected if the food material is exposed to both light and high temperature.

Nicotinic acid is probably the most stable of the B vitamins. Losses reported in the literature are generally explained by extraction into the processing liquid. An interesting effect of heat treatment is the liberation of bound and biologically unavailable nicotinic acid in cereals, a subject to be covered by a separate paper in this symposium.

VITAMIN B_6 AND PANTOTHENIC ACID

Schroeder[20] compiled data from many laboratories on vitamin B_6 and the pantothenic acid content of canned foods. From data on corresponding raw materials he calculated processing losses, a summary of which is referred in Table 7. The results were found so alarming that the author raised doubt on the adequacy of the American diet for these two vitamins. However, his calculations have been strongly criticised,[21] and much better retention is found in studies where the analyses are done on identical

TABLE 7
Losses of vitamin B_6 and pantothenic acid in canned food products[20]

Food	Varieties	Average % loss from canning	
		Vitamin B_6	Pantothenic acid
Fish and sea food	11	48·9	19·9
Meats and poultry	5	42·6	26·2
Dairy products	4	15·7	34·8
Vegetables, roots	4	63·1	46·1
Vegetables, legumes	5	77·4	77·8
Vegetables, green	7	57·1	56·4
Fruit and fruit juices	24	37·6	50·5

TABLE 8
Vitamin B_6 losses during aseptic canning as compared with conventional canning[22]

Product	Container	Processing method	Pyridoxine loss (%)
Strained lima beans	Tin	HTST	9·5
Strained lima beans	Glass	HTST	7·0
	Tin	Conventional canning	10·1
	Glass	Conventional canning	13·3
	Tin	HTST	4·1
Strained beef	Glass	HTST	1·7
	Tin	Conventional canning	2·9
	Glass	Conventional canning	5·8

material before and after processing. Table 8, as an example, gives the vitamin B_6 results of Everson et al.[22] in strained beans and strained beef sterilised by conventional canning and aseptic canning, respectively. Losses were small and approximately the same by both methods.

Losses of 30–50 % of vitamin B_6 and pantothenic acid have been reported during the cooking or frying of meat and vegetables. Leakage of tissue juice could account for most of the losses. By comparing cooking in a microwave oven with conventional cooking no significant difference has been found in the retention of vitamin B_6 as well as other B vitamins.

In-bottle sterilisation of milk or infant formulations destroys about 50 % of the vitamin B_6 content, pasteurisation or ultra-high temperature treatment about 10 %. Pyridoxal and pyridoxamine are supposed to be responsible for the major losses, since pyridoxine is more heat stable.[23] Excellent recoveries have been found in sterilised milk products fortified with pyridoxine hydrochloride, and similar results have been reported in bread and cereals made from enriched flour.

A phenomenon which has been too little studied is the conversion of one vitamin B_6 form to another during food processing. Since the three vitamin B_6 forms respond somewhat differently to Saccharomyces carlsbergensis, analyses with pyridoxine as the standard may not give the true vitamin B_6 activity of the food. Davies et al.[24] in a comparative study on sterilised milk found significantly higher retention values by chicken and rat bioassays than by microbiological assay.

Pantothenic acid, in contrast to other B-vitamins, has its optimum stability in the pH range 6–7. This is reflected in the processing losses referred to in Table 7, which are lower in canned foods of animal origin than in fruit products. No significant losses of pantothenic acid have been reported in the pasteurisation, sterilisation or drying of milk.

FOLIC ACID AND VITAMIN B_{12}

The stability of folic acid has been given particular attention because of the potential danger of a low dietary intake to pregnant women, infants and elderly people. The folic acid activity of food is, however, complicated by the presence of several forms having different thermal lability[25] and different bioavailability, and too little is known about the relation between microbiological assay and activity for man.

In bread making, about one-third of the natural folic acid is said to be destroyed, whereas only 10% of added folic acid is lost.[26] Sterilisation of milk when done in bottles destroys 50% of folic acid and nearly all vitamin B_{12}.[11] UHT-processing causes 10–20% loss of folic acid and 20% loss of vitamin B_{12}. Loss of folic acid can be reduced by adding ascorbic acid to the milk before processing.[25]

During the boiling of meat, up to 30% of vitamin B_{12} may be lost, but the total loss is retrieved in the liquid. Vitamin B_{12} is reasonably stable when autoclaved in non-alkaline medium, so therefore losses of no more than 10–20% would be expected during normal canning procedures.

BIOTIN, CHOLINE AND INOSITOL

Data published on biotin content of, for instance, Norwegian fish and fish products,[27] indicate that no important reduction of biotin occurs during canning. The same good stability is experienced in sterilisation and drying of milk. Losses reported in meat and vegetables during cooking may be ascribed to the general leaching of water-soluble ingredients into the liquid. Information is scarce with respect to choline and inositol stability in food, but it seems that these compounds are very little affected by thermal processing.

CONCLUSIONS

It seems that the vitamins C, B_1, B_6 and folic acid are the most vulnerable among the vitamins to which attention should be paid during food

A. Benterud

processing. Losses of pantothenic acid and vitamin B_{12} caused by processing are of less nutritional importance since a dietary shortage rarely occurs. Stability data can only be given within wide ranges of variability because experimental conditions and analytical methods differ from one laboratory to another. For some of the vitamins more investigation of the relationship between analytical data and biological value will be needed.

In general, major losses of vitamins during processing should be taken as evidence of faulty handling. With the advanced processing technology which is available today it is possible in most cases to retain a high percentage of vitamins as well as other nutrients. If fortification with synthetic vitamins is required to maintain a desired potency level, commercial qualities of vitamins with a good stability to heat can be added.

Industrial processing is only one of the links in the food-handling chain from the farm to the consumer. Greater losses of vitamins often occur in other links of the chain, such as storage of foods before and after processing, preparation of the food in the kitchen prior to eating and keeping food hot for an extended period of time before serving. In all stages of processing, information and education are necessary in order to reduce deleterious changes in the nutrient content of foods to a minimum.

REFERENCES

1. Harris, R. S. and von Loesecke, H. (1960). *Nutritional Evaluation of Food Processing*, Wiley, New York.
2. Bender, A. E. (1971). The fate of vitamins in food processing operations. Univ. Nottingham Residential Seminar on Vitamins, ed. by Mendel Stein, Churchill Livingstone, Edinburgh and London.
3. Borenstein, B. and Bunnell, R. H. (1966). Carotenoids, properties, occurrence and utilisation in foods. *Advances in Food Research*, **15**, 195.
4. Sweeney, J. P. and Marsh, A. C. (1971). Effect of processing on provitamin A in vegetables. *J. Amer. Dietetic Ass.*, **59**, 238.
5. Benterud, A. (1962). Stability of vitamin A in concentrates and vitaminised foodstuffs. Summary Report from Third Scandinavian Symposium on Fat Rancidity. *Acta Polytechn. Scand.*, **21**, 70.
6. Spencer, M. (1973). Chemical changes during cooking, processing and storage of food: Part 2. *Nutr. and Food Sci.*, No. 32, 11.
7. Bunnell, R. H., Keating, J., Quaresimo, A. and Parman, G. K. (1965). Alpha-tocopherol content of foods. *Amer. J. Clin. Nutr.*, **17**, 1.
8. Moore, T., Sharman, J. M. and Ward, R. J. (1957). Destruction of vitamin E in flour by chlorine dioxide. *J. Sci. Food Agric.*, **8**, 97.

9. Olliver, M. (1954). Ascorbic acid. In *The Vitamins*, ed. by Sebrell, W. H. and Harris, R. S., Academic Press, New York.

10. Zacharias, R. (1965). Ascorbic acid losses in preparation and processing of foods (German). In *Wissenschaftl. Veröffentl. Deutsch. Gesellsch. Ernährung*, **14**, 187.

11. Porter, J. W. G. and Thompson, S. Y. (1969). The effect of heat treatment on the nutritive quality of liquid milk, with particular reference to the UHT process. In *Dechema-Monographien*, **63**, 233.

12. Jadhav, S., Steele, L. and Hadzlyev, D. (1975). Vitamin C losses during production of dehydrated mashed potatoes. *Lebensmittel-Wissenschaft-u. Technol.*, **8**, 225.

13. Newmark, H. L., Osadca, M., Araujo, M., Gerewz, C. N. and Ritter, E. D. (1974). Stability of ascorbate in bacon. *Food Technol.*, **28**, 28.

14. Dwiwedi, B. K. and Arnold, R. G. (1973). Chemistry of thiamine degradation in food products and model systems: A review. *J. Agric. Food Chem.*, **21**, 54.

15. Farrer, K. T. H. (1955). The thermal destruction of vitamin B_1 in foods *Advances in Food Research*, **6**, 257.

16. Teixeira, A. A., Dixon, J. R., Zahradnik, J. W. and Zinsmeister, G. E. (1969). Computer optimisation of nutrient retention in the thermal processing of conduction-heated foods. *Food Technol.*, **23**, 845.

17. Everson, G. J., Chang, L., Leonard, S., Luh, B. S. and Simone, M. (1964). Aseptic canning of foods: II. Thiamine retention as influenced by processing method, storage time and temperature, and type of container. *Food Technol.*, **18**, 84.

18. Rognerud, G. (1972). Content of some nutrients in raw and prepared chicken (Norwegian). *Tidsskr. hermetikkind.*, **58**, 125

19. Beetner, G., Tsao, T., Frey, A. and Harper, J. (1974). Degradation of thiamin and riboflavin during extrusion processing. *J. Food Sci.*, **39**, 207.

20. Schroeder, H. A. (1971). Losses of vitamins and trace minerals resulting from processing and preservation of foods. *Amer. J. Clin. Nutr.*, **24**, 562.

21. Orr, L. M. and Watt, B. K. (1972). Letter to the editor. *Amer. J. Clin. Nutr.*, **25**, 647.

22. Everson, G. J., Chang, L., Leonard, S. Luh, B. S. and Simone, M. (1964). Aseptic canning of foods: III. Pyridoxine retention as influenced by processing method, storage time and temperature, and type of container. *Food Technol.*, **18**, 87.

23. Hassinen, J. B., Durbin, G. T. and Bernhart, F. W. (1954). The vitamin B_6 content in milk products. *J. Nutr.*, **53**, 249.

24. Davies, M. K., Gregor, M. E. and Henry, K. M. (1959). The effect of heat on the vitamin B_6 of milk: II. A comparison of biological and microbiological tests of evaporated milk. *J. Dairy Res.*, **26**, 215.

25. Ghitis, J. and Candanosa, C. (1966). The labile folate of milk. *Amer. J. Clin. Nutr.*, **18**, 452.

26. Keagy, P. M., Stokstad, E. L. R. and Fellers, D. A. (1973). Folacin stability during bread processing and family flour storage. *Cereal Chem.*, **52**, 348.

27. Brækkan, O. R. and Boge, G. (1960). Biotin in Norwegian Fish and Fish Products (Norwegian). *Tidsskr. hermetikkind.*, **46**, 467.

12

Inactivation of Enzymes during Thermal Processing

SVANTE SVENSSON

SIK, The Swedish Food Institute, Fack, S-400 21 Göteborg 16, Sweden

INTRODUCTION

A very high proportion of changes taking place in stored unprocessed food is caused by enzymes. Since in the living organism numerous enzymes are present and active in different metabolic processes, unprocessed foods contain enzymes that can drastically reduce their shelf life. This is especially pronounced for vegetables, where deteriorative enzymes can change the quality of, e.g. frozen peas from superior to inedible in 3 or 4 weeks, if the amount of active enzymes is not sufficiently reduced before freezing.

Some important enzymes which cause undesirable quality changes in foods are listed in Table 1. The enzymes are separated into four groups related to changes in flavour, colour, texture/consistency and nutritional value. Although the enzymes peroxidase and catalase are often reported to cause off-flavour, the reactions involved have not been conclusively identified. The hydrolysis of pectin to methanol and pectic acid by pectin methylesterase will sometimes give rise to a firmer texture by introducing more carboxyl groups in the pectin chain. Together with divalent ions such as calcium these carboxyl groups form bridges between different pectin chains making a gel structure.

Enzyme activity is destroyed by irreversible denaturation or by hydrolytic breakdown of the protein molecule. Denaturation can be brought about by such agencies as extremes of acidity or alkalinity, protein-modifying substances such as formaldehyde or configuration-changing compounds such as urea. In the food industry, however, heating is the most convenient method for enzyme inactivation. To minimise heat-induced quality changes in the food, the heat treatment is often designed as a high temperature short time (HTST) process. In this paper the influence of

TABLE 1
Enzymes related to food quality

Enzyme	Catalysed reaction	Quality defect
Flavour		
Lipolytic acyl hydrolase (lipase, esterase etc.)	Hydrolysis of lipids	Hydrolytic rancidity (soapy flavour)
Lipoxygenase	Oxidation of poly-unsaturated fatty acids	Oxidative rancidity ('green' flavour)
Peroxidase/catalase	?	'Off-flavour' (?)
Protease	Hydrolysis of proteins	Bitterness
Colour		
Polyphenol oxidase	Oxidation of phenols	Dark colour
Texture/consistency		
Amylase	Hydrolysis of starch	Softness/loss in viscosity
Pectin methylesterase	Hydrolysis of pectin to pectic acid and methanol	Softness/loss in viscosity
Polygalacturonase	Hydrolysis of α-1,4 glycosidic linkages in pectic acid	Softness/loss in viscosity
Nutritional value		
Ascorbic acid oxidase	Oxidation of L-ascorbic acid	Loss in vitamin C content
Thiaminase	Hydrolysis of thiamine	Loss in vitamin B_1 content

thermal processing on enzyme activity will be discussed. The examples are mainly drawn from our own investigations.

INFLUENCE OF TEMPERATURE ON ENZYME-CATALYSED REACTIONS

The influence of temperature on an enzyme-catalysed reaction is shown schematically in Fig. 1. As the temperature increases, enzyme activity increases in such a way that the reaction rate is approximately doubled for every 10 °C rise in temperature. This part of the curve can usually be expressed as a straight line in an Arrhenius plot and an apparent activation energy can be calculated. A doubling of the reaction rate for every 10 °C

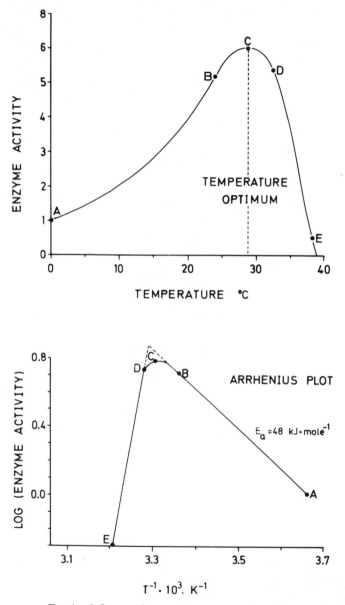

FIG. 1. Influence of temperature on enzyme activity.

increase in temperature corresponds to an activation energy of approximately 48 kJ/mol. On reaching a certain critical temperature the enzyme molecules begin to denature, leading to a change in catalytic activity of the molecule followed by a rapid decrease in the number of active molecules on further increase of the temperature. For some enzymes the 'denaturation part' of the curve can also be expressed as a straight line in an Arrhenius plot.

The optimum temperature is often quoted as a characteristic of a certain enzyme. However, this is not a satisfactory criterion for characterisation of the enzyme as it is not a fixed feature. The temperature optimum has been shown to be time- and pH-dependent and, therefore, is only applicable to specific conditions. It will increase in inverse proportion to the length of the reaction time used to measure the enzyme activity.

The great majority of enzymes show optimal activities within the temperature range 30–50 °C but there also exist enzymes such as microbial kinases which show[14] optimal activities at temperatures around 100 °C. It is, however, not obvious that the thermostability of an enzyme can be correlated with the temperature optimum. Pea lipoxygenase shows a relatively low temperature optimum, 30 °C, which is about 30 °C lower than what might be expected from thermostability data. The 'catalysing molecule' is often considerably more thermolabile than the 'resting molecule'.[1]

It should also be noted that in heat treatment processes intended to inactivate enzymes the temperature optima of these enzymes must always be passed. In some cases, where extremely high amounts of deteriorative enzymes are present, and when the temperature rise in the material is very slow, the enzymes can change the material during the heat treatment process itself.

THERMAL INACTIVATION OF ENZYMES

It is generally believed that thermostability of an enzyme is decided principally by its amino acid sequence and the specific conformation derived from the sequence.

Studies on the thermal inactivation of enzymes in solid foods are performed either with geometrically well defined samples of the tissue or with water extracts of the tissue. In the intact material the molecular conformation and in consequence the thermostability of the enzyme is influenced by substances in the near vicinity of the enzyme molecule and by the microstructure of the cell. Thermal inactivation data obtained from investigations on intact tissue are very valuable since these data include the influence of the micro-environment of the enzyme on thermostability. It is however, very difficult to obtain good HTST inactivation data in this way because of the disturbing effect of the non-uniform temperature distribution in the material during temperature increase. Therefore HTST inactivation data are usually obtained from studies on the thermal inactivation of enzymes in solution.[2]

Apparatus for Heat Treatment of Enzymes in Solution

A prerequisite for obtaining reliable inactivation data for enzymes in solution is to use properly designed heat treatment methods and apparatus. In our laboratory two different types of apparatus for reproducible heat treatment of enzymes in solution have been constructed. In the 'continuous flow apparatus' (Fig. 2) a cold enzyme solution is pumped through thin-walled glass tubes immersed in different thermostatted water baths. When passing from the cold part of the apparatus to the heat treatment part, the solution first enters a preheater, where the temperature of the solution is raised from 3 °C to the selected heat treatment temperature in 2·3 sec. This temperature is maintained for a certain time before cooling the solution.

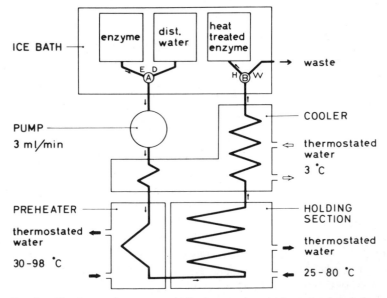

FIG. 2. 'Continuous-flow apparatus' for heat treatment of enzymes in solution.

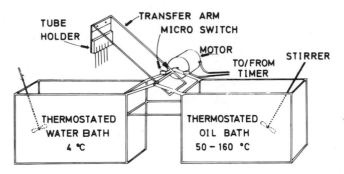

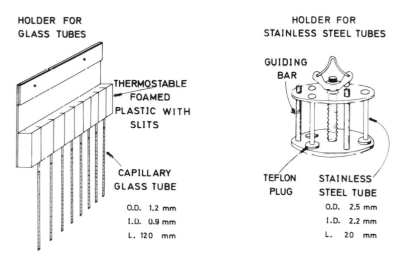

FIG. 3. 'Tube immersion apparatus' for heat treatment of enzymes in solution.

This type of apparatus performs highly reproducible thermal inactivation under isothermal conditions since the preheating time is constant irrespective of the selected heat treatment temperature.[3]

In the 'tube immersion apparatus' (Fig. 3) thin-walled melting point glass capillary tubes enclosing the enzyme solution are transferred from a cold thermostatted water bath to a hot thermostatted oil bath, held there for a selected time and then transferred back to the cold water bath. Since the capillary tubes are closed this apparatus is specially designed for heat treatment studies in the temperature range 80–160 °C. Temperature measurements with a thermocouple located in the centre of a capillary tube showed that a period of 6·5 sec was required for the content in the centre to reach 139·5 °C for an oil bath temperature of 140 °C. By using stainless steel tubes small cylinders of tissue can be bored out from, e.g. potato tubers, after which the tubes with the potato cylinders inside can be inserted between two Teflon seals in a special tube holder and heat treated in the same way as solutions in capillary glass tubes.[4]

Influence of Time and Temperature

Thermal inactivation of most enzymes follows first-order reaction kinetics. This is illustrated in Fig. 4 which shows the influence of heat treatment time on the inactivation of potato lipolytic acyl hydrolase at five different temperatures ranging from 67·7 to 72·8 °C. The thermal inactivation rate accelerates considerably as the heating temperature increases in this narrow temperature range. Hence a temperature increment of one degree approximately doubles the inactivation rate.[4]

Judging from inactivation data, the potato lipolytic acyl hydrolase is a homogeneous enzyme. This is, however, not the case with three other quality related potato enzymes—lipoxygenase, polyphenol oxidase and peroxidase. For these enzymes a representative inactivation curve consists of an initial, steep straight line, an intermediate curved portion, and a final straight line with a shallow slope. This type of curve is the result of two

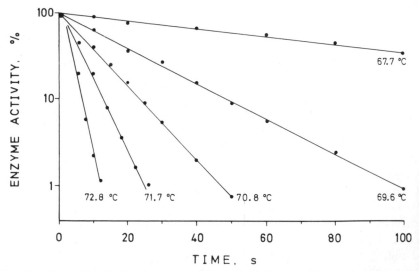

FIG. 4.   Thermal inactivation of potato lipolytic acyl hydrolase in buffer extract of potato as a function of time and temperature.

independent first-order reactions derived from the inactivation of two enzyme fractions with different thermostability. The amount of the most thermostable form can be estimated by extrapolating the inactivation curve to zero time, as shown for potato lipoxygenase in Fig. 5. The thermostable fractions constitute approximately 55, 45 and 7 % of the total lipoxygenase, polyphenol oxidase and peroxidase activities.[4]

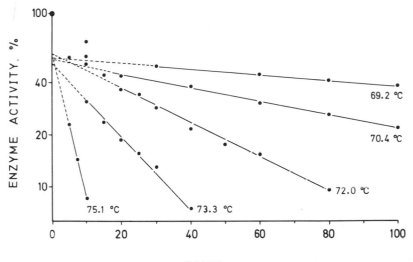

FIG. 5.   Thermal inactivation of potato lipoxygenase in buffer extract of potato as a function of time and temperature.

To illustrate the influence of temperature on enzyme inactivation, Arrhenius plots are usually used. Food scientists are however probably more familiar with a method of plotting thermal inactivation data which resembles that of death time curves for micro-organisms. In this case the heat treatment time required to reduce the enzyme activity to 10 % of the original value— the *D*-value—is plotted versus temperature. Such inactivation curves for the thermostable fractions of the four potato enzymes mentioned above are shown in Fig. 6. The slope of the peroxidase curve is considerably less steep than that of lipolytic acyl hydrolase. Generally enzymes with high thermostability show far less temperature dependence in the temperature range in question than do thermolabile enzymes. The temperature dependence of the inactivation is often expressed as the *Z*-value, which is the temperature increment required to reduce the

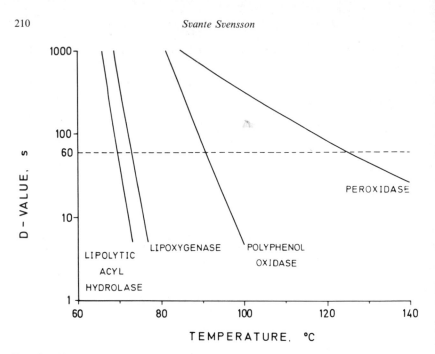

FIG. 6.   Thermal inactivation of the thermostable fraction of potato lipolytic acyl hydrolase, lipoxygenase, polyphenol oxidase and peroxidase as a function of temperature.

D-value to 10 % of the original one. Z-Values together with experimental activation energies for the four enzymes are listed in Table 2. As is shown there is a large difference between the Z-value of the thermostable enzyme peroxidase, (35 °C) and the relatively thermolabile enzyme lipolytic acyl hydrolase (3·1 °C). Compared to *Clostridium botulinum* spores (Z-value, 10 °C) only very thermostable enzymes show higher Z-values.[4]

TABLE 2

Experimental activation energy and Z-values for thermal inactivation of the thermostable fraction of potato enzymes

Enzyme	Thermostable fraction (%)	Activation energy (kJ/mol)	Z-Value (°C)
Lipolytic acyl hydrolase	100	725	3·1
Lipoxygenase	55	636	3·6
Polyphenol oxidase	45	313	7·8
Peroxidase	7	80	35

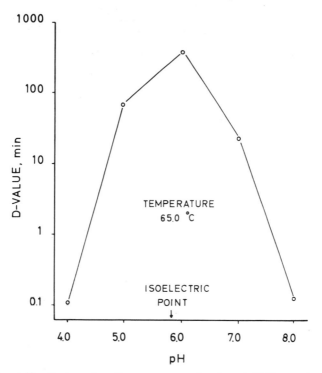

FIG. 7.   Thermal inactivation of pea lipoxygenase as a function of pH. Temperature 65·0 °C.

Influence of pH and Water Activity

The thermostability of an enzyme is greatly influenced by the hydrogen ion concentration. Enzymes are usually most stable at pH values close to their isoelectric point. The influence of pH on the thermal inactivation of pea lipoxygenase at 65 °C is shown in Fig. 7. A change in pH from, say, 6 to 4 will cause a large reduction in the D-value, from 400 to 0·1 min.[3]

Together with the hydrogen ion concentration, the relative humidity or the water activity of the food will greatly influence the thermostability of an enzyme. In the investigations by Rothe and Stöckel[5] and Acker,[6] it was shown that the thermostability of an enzyme is considerably increased by decreasing the amount of water in the food.

Thermostability of Different Enzymes

Thermostability data from the literature for some quality related enzymes are summarised in Fig. 8. The thermostability is expressed as the

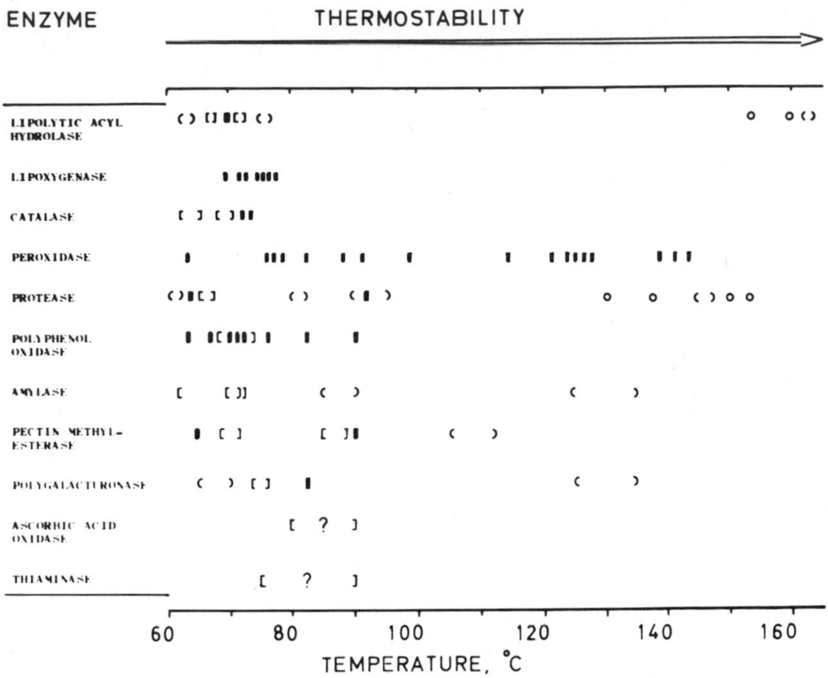

ENZYME   THERMOSTABILITY

FIG. 8. Thermostability data for some quality-related enzymes from the literature expressed as the temperature required to reduce the enzyme activity to 10 % of the original in one minute. Enzymes originating from vegetables or animals ▮, [ ] and micro-organisms ○, ( ). Experimentally found values ▮, ○. Extrapolated values [ ], ( ).

temperature required to reduce the enzyme activity to 10 % of the original value in one minute. Inactivation data of enzymes originating from food contaminating micro-organisms are included. Many investigations on the thermal inactivation of enzymes do not include the *D*-values or the one minute heat treatment time.

In those cases where plausible extrapolations could be done the 'thermostability temperatures' have been estimated and are shown as brackets in the figure. Regarding enzymes with a 'thermostability temperature' of 100 °C or more, only peroxidase originates from vegetables. All other enzymes, lipolytic acyl hydrolase, protease, amylase, pectin methylesterase and polygalacturonase, are microbial enzymes. The peroxidase enzyme is the only very thermostable plant enzyme, while several enzymes such as lysozyme, ribonuclease, acid phosphatase and phospholipase, which derive from animals, have been shown to be extremely thermostable.[4]

Reactivation of Enzymes

It is well known that, under certain conditions of limited heat treatment, enzymes can regain activity during storage, which results in quality changes in foods. Reactivation has been shown for such enzymes as trypsin, ribonuclease lysozyme and peroxidase of which the latter is the most extensively studied one. The rate and extent of reactivation vary from one enzyme to another and it is well known that the faster the temperature is raised to a given value the greater the extent of activity regained. This can be very troublesome when dealing with HTST heat treatment processes.[7]

Thermal Activation of Haemoproteins as non-enzymatic catalysts

Haemoprotein enzymes, such as catalase and peroxidase, can function as catalysts in two different ways. One is to act as an enzyme in the decomposition of a substrate, such as hydrogen peroxide, the other is to catalyse non-enzymatically the oxidation by molecular oxygen of unsaturated compounds, e.g. polyunsaturated fatty acids. In this case thermal denaturation of the enzyme molecule can result in a more than 10-fold increase in non-enzymatic lipid oxidation activity concomitant with loss of enzyme activity.[8, 9]

Calculations of Thermal Inactivation of Enzymes in Solid Food

Computer simulation studies on the effect of different time–temperature combinations on enzyme inactivation in solid food could be very valuable when designing new heat treatment processes. In a rather comprehensive study on the thermostability of pea lipoxygenase we calculated the enzyme activity distribution in heat treated peas from basic thermal inactivation data of the purified enzyme, enzyme distribution pattern and heat penetration data. In Fig. 9 calculated inactivation data are compared with experimentally found activities for lipoxygenase during hot water blanching of peas. Different combinations of temperature, 75–95 °C, and blanching time, 5–100 sec, were used. As a whole there is quite a good agreement between calculated and experimentally found enzyme activities, demonstrating that computer simulations can successfully be used when studying the effect of different process parameters in the blanching procedure. The great influence of temperature on the thermal inactivation of enzymes manifested by high activation energies for the inactivation process enable enzymes to be used as temperature indicators in food material.[10, 11]

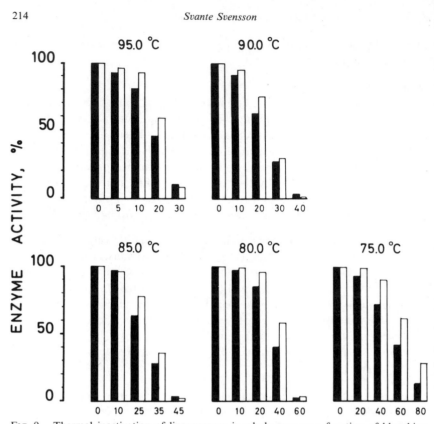

FIG. 9. Thermal inactivation of lipoxygenase in whole peas as a function of blanching temperature and time; calculated (open columns) and experimentally found (solid columns) lipoxygenase activities. The figures under the columns are the blanching times in seconds.

THE EFFECT OF DIFFERENT HEAT TREATMENT PROCESSES ON ENZYMES

In most industrial heat treatment processes enzymes are partially or completely inactivated (Table 3). In the processes aimed at killing micro-organisms, only extremely thermostable enzymes such as microbial lipase and protease survive the conventional sterilisation process, while a great number of enzymes withstand pasteurisation. The influence of temperature on the inactivation of potato peroxidase (thermostable fraction) and an extremely stable lipase produced by the psychrophilic micro-organism *Pseudomonas fluorescens* are shown in Fig. 10. The time–temperature area representing *F*-values between 2·7 and 10 is included in the figure. At

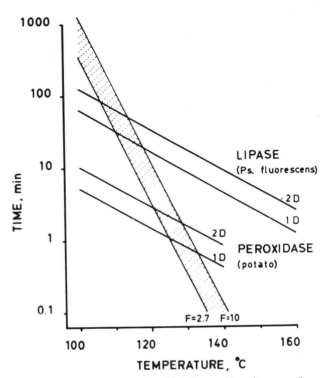

FIG. 10. Thermal inactivation of a microbial lipase (*Pseudomonas fluorescens*) and thermostable potato peroxidase as a function of temperature.

temperatures higher than 120 °C the sterilisation time is much less than that required for 90 % inactivation of the enzyme (the unit-*D*-curve) leading to an insufficient enzyme inactivation when using HTST sterilisation processes. This will give rise to a microbially stable but enzymatically unstable food product if contaminating microbial lipase is present in the unprocessed food. If deteriorative extremely thermostable enzymes having *Z*-values considerably higher than 10 °C are present in a food which is to be sterilised it is usually meaningless to use HTST processes. In those cases a combination of a low temperature, long time heat treatment process, causing sufficient enzyme inactivation, and a HTST sterilisation process could be successful.

The purpose of most pasteurisation processes is both the killing of micro-organisms and the inactivation of deteriorative enzymes. Often the inactivation of certain enzymes is used as a processing index for the verification that the food has passed the desired heat treatment process.

TABLE 3
Influence of different heat treatment processes on enzymes

Process	Influence on enzymes
Sterilisation	Extremely thermostable enzymes survive
Pasteurisation	Thermostable enzymes survive
Blanching	Quality-changing enzymes inactivated
Cooking/frying/baking	Most enzymes inactivated
Drying	Dependent on processing temperature
Extraction/concentration/distillation	Dependent on processing temperature

The blanching process is specially designed to inactivate enzymes causing adverse changes in frozen or dried vegetables during storage. It is common practice to let the inactivation of peroxidase—the most thermostable enzyme in vegetables—serve as an index of blanching adequacy although the quality-changing effect of this enzyme has not been conclusively verified. Usually green peas show a negative peroxidase test, which means a remaining activity of less than 1 % after a 3 min hot water blanching process, while Brussels sprouts demand a much more drastic heat treatment for complete enzyme inactivation. In an investigation by Böttcher[12, 13] it was shown that 'the relatively best quality' after frozen storage of different vegetables (peas, green beans, cauliflower and Brussels sprouts) was obtained when 1–10 %—depending on the vegetable—of the original peroxidase activity remained after hot water blanching. In my opinion, complete inactivation of the thermostable fraction of peroxidase during blanching is not essential in order to obtain high quality food products with a long shelf life.

In the cooking, frying and baking processes most enzymes are usually inactivated except for the very thermostable ones. Drying processes leave food products with varying degrees of enzyme activity depending on the temperatures used in the process. As mentioned above the thermostability of an enzyme is greatly increased as the water activity in the food decreases during the drying process.

In extraction, concentration and distillation processes, enzyme inactivation is entirely dependent on the processing temperatures used. If deteriorative enzymes are present in the material, a special heat treatment process is usually added to the original process in order to inactivate enzymes before processing.

As a whole the knowledge of quality-changing enzymes is restricted to a very limited number of enzymes headed by peroxidase. For future investigations I believe a more diversified view will be given by studies of correlations between enzyme activities, thermal inactivation data and sensory and instrumentally measured quality changes during storage of foods processed in different ways.

REFERENCES

1. Svensson, S. G. (1973). Thermal inactivation of the enzyme lipoxygenase from peas (*Pisum sativum* L.): (Swedish). Thesis, Göteborg. 1–43.
2. Park, K. H. (1976) Thermal inactivation and storage properties of industrially important enzymes. (German). Thesis. Karlsruhe. 1–90.
3. Svensson, S. G. and Eriksson, C. E. (1972). Thermal inactivation of lipoxygenase from peas (*Pisum sativum* L.): I. Time temperature relationships and pH dependence. *Lebensm.-Wiss. u. Technol.*, **5**, 118–23.
4. Svensson, S. G. (1976). To be published.
5. Rothe, M. and Stöckel, J. (1967). Activity changes with enzymes caused by heat: Part I. Lipase, acetyl esterase, peroxidase, and lipoxygenase of cereal products. (German). *Nahrung*, **11**, 741–9.
6. Acker, L. W. (1969). Water activity and enzyme activity. *Food Technol.*, **23**, 1257–8, 1261, 1264, and 1269–70.
7. Lu, A. T. and Whitaker, J. R. (1974). Some factors affecting rates of heat inactivation and reactivation of horseradish peroxidase. *J. Food Sci.*, **39**, 1173–8.
8. Eriksson, C. E., Olsson, P. A. and Svensson, S. G. (1971). Denatured haemoproteins as catalysts in lipid oxidation. *J. Amer. Oil Chem. Soc.*, **48**, 442–7.
9. Eriksson, C. E. and Vallentin, K. (1973). Thermal activation of peroxidase as a lipid oxidation catalyst. *J. Amer. Oil Chem. Soc.*, **50**, 264–8.
10. Svensson, S. G. and Eriksson, C. E. (1974). Thermal inactivation of lipoxygenase from peas (*Pisum sativum* L.): IV. Inactivation in whole peas. *Lebensm.-Wiss. u. Technol.*, **7**, 145–51.
11. Ohlsson, T. and Svensson, S. (1974). Calculation of heat transfer and enzyme inactivation in blanching of peas. *Proc. 4th Int. Congress Food Sci. and Technol.*, Madrid, September 22–27.
12. Böttcher, H. (1975). On the question of enzyme activity and quality of frozen vegetables: Part I. The remaining residual activity of peroxidase. (German). *Nahrung*, **19**. 173–9.
13. Böttcher, H. (1975). On the question of enzyme activity and quality of frozen vegetables: Part II. The effect on the quality of frozen vegetables. (German). *Nahrung*, **19**, 245–53.
14. Kobayashi, S.-H., Hubbell, H. R. and Orengo, A. (1974). A homogeneous thermostable deoxythymidine kinase from *Bacillus stearothermophilus*. *Biochem.*, **13**, 4537–43.

13

Thermal Degradation of Meat Components

Franz Ledl

*Institut für Pharmazie und Lebensmittelchemie der Universität München,
D-8 München 2, Sophiestrasse 10, Federal Republic of Germany*

Flavour components responsible for the specific aroma of boiled, roast or grilled meat are formed when meat is thermally treated. Various flavours are also obtained from the thermal conductor used, for example water or fat. The purpose of this paper is to discuss possible reactions with regard to the formation of aroma compounds which contribute to 'meatiness'.

Investigations of changes due to the influence of heat have been made according to three approaches:

(a) Isolation of volatile compounds.
(b) Analysis of precursor components.
(c) Systematic evaluation of model systems which produce volatile aroma compounds.

VOLATILE COMPOUNDS

These have been investigated by many groups with the aid of a GC–MS system. At present, literally more than 500 substances have been identified.[1] Quite a few of these have also been found in other thermally treated foods such as coffee, nuts and chocolate.

PRECURSOR COMPONENTS

It has been shown that the precursors of cooked meat flavour are water-soluble, of low molecular weight and, when partially purified, rather unstable, spontaneously undergoing reactions suspiciously suggestive of

Maillard browning.[2] Essentially these fractions comprise nucleotides, amino acids and sugars. Of the α-amino acids known to occur in meat, cysteine and its dimer cystine appear to yield the meatiest odours when subjected to thermal processing.

MODEL SYSTEMS

Many model reactions for obtaining meat aroma found in the patent literature are based on the knowledge and information obtained from studies of precursors.[3] Table 1 gives details of some patents for the

TABLE 1
Processed meat flavour patents

1. Reducing sugars or glyceraldehyde and cysteine in water at 120 °C.

2. Amino acid, ribose-5-phosphate and water; heat in an autoclave.

3. Hydrolysed vegetable protein (cysteine-free), thiamine, a mono- or polysaccharide, water and heat.

4. Protein hydrolysate, water, methionine, heat, and a saccharide, optionally in the presence of a carboxylic acid.

5. Reducing sugar, cysteine, fat at 70 °C for 24 h, or at 140 °C until water-free.

production of meat odour. Amino acids containing sulfur, e.g. methionine or cysteine, are substituted by other compounds with sulfur in the molecule, e.g. thiamine. Carbohydrates may also be present, as in patent 2, where ribose-5-phosphate as a carbohydrate is used. Patent 2 is one of the few patents which does not stress the presence of sulfur-containing substances. It is also very seldom that one finds precise information on temperature ranges and the time involved in thermal processing.

Figure 1 shows the products obtained from a model reaction between xylose and cysteine.[4]

I would like to direct this paper to a consideration of sulfur-containing compounds. Thiazoles form a large aroma-active group which have also been recently isolated from cooked meat[5] and from roasted coffee beans.[5] The 5- and 6-ring heterocycles with one or more S atoms in the ring are of particular interest because few of these have been obtained in the low volatile fraction of the cooked meat itself.[7] The aroma of these compounds in low concentration is desirable, but quite unpleasant in higher concentrations, as is also the case with the 1,3-dithiolanes. Faint traces of impurities can give a false impression of the aroma, as was the case with dimethyltrithiolane, for which the aroma of roasted meat was assumed. All of these substances with the exception of the 1,2-dithiane are also formed in the absence of carbohydrates when cysteine is heated in fat at higher. temperatures ($\sim 200\,°C$) for 5 min and then maintained at $150\,°C$ for 30

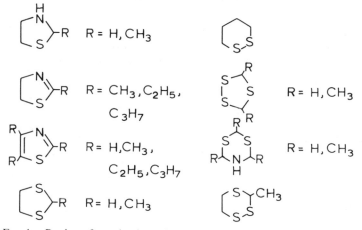

FIG. 1.   Products from the thermal reactions of a cystein–xylose mixture.[4]

min.[8] This provides further proof that cysteine is an important component of meat aroma. It is questionable whether the degradation of cysteine in the presence of carbohydrates follows a similar path.

Turning now to another type of model reaction, in which some volatile compounds obtained from cooked meat react with one another, Fig. 2 shows substances isolated from a reaction mixture of hydrogen sulfide, ammonia and saturated aliphatic aldehydes.[9] In this reaction, compounds are also formed which have already been identified in thermally processed meat, e.g. trimethyltrithiane and methylthioethanethiol. Since trimethyldithiazine and dimethyltrithiolane are formed, both of which have been isolated from the thermal degradation of cysteine, it may be assumed that hydrogen sulfide, ammonia and acetaldehyde are products of this process.

Figure 3 shows the products of the reaction of 4-hydroxy-5-methylfuran-3-one with hydrogen sulfide in an autoclave for 4 hours at $100\,°C$.[10] Researchers at Unilever have shown that the presence of a series of compounds with 1, 2 or 3 S atoms can be established. According to the

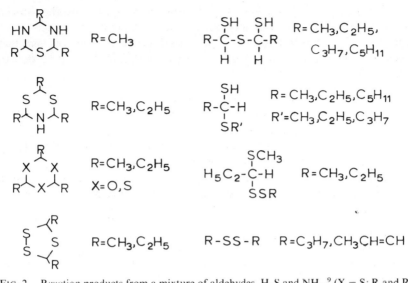

FIG. 2. Reaction products from a mixture of aldehydes, H_2S and NH_3.[9] (X = S; R and R' = CH_3.)

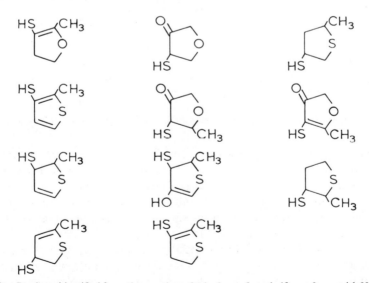

FIG. 3. Products identified from the reaction of 4-hydroxy-5-methylfuran-3-one with H_2S at 100°C for 4 h.[10]

Unilever results, the first row of Fig. 3 shows compounds with the roasted meat aroma, and the rest possess a meaty flavour. Although these compounds have been known for the past seven years, none of them had been identified in the cooked meat until recently. This could be due to various reasons:

(*a*)   The substances formed under the extreme conditions in an autoclave at 100 °C for 4 hours are not obtained when meat is cooked under normal conditions.

(*b*)   Even though the compounds are formed, their concentration levels are so low that identification is impossible.

(*c*)   More reactive substances than the hydroxyfuranone are present in meat which react with hydrogen sulfide to give different products.

Figure 3 has already demonstrated that hydrogen sulfide can react with ammonia and acetaldehyde at room temperature.

It must be stated at this point that any attempts to extrapolate from the chemistry of model systems to that of meat are difficult due to the obvious complexity of the latter. From all these investigations it is clear that studies of the aroma of thermally processed meat have not yet developed to the same extent as the studies of citrus and fruit flavours. Undoubtedly Maillard reactions, which obviously take place, as has been proved by the isolation of 4-hydroxy-5-methylfuran-3-one[11] and 2,3-dihydro-3,5-dihydroxy-6-methyl-4*H*-pyran-4-one[12] (Fig. 4), complicate the analysis.

FIG. 4.   The main products of the non-enzymatic browning of pentoses and hexoses.[11,12]

The furanone which is shown is formed when amines, amino acids or proteins react with pentoses and the pyranone is formed when the carbohydrate is a hexose or contains a hexose moiety.

A few examples of the reactions which take place in thermally treated meat have been given. It has been shown that, as with other thermally processed foods, a complex mixture of compounds is responsible for the aroma. It has been possible to show that sulfur plays an essential role in the formation of meat aroma, although the results are far from satisfactory. The non-enzymatic browning or Maillard reactions also play an important role in the thermal processing of meat. Further research on this subject could be useful and would perhaps provide more insight into the chemistry involved in the thermal degradation of meat components.

REFERENCES

1. Wilson, R. A. (1975). A review of thermally produced imitation meat flavours. *J. Agric. Food Chem.*, **23**, 1032–7.
2. Herz, K. O. and Chang, S. S. (1970). Meat flavour. *Advances in Food Res.*, **18**, 28–32, ed. by Chichester, C. O., Mrak, E. M. and Steward, G. F., Academic Press, New York and London.
3. Wilson, R. A. and Katz, J. (1974). Synthetic meat flavours. *The Flavour Industry*, **5**, 30–8.
4a. Ledl, F. and Severin, Th. (1973). Thermal decomposition of cysteine and xylose in tributyrine (German). *Chem. Mikrobiol. Technol. Lebensm.*, **2**, 155–60.
4b. Mulders, E. J. (1973). Volatile components from the non-enzymic browning reaction of the cysteine/cystine–ribose system. *Z. Lebensm. Unters. -Forsch.*, **152**, 193–201.
5. Wilson, R. A., Mussinan, C. J., Katz, J. and Sanderson, A. (1973). Isolation and identification of some sulfur chemicals present in pressure-cooked beef. *J. Agric. Food Chem.*, **21**, 873–6.
6. Vitzthum, O. G. and Werkhof, P. (1974). Oxazoles and thiazoles in coffee aroma. *J. Food Sci.*, **39**, 1210–5.
7. Brinkman, H. W. *et al.* (1972). Components contributing to beef flavour: Analysis of the headspace volatiles of beef broth. *J. Agric. Food Chem.*, **20**, 177–81.
8. Ledl, F. (1976). Thermal decomposition of cysteine, cystine, *N*-acetylcysteine-4-thiazolidine carboxylic acid and cysteine methyl ester in soya bean oil (German). *Z. Lebensm. Unters.-Forsch.*, **161**, 125–9.
9a. Ledl, F. (1975). Analysis of a synthetic onion aroma (German). *Z. Lebensm. Unters.-Forsch.*, **157**, 28–33.
9b. Boelens, M. *et al.* (1974). Organic sulfur compounds from fatty aldehydes, hydrogen sulfide, thiols, and ammonia as flavour constituents. *J. Agric. Food Chem.*, **22**, 1071–6.
10. Ouweland, G. A. M. and Peer, H. G. (1975). Components contributing to beef flavour: Volatile compounds produced by the reaction of 4-hydroxy-5-methyl-3(2*H*)-furanone and its thio analogue with hydrogen sulfide. *J. Agric. Food Chem.*, **23**, 501–5.
11. Tonsbeek, C. H. T., Plancken, A. J. and v. d. Weerdhof, T. (1968). Components contributing to beef flavour: Isolation of 4-hydroxy-5-methyl-3(2*H*)-furanone and its 2,5-dimethyl homologue from beef broth. *J. Agric. Food Chem.*, **16**, 1016–21.
12. Ledl, F., Schnell, W. and Severin, Th. (1976). Proof of 2,3-dihydro-3,5-dihydroxy-6-methyl-4*H*-pyranone in foods (German). *Z. Lebensm. Unters.-Forsch.*, **160**, 367–70.

14

The Effect of Temperature and Time on Emulsion Stability

LISBETH RYDHAG

The Swedish Institute for Surface Chemistry, Drotning Kristinas veg 45, S-11428 Stockholm, Sweden

INTRODUCTION

Emulsions are macrodispersed systems which consist of at least two non-miscible liquids or liquid crystals. Since emulsions are thermodynamically unstable systems, the phases making up an emulsion strive to achieve a reduced area of contact. If pure liquids such as water and oil are dispersed, a very unstable emulsion is formed which rapidly separates into a water layer and an oil layer. The stabilisation of such emulsions can be effected by adding amphiphilic substances which concentrate at the interfaces between oil and water. Different types of amphiphilic substances can be used for this purpose, e.g. surfactants or synthetic or natural polymers. Stabilisation can also be attained by the concentration of solid particles at the interface, the size of the particles being of the order of 500–1000 Å.

FLOCCULATION, COALESCENCE AND CREAMING

Two forms of instability are apparent in emulsions, i.e. flocculation and coalescence. In flocculation the emulsified droplets become associated in flocks without the destruction of the individual droplets, while in coalescence the walls between the droplets are destroyed, larger drops thus being formed (Fig. 1). When the dispersed phase is of lower density than the continuous phase, creaming takes place, which should not be confused with flocculation. Mulder and Walstra[1] present a thorough treatment of the subject of flocculation and coalescence.

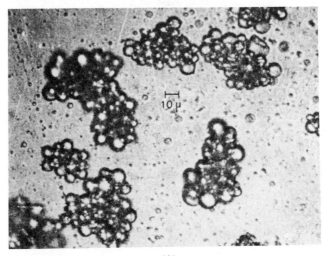

(A)

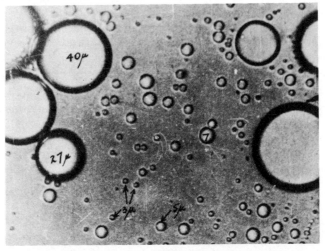

(B)

FIG. 1. Changes in emulsion stability: (A) flocculation and (B) coalescence.

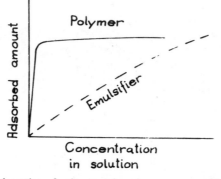

FIG. 2.   Maximum adsorption of polymers takes place even at very low concentrations of polymers; corresponding adsorption of surface-active substance does not take place below a concentration of a few per cent.

COALESCENCE INHIBITORS

Emulsions can be stabilised by building up around the dispersed phase an effective coalescence inhibitor which prevents the emulsion droplets from uniting, at contact, into larger units. Studies on the association conditions of surfactants with water and various organic phases have been carried out at The Swedish Institute for Surface Chemistry since the institute was founded in 1963. The importance of liquid crystalline structures in the stabilisation of different dispersion systems, including emulsions, has been investigated by Professor Friberg and his co-workers.[2, 3] Their work has shown that in emulsions multilayers of liquid crystalline phases can form protective membranes around the dispersed phase. This differs from previous theories concerning the stabilisation of emulsions by surfactants.[4]

Other mechanisms in emulsions can also lead to the formation of

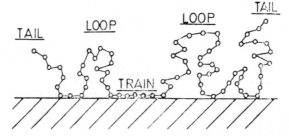

FIG. 3.   The extension of a polymer into the surrounding solution.

coalescence inhibitors. If polymers are adsorbed at the interface between water and oil phases, changes in conformation occur and the molecules acquire the energetically most favourable conformation (Figs. 2 and 3). Since, in addition, polymers have several segments adsorbed at the interface, the energy of activation for desorption of the polymers will be very high, and they can therefore be regarded as being irreversibly adsorbed at the interface. In food emulsions, polymers of biological origin are often present, e.g. proteins and polysaccharides.

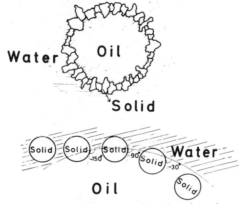

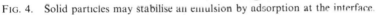

FIG. 4.   Solid particles may stabilise an emulsion by adsorption at the interface.

Coalescence inhibitors can further be built up by the concentration of solid particles at the interface[5] (Fig. 4). In this case the wetting conditions at the surface of the particles constitute the decisive factor. The best protection for the dispersed phase when the angle of contact θ is the same for the particle with the water phase and for the oil phase respectively, i.e. 90°. It is an empirical rule that the phase having the larger angle of contact with the particle is the one which forms the continuous phase.

EMULSIFIERS

It is characteristic of surfactants that their molecules are made up of a hydrophobic, or non-polar, part and a hydrophilic, or polar part (Fig. 5). Surfactants form colloidal solutions in water, i.e. their molecules associate to aggregates called micelles at the critical micelle concentration (CMC), as shown in Fig. 5. Substances of an organic nature can be solubilised in the interior of the spherical aggregates. At higher concentrations of surfactant, ordered structures are formed known as liquid crystalline phases, with

FIG. 5. Surface-active substances are amphiphilic molecules composed of hydrophilic and lipophilic parts.

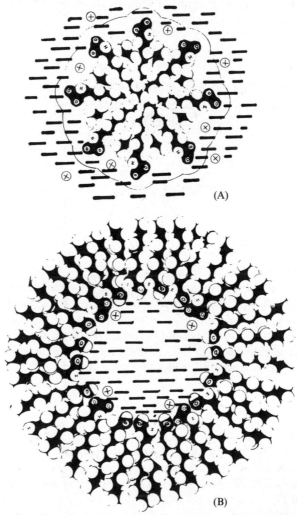

(A)

(B)

FIG. 6. Surface-active substances associate into micelles at the critical micelle concentration: (A) a normal micelle in a water solution; and (B) a reversed micelle in lipophilic solution.

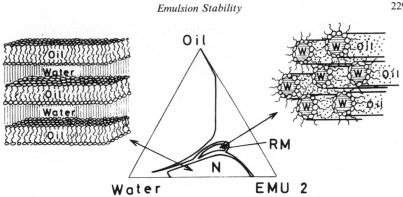

FIG. 7.   Surface-active substances form liquid crystalline phases when certain conditions according to concentration are fulfilled. (N = neat phase, M = middle phase.)

lamellar, hexagonal or cubic packing. Lyotropic liquid crystalline phases appear as separate phases in oil–water systems (Fig. 7) and have a high viscosity compared with the oil or water phases (Fig. 8).

Investigations at our institute have shown that the presence of liquid crystalline phases in emulsion systems gives an improved emulsion stability.

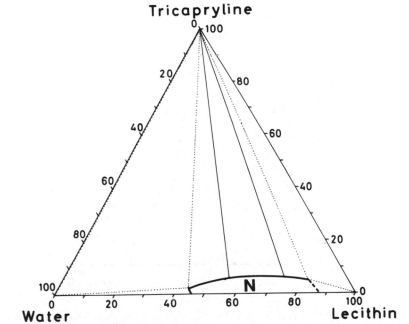

FIG. 8.   When the emulsifier, egg lecithin, is added to the water and oil phase a new phase appears in the system, i.e. a lamellar liquid crystalline phase (N).

Lisbeth Rydhag

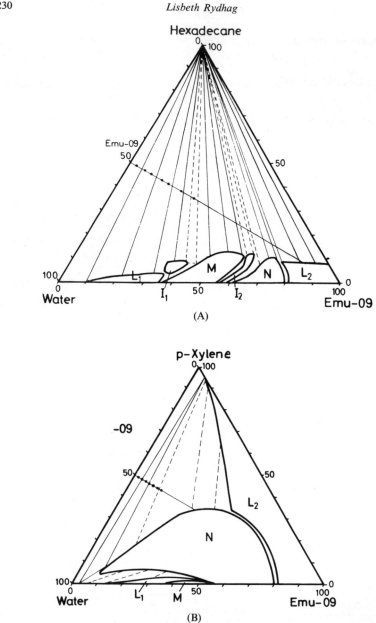

FIG. 9. The influence of an exchange of the hydrocarbon on the phase behaviour of a nonionic surfactant–water–hydrocarbon system: (A) aliphatic hydrocarbon; and (B) aromatic hydrocarbon.

Multilayers of the liquid crystalline phases concentrate at the interfaces between the dispersed and the continuous phases in emulsions and as a result they act as coalescence inhibitors at flocculation.

In flocculation and coalescence processes, the Van der Waals attraction potential plays a decisive part in the force with which two drops approach one another. Calculations made by Friberg *et al.*[3] show that the liquid crystalline phases reduce the Van der Waals energy in the coalescence process.

Choice of Emulsifier

The HLB value, or the hydrophilic–lipophilic balance[6] provides information concerning the solubility in the water or oil phase, and it has been used to make the optimum choice of emulsifier for a given emulsion system (Fig. 9). The HLB value can be calculated directly from the emulsifier's molecular structure but this does not take into consideration the fact that the interaction between different emulsifiers or between emulsifiers and the oil or water phase can give rise to structures which increase the emulsion stability.

The interchange between the polar and non-polar parts of a non-ionic emulsifier are very dependent on the temperature, changes of which result in changes in the solubility of the emulsifiers in oils and in water. The transition occurs over a narrow temperature interval, designated by Shinoda as the 'PIT' (Phase Inversion Temperature). Determinations of PIT in the presence of fluid water and oil phases make it possible to select with exactitude a suitable emulsifier for a specific emulsion system. Because the interfacial tension is at a minimum at PIT, optimum emulsifying can be achieved at this temperature, after which maximum stability can be obtained for the system 15–20 °C below PIT.[7] One advantage of the PIT method is that interaction between emulsifiers, water and oil phases[8] is taken into consideration.

EMULSIFIERS IN FOODS

Two types of emulsifier are mainly used in foods, i.e. soya bean lecithin, which consists of a mixture of surface-active phospholipids, and partially esterified glycerides (Fig. 10). The phospholipid constituents in soya bean lecithin—phosphatidyl—choline, phosphatidyl ethanolamine and phosphatidyl inositol all swell with water to give ordered structures of the liquid crystalline type.[9] Little work has been published on the association conditions in oil–water systems.

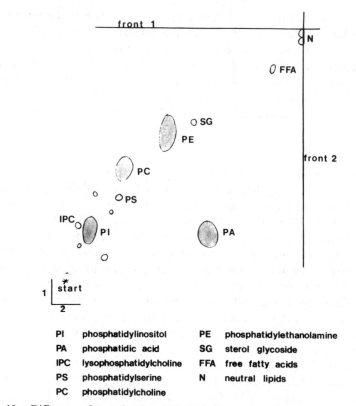

FIG. 10.   Different surface-active phospholipids from soya bean lecithin (commercial grade)
after two-dimensional TLC separation.

The polymorphism and liquid crystalline structure of the monoglycerides
has been thoroughly investigated by Larsson for monoglyceride–water
systems.[10] The monoglycerides, like the phospholipids, associate in the
presence of water to form liquid crystalline structures of different packing
types. In addition, the monoglyceride molecules assume different
crystalline forms, with different melting points, stability and capacity to
bind water.

FOOD EMULSIONS

Phospholipids and monoglyceride-type emulsifiers in foods associate with
both water and oil phases. In monoglycerides, in addition to liquid

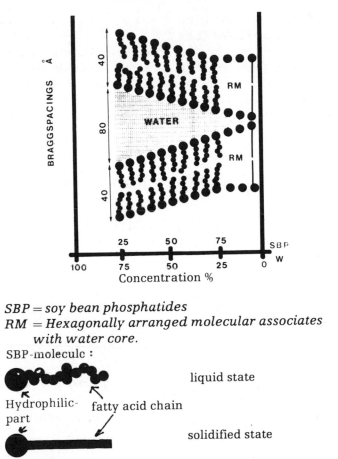

SBP = *soy bean phosphatides*
RM = *Hexagonally arranged molecular associates
with water core.*

SBP-molecule :

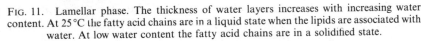

liquid state

Hydrophilic- fatty acid chain
part

solidified state

FIG. 11. Lamellar phase. The thickness of water layers increases with increasing water content. At 25 °C the fatty acid chains are in a liquid state when the lipids are associated with water. At low water content the fatty acid chains are in a solidified state.

crystalline forms, there also exists polymorphic forms, the formation of which is dependent on both temperature and time. A knowledge of the crystallisation properties of emulsifiers, and of the mixture of emulsifiers which may be employed, on association with oil or water at different temperatures, is of major importance in optimising the production of food emulsions.

The phase behaviour of well-defined egg lecithin as a function of temperature was published in 1967 by Small.[11] At water concentrations above 5%, only one lamellar phase was formed which extended over a temperature range of 20 to 220 °C.

The phase behaviour of soya bean phosphatides does not appear to be as dependent on temperature as that of the monoglycerides but changes can take place, depending on the composition of the phosphatides.

The phase behaviour of the monoglycerides in lipid–water systems can be exemplified by monolaurate, which can be packed in two different

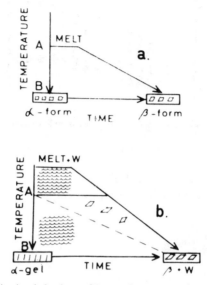

FIG. 12.   The crystallisation behaviour of 1-monolaurate as a function of temperature and time.[12]

crystalline forms, the metastable α-form and the stable β-form. The dependence, both on time and temperature, for the formation of the α- and β-forms is evident from Fig. 12.[12] The solubility of monoglycerides in, and crystallisation from, paraffin oil, is affected by heating and cooling processes (Fig. 13). On slow cooling, the monolaurate crystallises out in the β-form with only a small change in the melting point (Fig. 13, line A). On rapid cooling, on the other hand, a solid solution in oil of the monoglycerides' α-form is obtained (Fig. 13, line B). The behaviour of the monoglycerides with water was determined as a function of temperature (Fig. 14). The product, which was crystallised in the β-form, was hydrated at 38–40 °C. In the presence of water the softening temperature of the hydrocarbon chain was lowered, depending on the interaction of water

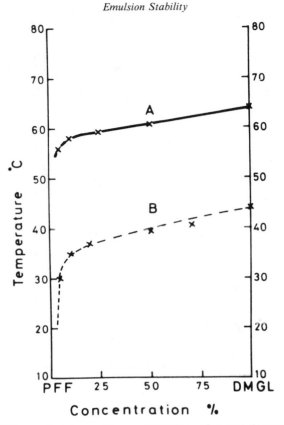

FIG. 13. Solubility and polymorphic crystallisation of 1-monolaurate in a liquid hydrocarbon.[12] A, melting curve of the β-form; B, melting curve of the α-form.

molecules with hydrogen bonds in the monoglyceride crystals. At low water concentrations a fluid isotropic phase was formed (I) and with amounts of water in excess of 20 % a lamellar liquid crystalline phase was obtained (N). The stability of the lamellar phase extends to temperatures around 70–80 °C. In the temperature range below this melting point the neat phase was metastable, with the hydrocarbon chains of the emulsifiers eventually assuming a β-packing, water being released from the lamellar phase. On rapid cooling lower than line B, Fig. 14, the liquid crystalline phase formed a gel (Fig. 15), this taking place because the hydrocarbon chains stiffen in the α-form of packing. The gel state did not affect the water binding because the hydrocarbon chains in this form of packing still have considerable freedom of rotation. The gel state is metastable, however, and reverts to the

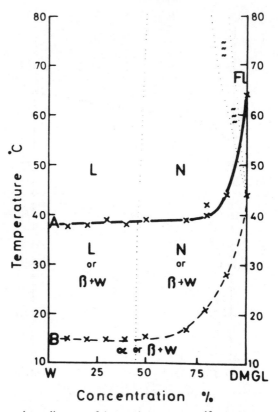

FIG. 14.   Binary phase diagram of 1-monolaurate–water.[12] Liquid crystalline phase: N = neat phase, Fl = fluid isotropic, and L = liposomes. A, melting curve of β-form; B, melting curve of α-form.

β-form after a certain time, water being released. The stability period for the gel state was dependent on temperature and the closer to the melting point of the α-form, the more rapid was the conversion to the β-form.

The effect of the heating process during the preparation of the emulsion was investigated in the system water–oil–monoglyceride. In the preparation of an emulsion with a monoglyceride monolaurate, the monoglyceride must be dissolved in a warm oil phase. When cold water is added, the metastable α-form crystallises at the interface of the water droplets. Controlled heating to a suitable temperature (ca. 40 °C) converts the crystalline monoglycerides to a liquid crystalline form which gives a neat phase at a DMG/water ratio of 1:1. Such a treatment gives a stable emulsion in which the water drops are protected by a multilayer of liquid crystalline phase. It is essential that the

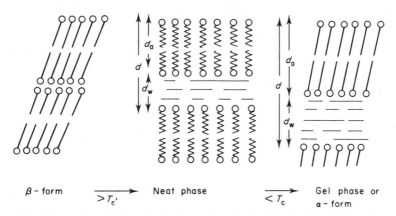

β- form $\xrightarrow{> T_{c'}}$ Neat phase $\xrightarrow{< T_c}$ Gel phase or α - form

FIG. 15. Schematic model showing the formation of a lamellar liquid crystalline phase and the α-crystalline gel phase.

temperature of the emulsion is adjusted to the range where the neat phase can exist, between 40 and 80 °C. If the temperature is allowed to fall, the emulsion will remain stable only during the time the monoglyceride retains its α-form, but as soon as conversion to the stable β-form begins, the emulsion will break below 40 °C.

Those temperature boundaries which are of importance for the emulsion stability can be defined on the basis of a few analyses in the two systems:

(a) Emulsifier–oil: determination of the crystallisation temperature of the α-form in the oil phase.

(b) Emulsifier–water: determination of the transition temperature of the neat phase to α-form, or vice versa.

CONCLUSIONS

The properties of fat emulsions depend to a high degree on the temperature during the pretreatment and mixing of the ingredients. This can be due to the temperature dependent changes involving the crystalline and the liquid crystalline phases which occur in fats and in lipid emulsifiers. In emulsions, the oil, water and emulsifiers are often associated with each other in the form of liquid crystalline structures at the interface. The presence of a liquid crystalline phase is known to increase the emulsion stability, probably due to a reduction of the available energy for coalescence of the dispersed droplets in the emulsion system.

REFERENCES

1. Mulder, H. and Walstra, P. (1974). Stability of the Fat Dispersion. In *The Milk Fat Globule*, Centre for Agricultural Publishing and Documentation, Wageningen, The Netherlands.
2. Friberg, S. and Rydhag, L. (1971). The system: water–*p*-xylene-l-amino-octane–octanoic acid: II. The stability of emulsions in different regions. *Kolloid-Z. und Z. Polymere*, **224**, 233.
3. Friberg, S., Jansson, P-O. and Cederberg, E. (1976). Surfactant Association Structures and Emulsion Stability. *J. Coll. Interface Sci.*, **55**, 614.
4. Becher, P. (1966). *Emulsions, Theory and Practice*, Reinhold Publ. Co., New York.
5. Schulman, J. H. and Leja, J. (1954). Control of contact angles at the oil–water–solid interfaces: emulsions stabilised by solid particles (BaSO$_4$). *Trans. Farad. Soc.*, **5**, 598.
6. Griffin, W. C. (1949). Classification of surface-active agents by 'HLB'. *J. Soc. Cosmetic Chemists*, **1**, 311.
7. Saito, H. and Shinoda, K. (1970). The stability of w/o type emulsions as a function of temperature and of the hydrophilic chain length of the emulsifier. *J. Coll. Interface Sci.*, **32**, 647.
8. Friberg, S. (1976). Emulsion Stability. In *Food Emulsions* (Food Science Series, Vol. 5), Marcel Dekker, New York.
9. Luzzati, V. and Tardieu, A. (1974). Lipid phases: structure and structural transitions. *Amer. Rev. Physic. Chem.*, **25**, 79.
10. Larsson, K. (1976). Crystal and Liquid Crystal Structures. In *Food Emulsions* (Food Science Series, Vol. 5), Marcel Dekker, New York.
11. Small, D. M. (1967). Phase equilibria and structure of dry and hydrated egg lecithin. *J. Lipid Res.*, **8**, 551.
12. Wilton, I. and Friberg, S. (1971). The relationship between the stability of fat emulsions and crystallisation at the interfaces. 6th Scandinavian Lipid Symposium, Grenå, Denmark.

15

Fish Products—Heat Penetration

K. H. SKRAMSTAD

*The Norwegian Canning Research Laboratory, Alex Kiellandsg. 2, N-4000
Stavanger, Norway*

The Research Laboratory of the Norwegian Canning Industry was founded
in 1931—and one of its most important tasks has always been to produce
reliable processing data for the canning industry. As early as the beginning
of the 1940s, thermal death time studies (TDT) were performed and heat
penetration measured in various products.[1-4] The method used—named
the general method—was that of Bigelow *et al.* later improved by Ball[5] and
Schultz and Olson in 1940.[6] The test organism used was a very heat resistant
strain of *Clostridium sporogenes*—isolated and described by Aschehoug
and Jansen.[7] Further studies on spore-forming anaerobes as spoiling agents
in canned foods were undertaken.[8]

The evaluated data relevant to the processing industry were later
collected in a large 'recipe' collection,[9] intended as a guiding manual for the
canning industry. This publication describes the industrial production of
almost 70 different canned products of fish, meat and vegetables. Times and
temperatures for safe processing are listed, together with recommended
retort pressures for different can sizes. Other data were given, for instance
when the products are filled hot, initial temperatures (IT) are given. A great
deal of work was performed through the 1950s to complete the publication
of the recipe collection. Later on in the 1960s the introduction of a German
agitating pilot retort called for further work and the evaluation of safe
processes for various meat products[10] and, for instance, fish meat balls in
brine.[11] A new edition of the 'Norwegian Canners' Manual' was published
in 1969,[12] including in this edition process and retort data for different
canned vegetables, meat and fish products.

Established scheduled processes were evaluated as already mentioned by
carrying out TDT studies on spores of the particular strain of *Clostridium
sporogenes* as described by Aschehoug and Jansen.[7] This organism had an

F-value of 6·0 min and *Z*-values of 10 °C as determined by the TDT tube method in phosphate buffer at pH 7·1.

For the various canned products the TDT can method was used. The cans were filled with exactly similar products as used in the industrial packs—the same pH, salt concentration and same environmental conditions—except for the solid fish, meat etc., which were chopped to fit the small can volume (22 cm^3). The same method as described by the National Canners Association (USA) has been used.[13] Finally, the TDT curves are drawn ($F_{121·1}$) and *Z*-values calculated for various products.

Heat penetration measurements are performed routinely for every product variant. Today we are using a German 'Labor Rotorzwerg' from H. Stock, Neumunster, constructed to simulate a still or agitating retort for pressure processing with steam or water as the heating medium. The retort is equipped with copper–constantan thermocouples and an automatic potentiometric recorder. There is also included an F_0-value computer (from ELLAB, Denmark) based on the lethal rates of *Clostridium botulinum* as worked out by the National Canners Association. The thermocouples are placed in the middle of the cans. and the temperature curves are automatically registered and calculated.

Today we have thus dropped the use of *F*-values from the aforementioned *C. sporogenes* strain, and instead we use international lethal values of *C. botulinum* to get more uniform and comparable process values.

Process calculation for low-acid products (pH > 4·5) today can also be achieved by graphical method—by summing the lethal rates from tables at every single minute during the sterilisation.

To establish the scheduled process for a product we first have to take into consideration factors influencing thermal resistance—and we often have to confirm the results with incubation tests. Factors influencing thermal resistance have to be constant and controlled in the industrial packed products, i.e. under similar conditions as in the TDT can tests. The pH and the water activity have to be constant—the oil fill, the fat content, the starch content have to meet the regulations etc. The research laboratory has specified regulations for various products, and the official Norwegian quality control institute keep regular control on the actual products to ascertain that the regulations are kept.

The sheduled process contains separate provisions for the various operations of the process (see Fig. 1). Specified data are laid down for:

    (*a*)   Temperature and pressure in the over-kettle (stock kettle) or steam boiler.

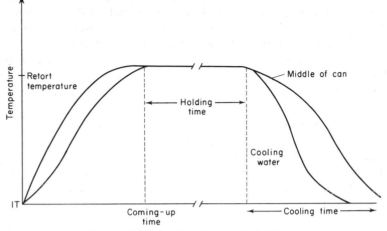

FIG. 1. Specified data for the scheduled process.

(*b*)   Coming-up time, which must be specified within strict time intervals (too short a time may imply danger for underprocessing).

(*c*)   Holding time.

(*d*)   Retort temperature and total pressure.

(*e*)   Time for hot water to return back to the stock kettle.

(*f*)   Cooling time and pressure conditions (pressure cooling).

The scheduled process is prepared with close attention to the actual conditions in the retorts.

Incubation tests and microbiological sterility tests are made on request—whenever assistance for a check-up is necessary. If a plant alters its routine processing conditions, requires new equipment or machinery, or if they adopt new products or variants on the older ones a thorough check-up has to be undertaken.

The official quality control institute is responsible for carrying out incubation and sterility tests on samples drawn from regular production. If necessary, test trials are performed in the pilot plant of the canning laboratory, in connection with process calculations. The managers of the plants make inquiries to the canning laboratory whenever they need instructions for converting process data from one can size to another, converting from one retort to another, changing the IT etc.

As a typical example of an important product sardines will be mentioned. This product consists of either smoked sprats or small herrings packed in oil or sauces in small ($\frac{1}{4}$ dingley) aluminium cans (112 ml capacity and ca. 20 mm high). Since the first heat penetration measurement in the 1930s, calculations from time to time have always shown that *F*-values for sardines are far above the safe *C. botulinum* limit, thus an over-processing of sardines generally takes place. Canned products usually lose quality when they are over-processed. An example to the contrary is the Norwegian sardine, where over-processing is necessary to achieve soft bones (backbones, fins). The process is safe even in respect of a more heat resistant spore-former, i.e. *C. sporogenes*.

A scheduled process of 112 °C for 60 min was formerly evaluated with a *C. sporogenes* strain, as mentioned above. On this basis we got a $F_{121\cdot1}$-value of 8 min and $Z = 11$ °C. As outlined above, both TDT tube and TDT can methods were used, with 'heavy' inoculums of spores in smears of sardines in oil. The moment was observed when no survivors were detected in the cans at three or four different temperatures as a function of time. Thus, the evaluated *F C. sporogenes* value is not related to the *D*-value of the bacteria as in the 12 *D*-concept. On the other hand, the higher resistance of the *C. sporogenes* strain and the heavy inoculums (10^{10} spores/g) makes the process (112 °C, 60 min) safe enough.

Now, to be in line with other countries, for instance the FDA regulations

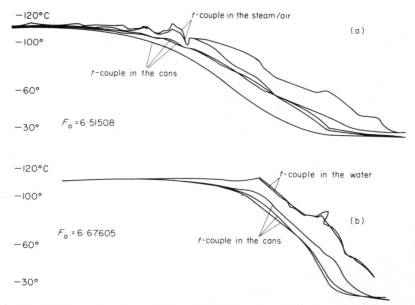

FIG. 2. Coming-up time for sardines in oil in (a) a steam/air-filled autoclave, $F_0 = 6\cdot51$ min; (b) a water-filled retort, $F_0 = 6\cdot67$ min.

in the USA, we use the lethal rates of *C. botulinum*, as worked out by the National Canners Association. With these lethal rates applied to heat penetration measurements in the cans, we state an F_0-value of 6·0 min for sardines. The basis for this value is the equation:

$$t = F_0 \log^{-1}\left(\frac{121 - T}{Z}\right)$$

where t = time in minutes; $F_0 = 6\cdot0$ min; T = temperature = 112°C; Z = slope = 10°C. Usually $F_0 = 4\cdot0$ is regarded as sufficient.

The stated F_0-value of 6·0 min is thus a practical value to be used in order also to get the bones softened. This goal is achieved with a minimum heat process of 50 min at 112°C. In order to fulfil this minimum process for every can in a retort batch—and due to small variations in the retort during processing—the official process for the industry is 112°C in 60 min, as recommended by the research laboratory of the Norwegian canning industry.

Laboratory examination of heat penetration in water-filled and

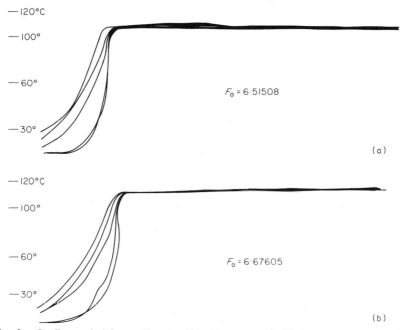

FIG. 3. Cooling period for sardines in oil in (a) a steam/air-filled retort, $F_0 = 6\cdot51$ min; (b) a water-filled retort, $F_0 = 6\cdot67$ min.

steam/air-filled retorts have been performed. The heat penetration data are shown in Figs. 2 and 3. Without describing any further the manner of operation of the two retorts, the F_0-value of the water-filled retort is 6·7 min and of the steam/air-filled retort 6·5 min. The heat transfer into the centre of the cans is very good in both cases, and the difference in F_0-value is negligible.

The top deviation in the curves for the steam/air retort—when the temperature reaches 112 °C—is due to inlet of air (not preheated) to attain the necessary counterpressure during processing of the *aluminium* cans. These top deviations are very small and do not affect the processing by more than 0·2 min in the difference in F_0-value (maximum), proving the performance of two types of retort are approximately equivalent. In the industrial retorts the air inlet has the same temperature as the steam.

The influence of initial temperature (IT) on the thermal processing of sardines is negligible. The filled and sealed cans are conveyed in channels of fresh tap-water or in pipes, and have a short delay in the retort baskets completely under fresh water before the thermal processing. The conveyor water temperature varies from 10 °C in the winter to 20 °C during the

TABLE 1
Example of nutritional labelling, as used in the USA

Norway Brisling sardines in olive oil, 2-layer
Net weight: $3\frac{3}{4}$ oz. (106 g).
Ingredients: smoked brisling, olive oil, salt.
Name and address of the manufacturer or distributor

NUTRITIONAL INFORMATION
Serving size: $3\frac{3}{4}$ oz. (106 g).
Contains one serving

Calories:	460	Excluding liquid	260
Protein:	19 g	Excluding liquid	19 g
Carbohydrate:	1 g	Excluding liquid	1 g
Fat:	42 g	Excluding liquid	20 g

Percent of US recommended daily allowances

Protein:	40	Calcium:	30
Vitamin A:	4	Iron:	10
Vitamin C:	0	Vitamin D:	100
Thiamine:	0	Vitamin B_{12}:	100
Riboflavin:	10	Phosphorus:	30
Niacin:	30	Magnesium:	10

---

summer time, i.e an IT difference of approximately 10 °C. Experiments are performed with IT = 10 °C and IT = 20 °C, and the lethal values are summarised in the two cases, resulting in a difference in F_0-values of less than 0·1 min. This is of course negligible in relation to the heat processing of canned sardines where F_0 equals 6·0 min.

As we have been talking about destruction of vitamins during heat processing I will mention nutritional labelling used in the USA. This shows (Table 1) that the product is high in vitamin D and low in thiamine and vitamins A and C.

REFERENCES

1. Lunde, G. and Kringstad, H. (1940). Varmegjennomtrenging og sterilisering av hermetikk I. *Tidsskr. f. Herm. ind.*, p. 185.
2. Ringstad, H., Vesterhus, R. and Aschehoug, V. (1941). Varmegjennomtrenging og sterilisering av hermetikk II—Sterilisering av fiskeboller. *Tidsskr. f. Herm. ind.*, p. 177.
3. Vesterhus, R., Aschehoug, V. and Kringstad, H. (1942). Varmegjennomtrenging og sterilisering av hermetikk III—Sterilisering av erter og spinat. *Tidsskr. f. Herm. ind.*, p. 189.
4. Jansen, E., Aschehoug, V. and Kringstad, H. (1943). Varmegjennomtrenging og sterilisering av hermetikk IV—Sterilisering av kjøtt i egen kraft, kjøttkaker i brun saus, kjøttpudding og blodpudding. *Tidsskr. f. Herm. ind.*, p. 256.
5. Ball, C. O. Mathematical methods and problems. 1928. Univ. Calif. Public Health, I, p. 230.
6. Schultz, O. T. and Olson, F. C. W. (1940). *J. Food Res.*, **5**, 399.
7. Aschehoug, V. and Jansen, E. (1950). Putrefactive anaerobes as spoilage agents in canned Foods. *J. Food Res.*, **15**, 62.
8. Aschehoug, V. (1947). Spore-forming anaerobes as spoiling agents in canned foods 4th Int. Congress of Microbiology, Copenhagen, July 20–26. Report of Proceedings.
9. Hermetikkindustriens Laboratorium. (1950). Oppskriftsamling, Stavanger, September.
10. Skramstad, K. H. (1962). Rapport over forsøk med rotasjons-sterilisering, Hermetikkindustriens Laboratorium, Januar: I, Høns i gele; II, Frankfurter-pølser: III, Leverpostei.
11. Skramstad, K. H. (1969). Sirkulaere no. 4. Rotasjonssterilisering av fiskeboller i lake.
12. Konservindustriens Oppslagsbok. (1969). Steriliseringsbetingelser for en del hermetiske produkter, *Tidsskr. f. Herm. ind.*, p. 91.
13. Laboratory Manual for Food Canners and Processors, 1968, Avi Publ., Westport, Conn.

16

Thermal Processing, Evaporation and Drying of Fish-Meal Products

GEIR E. SØBSTAD

Norwegian Herring Oil and Meal Industry Research Institute, 5033
Fyllingsdalen, Bergen, Norway

The fish-meal industry in Norway handles approximately two-thirds (ca. 2×10^6 tonnes) of the fish landed every year. Small pelagic fishes are the main raw material which is boiled and pressed to separate the lipids from the protein material. After pressing we obtain a solid fraction (the press cake) and a liquid phase carrying lipids, water and some protein material.

The flowsheet from the fish-meal process indicates that 20–25% of the fish protein is related to the press liquid. Under special conditions this may rise to 45% and it is obvious that this liquid is a resource of proteins and amino acids which should be taken care of. Today this is done in the fish-meal plants by the evaporation system where the press liquid is separated from the lipids and concentrated from 6% dry matter (d.m.) to about 30% d.m. Experiments of evaporation of fish solubles started in 1949 and the first evaporation plants were operated under pressure, which made it possible to produce concentrates of up to 70% d.m. This was probably due to the hydrolytic action which these conditions impose on the proteins (Fig. 1).

Today most plants operate as combined multiple effect pressure/vacuum systems as it is a more practical way of concentrating the 'fish solubles', but these conditions will not allow a concentration to much more than 30%. Beyond that concentration the flow properties of the fish solubles are greatly affected. The viscosity of the sticky water which carries these proteins, partly as soluble proteins and partly as suspended materials emulsified with fat, depends on the amount of suspended materials, soluble materials, pH, temperature and probably the stoichiometry of these components. The content of collagen may also contribute to the viscosity under certain conditions.

If the heat treatment in the cooker has been incomplete, there will be a

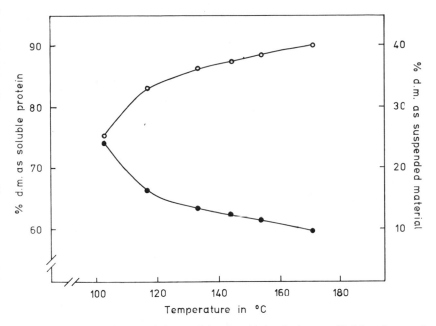

FIG. 1.   Hydrolysis of suspended materials after 4 h incubation at pH 6·5 and normal pressure. Incubation temperatures are varied: (○) soluble protein; (●) suspended material.

precipitation of proteins in the first stage, which is operated under pressure and a temperature of 110 °C, resulting in a rise in the level of suspended materials and a steep rise in the viscosity. The succeeding stages are operated under vacuum and lower temperatures (80 °C and 60 °C), but even these conditions have a proteolytic action on the proteins. Thus, it is demonstrated that there is a change in the pattern of soluble proteins from higher to lower molecular weight (Fig. 2). However, these reactions do not prevent a rise in the viscosity which can be expressed by a relation

$$(\eta)_{T,\,pH} = \eta_0\,(K:M^d + a\phi + b\phi^2 + c\phi^3) \qquad (1)$$

where ϕ = suspended materials; M = mean molecular weight of soluble; a, b, c, K = constants. This behaviour is shown in Fig. 3 for three different concentrates.

These curves show a reasonable fit to the mathematical relation, but in some cases there will be an enormous rise in the viscosity at moderate d.m. contents, as for instance 5000 cp at 22–25 % d.m. Such conditions may be a serious problem to the industry during certain periods and this seasonal

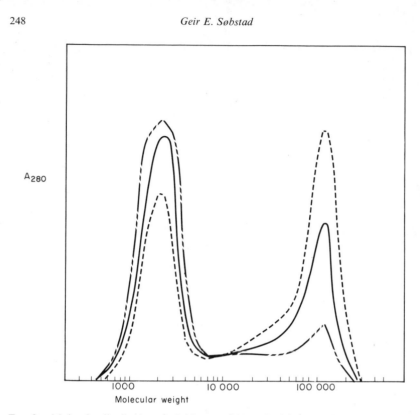

A_{280}

Molecular weight

FIG. 2.   Molecular distribution of soluble part of, (— — — —) sticky water, (————) concentrate
and (— · · · —) concentrate entering the driers.

variation turned our attention to collagen. Chemical analysis on such high-
viscosity fish solubles could not, however, confirm this theory.[1] Thus, it is
still an open question how these high viscosity concentrates arise and how
they can be avoided.

Laksesvela[2] tested the PER-value of different mixtures of concentrated
fish solubles and press cake meal and found that a mixture of 20% (w/w)
solubles to 80% (w/w) meal was optimal as far as the PER-value was
concerned. This ratio corresponds to the composition of whole meal.
Mixtures with a higher percentage of solubles were limiting in these essential
amino acids, due to the fact that the solubles are low in these.

The choice of driers in this industry is based on the desire to conduct the
dehydration operation economically and efficiently. The industrial driers in
use are therefore of two main types, namely the hot-air rotary drier and the
steam drier. In the first there is a stream of hot air passing over the meal and

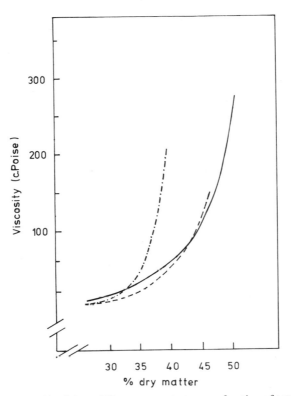

FIG. 3.   Viscosity of three different concentrates as a function of total dry matter.

drying the materials; in the steam drier the heat is administered through rotary heating surfaces and the water vapour is soaked off. Of these two, the hot-air drier has been more thoroughly investigated, and the discussion will be limited to this type. The nutritional value of the meal is of similar magnitude, whether it is dried in either type of drier. In fact, there are no indications from users of fish-meal that indicate any deterioration of feed quality during dehydration. The investigation of Myklestad[3] therefore aimed at measuring the physical conditions in the drier and calculating the most efficient operating conditions.

During operation, the solids (the press cake) leave the press with a moisture content of 55–60 % and 4–6 % lipid in addition to the phospholipids (∼ 3 %). This is mixed with the concentrate before entering the drier. The main effect in the material transport through the drier is due to the interaction of the horizontal air flow with the cascading materials.[4] In addition to this

Geir E. Søbstad

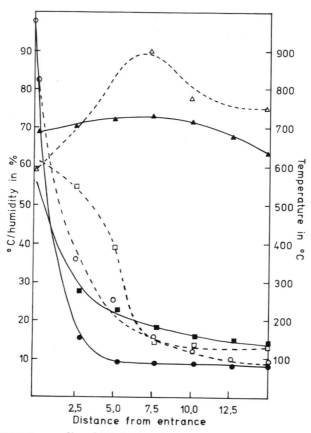

FIG. 4. Temperature profiles of (○) the hot air, (Δ) the meal and (□) the humidity of the meal. Dotted lines indicate experiments of series one.[5]

momentum transfer, there is a heat transfer from the hot air to the particles. This provides the energy for the evaporation of excess water from the solids. Some typical results from the work of Myklestad[3] and Midtsaeter[5] are shown in Fig. 4, which is a graphical representation of the temperature of the air, the temperature of the meal and the moisture content of the meal as it passes through the drier. The hot air enters with a temperature of ∼ 600 °C and this temperature falls rapidly in the first part of the drier as heat is transferred to the solids and water is evaporated and leaves with the air.

In the first experiments, the evaporation seemed to proceed through two different phases, each of them with an approximately constant drying rate.

In the second series of experiments it was not possible to distinguish between these two phases and there was a rapid decrease in the water content of the material. In these experiments the temperature of the meal rose to a maximum and then decreased towards the end of the drier. This temperature rise is, of course, an effect of the heat transfer from the hot air, but whether the temperature rise gives a correct picture of the heat gained by the solids or if the measurements are faulty *per se* is an open question. Anyway, as a first order approximation we can write

$$\int Gc_p \, dT = \int (c_p m_1 + c_f m_2 + c_w m_3) \, dT + \int \lambda \left(\frac{dw}{dt}\right) dt + \int c_p' \left(\frac{dw}{dt}\right) dt \, dT$$

$$(2)$$

Thus, putting numbers into the equation will confirm that the heat transferred to the particles in the cascading period is mainly used to evaporate the water, but some of the energy is consumed by the water vapour and some is consumed by the dried particles and this energy may be dissipated through different mechanisms. This energy is indicated by the first term on the right hand side of the Eqn. (1). As mentioned earlier, 33 % of the solids entering the drier is protein and 6 % lipid. It is quite clear that the polyunsaturated fatty acids (PUFA) in the lipid fraction are easily activated during this heat treatment, as the activation energies of these compounds are in the range 3–17 kcal/mol, depending on the content of catalytic compounds.[6]

In former days, dried meal could exert a rapid temperature rise in the first period of storage, due to lipid oxidation. The most vigorous reaction might even lead to ignition of the meal. Today, the meal is cooled down by recirculation and stored in bags which restrict the diffusion of oxygen into the meal. The oxidation of the activated fatty acids will therefore proceed slowly.

To us it is important to know and to control the reactions taking part during the heating of the meal, and there are indeed many. On the one hand there is oxidation of the lipids giving rise to different low molecular compounds, like aldehydes, ketones, epoxides and others which may react with one another to produce polymers of lipids, through direct association between free alkoxy and alkyl radicals.

$$-CH{=}CH- + HOOR \rightarrow -\overset{+}{C}HCH-O-OR \rightarrow -CH-CHOH$$
$$\quad\quad\quad\quad\quad\quad\quad\quad\quad\quad | \quad\quad\quad\quad\quad |$$
$$\quad\quad\quad\quad\quad\quad\quad\quad\quad\quad H \quad\quad\quad\quad OR$$

It is, however, believed that other types of polymers may form through C–C bonds. Such compounds may contain cyclic structures and add a dark brown colour to the lipid fraction.[7]

On the other hand, the loss of rehydration and textural properties in the product is of great interest to us for certain purposes, and causes us to speculate on the possible interactions between lipid oxidation products and other compounds, e.g. proteins. In fact, reactions between functional groups on proteins and reactive intermediates from lipid oxidation were described in 1953 by Koch.[8] Such interactions might be expected to give rise to changes in the solubility of the proteins and cross-linkages in the molecules, with resultant changes in the properties of the dehydrated product. Such deleterious reactions can sometimes be demonstrated by the formation of coloured compounds in the product. The browning reaction observed by dehydration of fish-meal may be the sum of several different paths of reactions including oxidative reactions in natural pigments (dark muscle, other pigments) and Maillard reactions, mainly due to reactions between ribose or decomposition products of ribose with lysine and other amino acids. Such reactions have been demonstrated in model systems by several authors.

Pokorný et al.[9] studied mixtures of polyunsaturated esters of marine oils and casein. They found a quick turnover of peroxides, but the browning proceeded slowly and neither depended on the concentration of peroxide nor the concentration of carbonyl compounds. The formation of this pigment was therefore believed to proceed through the formation of a colourless intermediate. In 1973 the same author studied the oxidation of phosphatidyl ethanolamine as a model for the reaction of phospholipids on heating.[10] There was found to be a clear connection between the total amount of peroxide decomposed, on one hand, and the total amount of brown pigment formed and, on the other hand, the total amount of primary amino groups blocked.

Dividek and Jiroušová[11] heated n-hexanal and glycine in an aqueous solution and obtained both soluble and insoluble brown pigments. The insoluble pigment contained nearly 0·6% nitrogen. Addition of antioxidant (butylated hydroxyanisole) had an inhibitory effect on the autoxidation of unsaturated fatty acids in a mixture with model proteins.[12] There were formed some brown pigments, most probably because there were some carbonyl lipid oxidation products in the mixture before the antioxidant was added. Under certain conditions such brown pigments may even act as antioxidants.[13] Brown pigments produced by the condensation of glycine and glyoxal or dehydroascorbic acid (or ethylamine and glyoxal) were demonstrated to increase the shelf life of heated fatty foodstuff to a considerable degree. These model experiments demonstrate that reactions between lipids and proteins are very complex and that a number of chemical and structural changes take place in the tissue. The connections between the reaction medium and the path of reactions are but partly known. Most probable are carbonylamine reactions between aldehyde and the ε-amino group of lysine, arginine, histidine and terminal α-amino groups of other amino acids. Such reactions may reduce the biological availability of the amino acids. Furthermore, inter- and intra-molecular cross-linkages may reduce the availability of alanine, glycine, isoleucine, leucine, phenylalanine and valine. There is also reason to believe that proteins with such modified amino acids are poorly hydrolysed in the digestive tract.[14] There are also indications that racemisation of amino acids can occur in proteins during heating, even during the practical heat treatment of fish-meal.[15]

On the other hand, it is an open question whether lipid–protein interactions of a non-covalent type play any part in protein hydrolysis *in vivo*. Anyway, there are several indications that some reactions will have a more serious effect on the nutritive value of the dehydrated fish proteins than others. Here, the digestive system of the animal will play an important part. Thus, Finot[16] reported that ε-deoxyfructosyl lysine was hydrolysed by rats, giving partly free lysine and partly pyrodisine and furosine. This gave a higher availability of the lysine than expected. Such species differences are also reflected in the biological testing of fish-meal produced under extreme conditions (Table 1).[3] While, in experiments with chickens, only fish-meal processed under extreme temperatures had a lower PER-value compared to the control sample, there was no significant difference observed in experiments with rats. Fish-meal processed under 'normal' conditions is considered to have proteins of high biological availability and this is true as far as the traditional consumers are concerned. However, in the diet of, say, young calves, the effects of dehydration may become significant.

This is the conclusion of a series of experiments carried out at SSF. The aim was to improve the digestibility of fish protein concentrates in the diet of young calves. The fish protein concentrates were produced in a pilot plant and the processing conditions are varied from one sample to the other, especially the heat treatments (e.g. the boiling and the drying). Considering the digestive functions in the animals, it is of importance to retain the feed in the ventriculus to have a controlled emptying in small portions into the duodenum. This means that we should have acceptable swelling properties in the meal to obtain a reasonably good 'coagulation' in the ventriculus.

TABLE 1
Biological evaluation of capelin meal[a]

Meal no.	Process	Max. air temperature (°C)	Chickens				Rats			
			NPU %	PER	%	Growth g/24 h	D_a (%)	Bal (%)	NPU_{calc} (%)
13–1	Freeze-dried	20	60·6	100	2·80	100	1·89	79·8	53·0	73·3
13–7	Factory product	900	59·8	99	2·18	78	1·75	80·3	55·0	73·3
19–1	Freeze-dried	20	68·1	100	2·75	100	1·80	81·4	57·9	79·6
19–7	Factory product	1065	64·9	95	2·50	91	1·77	80·2	55·4	76·9
20–1	Freeze-dried	20	69·4	100	2·92	100				
20–7	Factory product	790	65·5	94	2·43	83				

[a] NPU = net protein utility; PER = protein efficiency ratio; D_a = apparent digestibility; Bal = nitrogen balance.

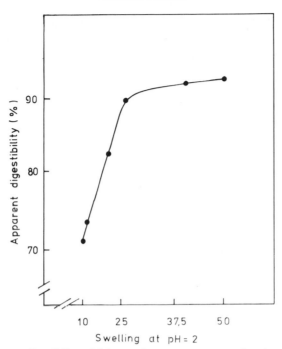

FIG. 5. Apparent digestibility of fish protein concentrates as a function of the swelling properties.

In Fig. 5 the apparent digestibility of fish proteins is plotted as a function of the swelling properties of the proteins. The calves were less than one week of age at the start of the experiment: it should be stressed that the digestibility improves with the age of the animal. However, in these experiments it is obviously true that the drying operation may have a deleterious effect on the proteins, because unfavourable drying conditions may destroy the functional properties (e.g. swelling, water binding). Nesse et al.[17] investigated the connection between the drying conditions and the rehydratisation index of the product. The experiments were conducted in a pilot plant drier where the proteins were dried in a stream of hot air on a stationary bed. Both temperature and humidity were recorded and the rehydratisation index measured after the drying. The data were collected on a data logger and processed in a computer. From these observations they concluded that the deteriorative effects of drying were best expressed by the integral.

$$\int W\theta_f \, dt \tag{3}$$

where W is the water content of the sample; θ_f, the temperature, and t, the time (Fig. 6). Furthermore, there were indications that the proteins were most susceptible to damage (e.g. loss in rehydrative properties) with a water content in the region 30–60 % (w/w).

While passing through this region the physical conditions in the drier (e.g. temperature, humidity) were decisive to the results. In these experiments only lean fish was used as raw material and the influence of

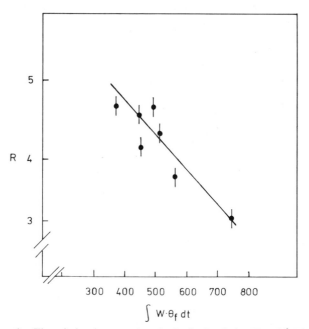

FIG. 6.   The relation between the rehydratisation index R and $\int W\theta_f \, dt$.

lipids was therefore not investigated. In experiments performed at our institute the level of rest lipids was measured in a series of fish-meal together with the functional properties of the meal (e.g. swelling). As far as the experiments permit any conclusions to be drawn, there is evidence that the rest lipids are deleterious to these functional properties measured. These relations are subject to further investigations because they may prove to be important to other properties of the fish-meal as well. This is especially interesting when we compare these results with those given in Fig. 5, describing the relation between the digestibility of the fish-meal and the swelling properties of the meal.

Even if there are a large number of possible reactions which may take part in the meal during the dehydration and even if it is difficult to find significant methods to measure these changes, we think that more work should be done to shed more light on the drying operation, not only because we want to retain the functional properties of the native fish, but also because this will enable us to obtain even better control of the processing and give us the possibility of manufacturing new and more refined products, which are of importance to new consumers of dried fish products.

REFERENCES

1. Tidemann, E. (1975). Thermal dehydration (Norwegian). SSF Report no. B 240.
2. Laksesvela, B. (1961). Extensive Examination of Dietary Amino Acid Balance (Norwegian). Universitetsforlaget, Oslo.
3. Myklestad, O. (1973). Hot-air driers in the fish-meal industry. Measurements of operation conditions and their influence on heat economy and quality of the product (Norwegian). SSF report no. D 205.
4. Saeman, W. C. (1962). Air–solid interactions in rotary driers and coolers. *Chem. Eng. Progr.*, **58**, 49–56.
5. Midtsaeter, J. (1973). Study of hot-air driers at Knarrevik herring meal factory (Norwegian). SSF report no. D 200.
6. Tappel, A. L. (1962). Haematin compounds and lipoxidase as biocatalysts. In *Symposium on foods: Lipids and their Oxidation*, ed. by Shultz, H. W., Day, E. A. and Sinnhuber, R. O., Avi Publ. Co., Westport, Conn.
7. Keeney, M. (1962). Secondary degradation products. In *Symposium on Foods: Lipids and their Oxidation*, ed. by Shultz, H. W., Day, E. A. and Sinnhuber, R. O., Avi Publ. Co., Westport, Conn.
8. Koch, R. B. (1962). Dehydrated food as model systems. In *Symposium on Foods: Lipids and their Oxidation*, ed. by Shultz, H. W., Day, E. A. and Sinnhuber, R. O., Avi Publ. Co. Westport, Conn.
9. Pokorný, J., El-Zearny, B. A. and Janiček, G. (1973). Browning produced by oxidised polyunsaturated lipids on storage with proteins in presence of water. *Z. Lebensm. Unters.-Forsch.*, **151**, 157–61.
10. Pokorný, J., Tài, P. and Janiček, G. (1973). Autoxidation and browning reactions of phosphatidyl ethanolamine. *Z. Lebensm. Unters.-Forsch.*, **153**, 322–5.
11. Davidek, J. and Jirioušová, J. (1975). Formation of volatile compounds and brown pigments in the model system n-Hexane–glycine. *Z. Lebensm. Unters.-Forsch.*, **159**, 153–9.
12. El-Zearny, B. A., Janiček, G. and Pokorný, J. (1975). Effect of an anti-oxidant on the discoloration of lipid–protein mixtures. *Z. Lebensm. Unters.-Forsch.*, **158**, 93–6.

13. El-Zearny, B. A., Pokorný, J., Velišek, J., Davidek, J. and Davidek, G. (1973). Antioxidant activity of some brown pigments in lipids. *Z. Lebensm. Unters.-Forsch.*, **153**, 316–21.
14. Opstvedt, J. (1975). The influence of lipid oxidation on protein quality with special reference to fish protein concentrates (Norwegian). Thesis. Norwegian Agricultural College, Ås, Norway.
15. Bjarnarson, J. and Carpenter, K. J. (1970). Mechanisms of heat damage in proteins. *Brit. J. Nutr.*, **24**, 313–29.
16. Finot, P. A. (1973). Non-enzymic Browning. In *Proteins in Human Nutrition*, ed. by Porter, J. W. G. and Rolls, B. A., Academic Press, London.
17. Nesse, N., Halvorsen, K. and Vassbotn', T. (1972). FPC with rehydrative properties (Norwegian). SINTEF Report no. B 2572.

17

Milk Products—Pasteurisation and Sterilisation

E. -G. Samuelsson

The Royal Veterinary and Agricultural University, Department of Dairy Technology, Bülowsvej 13, DK-1870 Copenhagen V, Denmark

The dairy industry is a highly specialised industry which processes milk and milk products and some fruit juices. Except milk for some special kinds of cheese, e.g. Emmenthaler cheese, all milk is always given some heat treatment, which we may call either pasteurisation or sterilisation. The heat treatment influences the number of different micro-organisms, but also the chemical and physico-chemical properties of the milk. The effect of the heat treatment depends on the technique for heat treatment, the temperature, and the holding time. Modern process-technology for heat treatment has favoured a high temperature for short time but heat treatments for up to 20 min are still common practice within the dairy industry, i.e. for cultured dairy products and, of course, for autoclave-sterilised evaporated milk.

PASTEURISATION

Originally, pasteurisation was used to protect the consumer against all pathogenic micro-organisms and to prolong the keeping quality. It is still accepted that *Mycobacterium tuberculosis* is the most resistant pathogenic bacterium and the pasteurisation process was adjusted to guarantee that all milk is free from *M. tuberculosis*. Therefore, today, milk for consumption is pasteurised at 71–2 °C (minimum) for 15 sec. Figure 1 shows the thermal death curve for *M. tuberculosis* and the inactivation curves for the phosphatase and the peroxidase enzymes which are used for control if the milk is pasteurised too low or too high.

In addition to destroying pathogenic bacteria, normal pasteurisation (72 °C for 15 sec) also reduces the total number of bacteria, and this effect was from the beginning found to be about 99·9 %, which improves the keeping

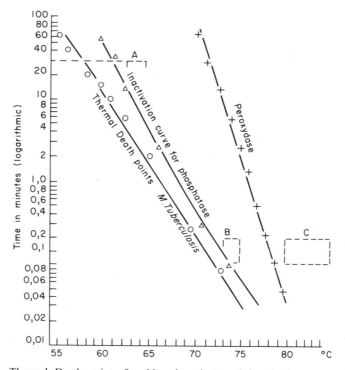

FIG. 1. Thermal Death points for *M. tuberculosis* and inactivation curves for the phosphatase enzyme and peroxidase enzyme (Storch's enzyme) in milk. A, LTLT 63–5 °C for 30 min; B, low pasteurisation, 71–3 °C for 6–12 sec; C, high pasteurisation (HTST), 80–6 °C for 6–12 sec.

quality to a great extent. Of course this effect is more pronounced at higher temperatures, but not more than 20 years ago the creaming of milk was an important quality factor and even more important than the keeping quality. Creaming is highly temperature dependent and is almost lost at 76 °C for 15 sec, so the pasteurisation temperature was held as close as possible to a negative phosphatase test. The reason is that agglutinine, which is responsible for the creaming, is inactivated above 70 °C for 15 sec (Fig. 2).

Today all milk for consumption is homogenised, and the creaming criterion is ruled out as a quality factor. This permits an increase of the pasteurisation temperature up to 77–8 °C for 15 sec. This is common practice today, for two reasons:

(*a*)   Milk production has undergone important changes, e.g. better hygiene and delivery every second day. This has changed the

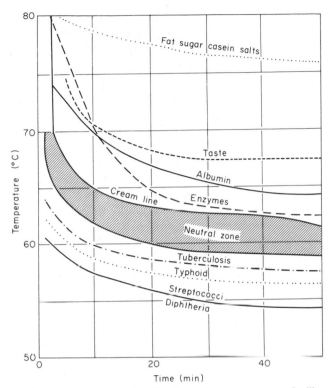

Fig. 2. Time–temperature relationships for various components of milk products.

microflora, which are reduced in number but to a more psychrotrophic flora, and also to a more thermo-resistant flora. Therefore we now seldom reach a pasteurisation effect of 99·9 % at 72 °C for 15 sec, or a negative phosphatase test. To obtain this we have to increase the temperature to about 76–8 °C for 15 sec.

(b) Modern marketing of liquid dairy products requires a prolonged keeping quality, and this is also obtained by an increased pasteurisation temperature, but is limited by some negative effects because of chemical changes in the milk.

The reason why, for example, milk for consumption is not pasteurised to an even higher degree is the development of cooked flavour, which is a consequence of heat denaturation of β-lactoglobulin, involving liberation of some sulphur-containing substances, e.g. hydrogen sulphide.

Thus modern dairy technology means the pasteurisation of milk for consumption at 77 °C for 15–20 sec. It should be mentioned that milk for cheese-making is never pasteurised to more than 71–2 °C for 15 sec, because rennetting is highly dependent on heat treatment.

For cream, especially cream for butter-making but also for both coffee cream and whipping cream, pasteurisation of up to 95–105 °C for 15–20 sec is often used. In this case the cooked flavour is accepted, and the high temperatures are used for increasing the keeping quality as much as possible for these relatively expensive dairy products—but also for another reason. The psychrotrophic flora, which modern production technology has introduced to the dairy industry, may produce very thermostable lipolytic and proteolytic enzyme systems in the milk. These may be concentrated in cream and must be inactivated. This is the reason for the extreme pasteurisation conditions. It should also be mentioned here that culturing the cream eliminates the cooked flavour, and cultured cream and butter have no cooked flavour.

The positive effects of an increased pasteurisation temperature should also be mentioned. The most significant heat-induced changes of technological importance will be summarised.

Heat treatment promotes denaturation phenomena in the milk proteins. The β-lactoglobulin is especially sensitive and starts to show the development of SH groups and formation of H_2S at a temperature of 75 °C for 15 sec. As mentioned, this is the reason for a cooked flavour, but it also has a positive effect because of the antioxidant effect of the SH groups, which is of great value for the keeping quality and oxidative stability of cream and butter.

Heat treatment is also responsible for the degradation of milk proteins into free amino acids and peptides. These stimulate the growth of the starter culture, which is responsible for the quality, especially the flavour, of butter and cultured milk products. On the other hand a high pasteurisation temperature makes the milk sensitive to reinfection.

Milk, as a colloidal system is sensitive to heat treatment. The serum proteins denature if pasteurised at 80–5 °C for 20 sec, precipitate and associate with the casein micelles. At the same time the distribution of calcium is changed, and insoluble calcium phosphates are formed. The calcium dissociates from the calcium–caseinate system, and therefore the particle size of the casein distribution is reduced, and small caseinate particles are formed. These changes are of great importance in connection with the viscosity and stability of cultured dairy products, e.g. yoghurt. Viscosity and stability are a question of water-binding. Because cultured dairy products may be looked on as an extremely voluminous, caseinate, serum-protein precipitate, the viscosity and stability may depend on the amount of serum which can be capillary-bound and immobilised. Heat treatment reduces the particle size of the casein particles and the serum proteins become associated. Both increase the amount of capillary-bound and immobilised water, and therefore the viscosity and stability of cultured dairy products are increased.

Normal milk is often stabilised before sterilisation; this may mean a heat treatment at 80 °C for 10 min. The serum proteins denature and associate with the caseinate particles. When the milk afterwards is sterilised the 'burning on' of the serum proteins during the sterilisation process is eliminated. In other cases the running time of the sterilisers can be very limited.

Preheating is, in principle, also used in connection with the production of evaporated milk to control the viscosity of the product after autoclave sterilisation.

STERILISATION

In the strict sense sterilisation means the complete reduction of all micro-organisms, but we talk of potential and commercial sterility or aseptic products. Because of interaction between acidity and heat treatment, cultured dairy products can be made aseptic by treatment at 60 °C for a few minutes or at 50–55 °C for 30 min. On the other hand, heating to a higher temperature (75–80 °C) gives products with a sandy texture. Some stabiliser must also be added—and after this it is difficult to say if the product can be considered to be as a cultured dairy product.

Autoclave sterilisation means heat treatment at 115–120 °C for 15–20 min, and is used for making evaporated (2:1) milk. Much coffee cream is also autoclave-sterilised. The process is well known and gives excellent results, especially the continuous hydrostatic autoclaves. However, the products get a cooked flavour, sometimes a browny yellow colour, the process is relatively expensive and the emballage costs are high.

Today, UHT process for the sterilisation of milk products is of the greatest interest. More than 20% of the milk for consumption in West Germany is UHT-treated milk. The process means heat treatment at 135–55 °C for 2·4–7 sec. The reason for the success of this treatment is illustrated in Fig. 3. The figure shows that at high temperatures for a short

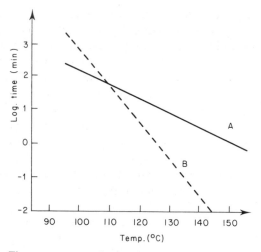

FIG. 3.   Time–temperature diagram for A, browning and B, sterilisation.

time sterility is reached prior to the development of important chemical reactions.

To control the thermal death rate for the microflora of milk and the efficiency of different equipment for sterilisation, Dr T. E. Galesloot has introduced the following definition:

$$\text{sterilisation effect (SE)} = \log\left(\frac{\text{number of spores before sterilisation}}{\text{number of spores after sterilisation}}\right)$$

According to Galesloot, SE should be greater than 8. If we assume that milk has 10 spores per ml, and SE = 8, the number of spores after sterilisation is 10^{-7} or 1 in 10 000 litres. (SE is checked by inoculation of *B. stearothermophilus*.)

UHT-sterilised milk has the same appearance as normal milk but still has a more or less pronounced cooked flavour, and what we call a 'chalky' flavour. Cooked flavour can be eliminated, e.g. by addition of a suitable oxidising substance or a substance that reacts with the sulphur compounds in some other way, but also during the process by using as optimal and gentle a temperature treatment as possible when heating the milk. This arrangement also reduces the chalky flavour. The reasons for a chalky flavour are not fully understood, but it may be the result of changes in the casein dispersion and/or the minerals, e.g. phosphates, associated with casein micelles.

Sediment formation is another problem in connection with UHT processing and depends on changes of the casein phase. Reduction of the particle size by addition of sodium citrate or sodium bicarbonate or by increasing pH, and as mentioned preheating the milk and placing the homogeniser in the end of the sterilisation process, all contribute to minimise sediment formation.

A serious problem for UHT-sterilised dairy products is sweet coagulation. The reason is the existence of very thermostable exo-and endo-proteolytical enzymes. It has also been shown that the psychrotrophic flora of the milk is especially important in this connection. Autoclave-sterilised milk never gives sweet-coagulation problems, and therefore a temperature in excess of that which is necessary for sterilisation was recommended to control this fault. Therefore milk for sterilisation must be protected against infection especially from psychotrophic flora.

The risk of coagulation and increased viscosity is even worse for UHT-sterilised concentrated milk, e.g. evaporated milk. Preheating has been suggested as a method to control the problem—85 °C for 30 min and 120 °C for 4 min before concentration have been proposed. Placing the homogeniser at the end of the process and the addition of sodium phosphate are also mentioned as possibilities.

Thus within the dairy industry heat treatment is an important aspect of process technology. Pasteurisation is used to control the microflora of the milk but also to produce important chemical and physico-chemical properties of many dairy products. Sterilisation makes dairy products keep for ever, from a bacteriological point of view, but creates many problems as a consequence of the necessary heat treatment. The goal is sterilised dairy products which are identical with our well-known dairy products, at least from an organoleptical and nutritive point of view. The interest today is focused on different types of UHT treatment, together with aseptic filling.

REFERENCES

1. IDF Monograph on UHT Milk. (1972). Annual Bulletin, Part V, Secretariat General, 1040 Bruxelles.
2. Bengtsson, K. (1973). Smakförändringar och koaguleringsfenomen i UHT-behandlad mjölk. *Avd. för livsmedelsteknologi*, Kemicentrum, Lund.
3. Stordal, C. P. (1975). Procesteknologiske aspekter på syrnede maelkeprodukters stabilitet og viskositet, specielt med henblik på fremstillingen af mager tykmaelk og yoghurt, Dairy Dept., Royal Vet. and Agric. Univ., Copenhagen.

4. Samuelsson, E. -G. and Holm, S. (1966). Technological principles for UHT-treatment of milk. XVII Int. Dairy Congress, B. 57–65.
5. Samuelsson, E. -G. (1971). US Pat. 3,623,984.
6. Persson, T. and von Sydow, E. (1974). Kan man undvika konservsmak? *Livsmedelsteknik*, no. 4, SIK-Rapport no. 362.

18

The Influence of Evaporation and Drying on Milk

K. KUSSENDRAGER AND J. SCHUT

*Research Laboratory of the Zuid Nederlandse Melkindustrie B.V.,
D.M.V., P.O. Box 13, Veghel, The Netherlands*

INTRODUCTION

From the various dairy products summarised in Fig. 1, condensed milk, milk powder and whey powder are of the main interest for our discussion. Unlike the consumption of liquid milk and milk products, the world's production of milk, butter[1] and cheese are increasing,[2] (Fig. 2). This results in an overall increase of milk by-products, such as skim milk and whey. The surplus of skim milk gets even more pronounced, since increasing prices have forced the manufacturers of calf-starters to partly substitute whey powder for skim milk powder. Furthermore, as rejection of whey is no longer allowed, the production of dried whey products is also sharply increasing,[2] which justifies the expectation that more and more skim milk and whey in any suitable form will become available for human consumption.[2, 3]

THE USE AND APPLICATION OF MILK BY-PRODUCTS

In considering the influence of evaporation and drying on milk by-products, the question arises as to whether this influence is favourable or not, and this can only be answered in the light of the application and function of the products. In Table 1, the main function of milk by-products in various human foods is summarised. Since the supply of calories and proteins in developed countries is, generally speaking, more than sufficient, technological, sensory and economical functions will in many cases prevail. For developing countries, the biological function has always been and still is strongly emphasised, although recent ideas point to a first need for food (energy) rather than for protein.[4−7] This would mean that for, say, Europe

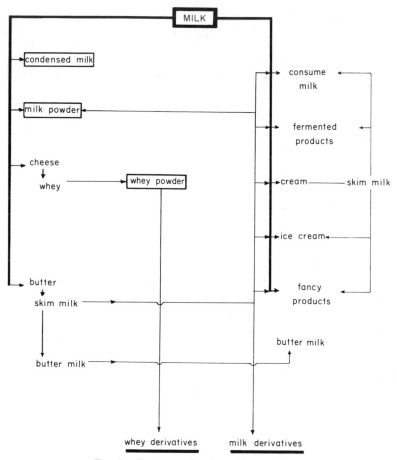

FIG. 1.   Summary of various dairy products.

the first role of milk by-products would be that of a food ingredient that enables the production of attractive, high-energy, well-balanced foods at an acceptable price.

THE INFLUENCE OF EVAPORATION AND DRYING ON THE SUITABILITY FOR APPLICATION OF MILK BY-PRODUCTS

This influence is mainly determined by the effect of temperature and time. Beyond a certain temperature/time combination κ-casein reacts with β-

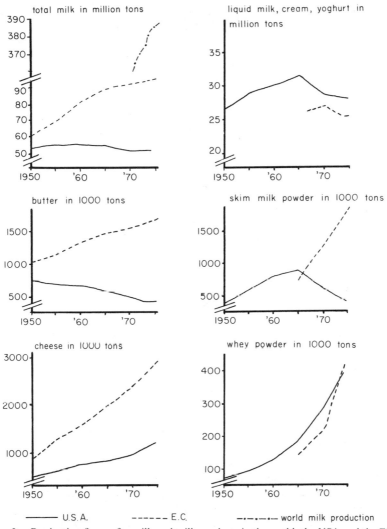

FIG. 2.   Production figures for milk and milk products in the world, the USA and the EEC.

lactoglobulin. This and other reactions create changes in the milk protein which are reflected in the suitability of milk by-products in various foods. For example, for the production of reconstituted milk, ice cream, cheese, desserts, etc. low-heat milk powders are required, whereas the bakery industry needs high-heat milk powder in order to prevent slackness of doughs and reduction of loaf volume. In the confectionery industry

TABLE 1
Main reasons for the use of milk by-products[a]

Food	Main reason for use				
	Basic ingredient	Biological function	Technological function	Sensoric function	Economic function
Reconstitution	×	×			
Infant foods	×	× ○			
Dietary foods		× ○			
Breakfast cereals		× ○			
Soft drinks		○			
Ice cream	×				○
Margarine			×		○
Bakery products				×	○
Confectionery			×	×	○
Dairy products	×		×		
Meat products			×		× ○
Soups and sauces			×	×	○

[a] × = milk solids, ○ = whey solids.

medium-heat milk powders and condensed milk are preferred for some specialities.[8] And since high drying temperatures diminish solubility, many food industries can exclusively use spray-dried powders. As milk is progressively heated, sulfides liberated from sulfhydryl radicals cause cooked flavour and odour, along with Maillard reaction products, which are also partly responsible for changes in colour. Whether the organoleptic changes are favourable or not, again depends on product application.

Heating of milk may lead to losses of available lysine as a consequence of a step in the Maillard reaction in which the ε-lysine amino group reacts with carbonyl groups of reducing sugars.[9-11] Furthermore, losses of vitamins can occur during heating. The above-mentioned losses, however, have been reported to be mainly limited to a few water-soluble vitamins[12] and a negligible reduction of available lysine, provided severe heating is avoided,[13, 14] i.e. falling film evaporation and spray-drying for the production of sweetened condensed milk and powders. So, these milk by-products, probably enriched with some vitamins, seem to be suitable for reconstitution purposes. It needs to be asked, however, whether the heat processing of many foods in which milk by-products are used as an ingredient does not do more harm to the biological value and other properties than do appropriate processes to milk by-products themselves.

This question seems to be particularly justified for bakery products and canned foods.

MILK COMPONENTS

In this respect, isolated milk components such as casein and their derivatives, the caseinates, should be mentioned (Fig. 3). The absence of reducing sugars enables the production of caseinates with extremely low bacterial and spore counts, without a noticeable effect on the biological value.[15-17] Sodium and potassium caseinate are very effective emulsifiers,

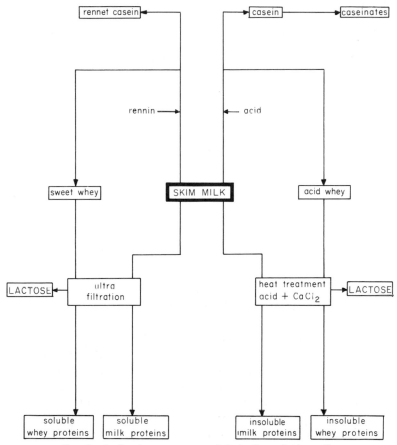

FIG. 3.   Isolated components and derivatives of skim milk.

stabilisers and foaming agents, and their use in foods enables processing without the development of often unwanted changes in colour, taste and flavour. Calcium caseinate is used for protein enrichment and as an essential ingredient in foods and special dietary formulations, e.g. in diets for lactose-intolerant patients. Even imitation milk with sodium and calcium caseinate as a protein source has been found reasonably acceptable.[2] Losses in biological value during processing may be diminished if the formulae are carefully chosen. For example, biscuits enriched with calcium caseinate and saccharose as a source of sugar are preferred to those enriched with skim milk powder.[17, 18]

At present, whey protein concentrates and isolates produced by modern techniques such as reversed osmosis and ultrafiltration enjoy a great interest for both technological and biological reasons.

Summarising, one could say that there is a definite influence of heat treatment on milk by-products, but also that in the quality evaluation of these products attention should possibly be directed more to their technological and sensory properties. In this respect emphasis is put on the application of milk components. The study of losses in biological value, the possible development of toxic components and other unwanted changes due to heating, remain of great importance in order to control production processes. In our laboratory investigations have been started to develop an easy method to follow the reactions induced by the heating of milk products.

EXPERIMENTAL DTA TECHNIQUES

Basic Method

Since it is known that interactions between protein and sugar can cause rather severe chemical changes in the reacting components, it might be possible to get an impression of the occurrence and of the progress of these complex reactions using differential thermal analysis under suitable conditions. Attempts have been made to follow these reactions in model systems consisting of mixtures of skim milk or caseinate with sugars.

Differential thermal analysis (DTA) enables measurement of qualitative and also, with the modern equipment, quantitative heat effects in samples subjected to a given temperature programme. Physical and chemical transitions accompanied by heat changes can be followed. We used the Mettler TA 2000 system in which two cups, one filled with the sample under investigation and one with a reference, are placed on top of a thermocouple detector in a small oven. During a given temperature programme the temperature differences between sample and reference, representing the DTA signal, are recorded continuously.

The surface area of the DTA curve is a quantitative measure of the thermal changes. After some preliminary work we chose the isothermal programme at 110 °C to study the milk protein–sugar interactions. Sealed cups were used to prevent unwanted evaporation effects, with sample weights of 7–10 mg, and using air as reference.

To account for thermal changes which occur just before the system reaches 110 °C, the curve of the thermally inert material Al_2O_3 against air has been used as a blank.[19]

Preparation of Samples

Caseinate and skim milk samples were prepared as shown in Fig. 4. Caseinate solutions were prepared from fresh skim milk by isolectric precipitation at pH 4·6, and subsequent washing and neutralisation to pH

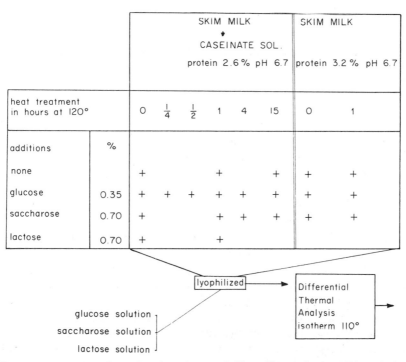

		SKIM MILK ↓ CASEINATE SOL. protein 2.6% pH 6.7						SKIM MILK protein 3.2% pH 6.7	
heat treatment in hours at 120°		O	$\frac{1}{4}$	$\frac{1}{2}$	1	4	15	O	1
additions	%								
none		+			+		+	+	+
glucose	0.35	+	+	+	+	+	+	+	+
saccharose	0.70	+			+	+	+	+	+
lactose	0.70	+			+				

lyophilized ⟶ Differential Thermal Analysis isotherm 110°

glucose solution ⎫
saccharose solution ⎬
lactose solution ⎭

FIG. 4.   Preparation of samples of caseinate and skim milk samples for DTA analysis.

caseinate–glucose mixtures, lyophilized

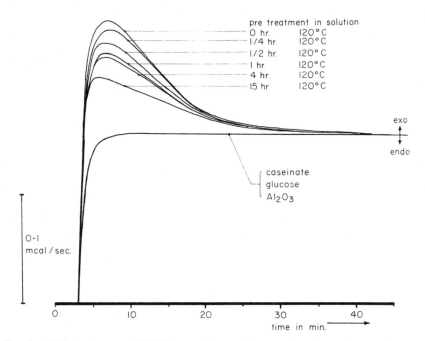

FIG. 5.   DTA isotherms at 110 °C from caseinates, glucose and caseinate–glucose mixtures.

6·7. Reducing and non-reducing sugars were added as tabulated in Fig. 4. The solutions were lyophilised directly or after heating for different times at 120 °C and subsequently subjected to DTA. The caseinate samples had moisture contents ranging from 5–8 %, the skim milk samples from 10–11 %.

Results and Discussion

Variations in moisture content within the ranges mentioned above did not significantly influence the curves of the samples investigated. Figure 5 shows the DTA results from caseinates, glucose and the caseinate–glucose mixtures, respectively. Caseinate and glucose followed the curve of the Al_2O_3 control. The mixtures, however, showed exothermal effects, the extent of which appeared to depend on the time of preheating in solution. Also a colour change to ochre/brown was observed. This could suggest that the heat effects are due to protein–glucose interactions and possibly represent the overall reaction enthalpy of several stages of the Maillard

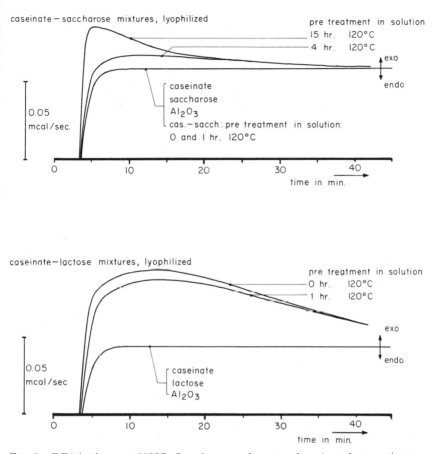

FIG. 6. DTA isotherms at 110 °C of caseinate–saccharose and caseinate–lactose mixtures, lyophilised.

reaction, initiated by amino groups of the protein and the carbonyl groups of the reducing sugar molecule. This suggestion is supported by the lower heat effects found for the preheated mixtures. Moreover, these enthalpy changes decreased with increasing time of preheating in solution. Another fact in support of this suggestion can be found in the thermograms for caseinate–saccharose mixtures, shown in Fig. 6.

The results elucidate the absence of heat effects in the samples which were not preheated and those preheated for one hour, for which almost no change in colour occurred. The exothermal effects observed after preheating the solutions during 4 and 15 h, respectively, can be explained

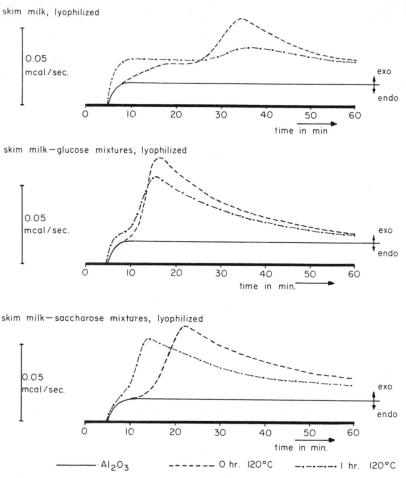

FIG. 7.   DTA isotherms at 110 °C of skim milk with and without added sugars.

partly by hydrolysis of the saccharose molecule, which has been confirmed by chemical analysis.

The caseinate–lactose mixtures in turn showed considerable exothermal effects. Preheating this solution for one hour caused a decrease in the heat quantities liberated. Both mixtures showed colour changes from white to ochre after DTA treatment. The same suggestion as given for the caseinate–glucose mixtures could also apply to these mixtures. The DTA results of the skim milk samples with and without added sugars are shown in Fig. 7. Exothermal effects were also observed in this case, accompanied by

DTA Results
lyophilized samples

DTA   : isotherm 110°C, closed cups
sample size : 7–10 mg
heat effects : in cal/g protein (with Al_2O_3 as a reference)

	CASEINATE						SKIM MILK	
heat treatment in hours at 120°C	0	$\frac{1}{4}$	$\frac{1}{2}$	1	4	15	0	1
additions :								
none	0			0		0	20.0	16.5
glucose	13.8	13.3	12.2	10.5	10.2	7.8	20.2	16.7
saccharose	0	0	0	0	1.9	4.7	21.3	17.6
lactose	17.0			15.4				
	ΔH after 40 min 110°C.						ΔH after 60 min 110°C	

	ΔH in cal/g sugar 60 min 110°C.
glucose	0
saccharose	0
lactose	0

FIG. 8.   Summary of DTA results for caseinate and skim milk with and without added sugars.

considerable colour changes. Preheating for one hour decreases the heat effects but also influences the shape of the curve. When glucose was added the mixture was thermally more reactive in the first DTA period, and competition between glucose and lactose in the reaction with protein may explain this phenomenon.

The skim milk–saccharose mixtures showed unexpected results, since the curves were significantly different from those of skim milk alone. This phenomenon may result from the participation of saccharose in the protein–lactose reaction system, probably as a result of inversion during the DTA run or directly. Also synergistic and catalytic effects must not be excluded.

Figure 8 summarises the enthalpy changes calculated from the DTA curves shown. Although investigations are far from complete, it seems that enthalpy effects resulting from protein–sugar reactions could be a measure of the extent to which these reactions have taken place. In this respect further studies must be undertaken to establish whether DTA can make a valuable contribution to studies of this reaction system.

REFERENCES

1. Richarts, E. and Münz, A. (1976). The agriculture market 1975, W. Germany, E.E.C. and world market: Milk and milk products. (German). Zentral Markt und Preisberichtstelle, 55 Bonn Bad Godesberg 1. W. Germany.
2. Winkelmann, F. (1974). Imitation and imitation milk products, FAO, Rome.
3. USDA. Dairy Situation, USDA, no. 332.
4. McLaren, D. S. and Pellet, P. L. (1970). Nutrition in the Middle East. *World Rev. Nutr. Diet.*, **12,** 43.
5. Sukhatme, P. V. (1970). Incidence of protein deficiency in relation to different diets in India. *Brit. J. Nutr.*, **24,** 477.
6. Sukhatme, P. V. (1975). *The Man/Food equation*, ed. by Bourne, A., Academic Press, London.
7. Crawford, M. A. (1975). Meat as a source of lipids. 21st Meeting of Meat Research Workers, Berne, Switzerland.
8. Webb, B. H. and Whittier, E. O. (1970). *By-products from milk*, Avi Publ. Co., Westport, Conn.
9. Hodge, J. E. (1953). Dehydrated foods: Chemistry of browning reactions in model systems. *J. Agric. Food Chem.*, **1,** 928.
10. Schwartz, H. M. and Lea, C. H. (1952). Reaction between proteins and reducing sugars in the 'dry' state: Relative reactivity of the α- and ε-amino groups in insulin. *Biochem. J.*, **50,** 713.
11. Hannan, R. S. and Lea, C. H. (1952). Studies of the reaction between proteins and reducing sugars in the dry state: VI. The reactivity of the terminal amino groups of lysine in model systems. *Biochim. Biophys. Acta*, **9,** 293.
12. Ashton, W. H. (1972). The components of milk, their nutritive value and the effects of processing, part 2, *Dairy Ind.*, Nov., 602.
13. Bender, A. E. (1971). Processing damage to protein foods. 18th PAG Meeting, Rome.
14. Bruel, A. S. M. van den, Jenneskens, P. J. and Mol, J. J. (1971). Availability of lysine in skim milk powders processed under various conditions. *Neth. Milk Dairy J.*, **25,** 19.
15. Osner, R. C. and Johnson, R. M. (1974). Nutritional and chemical changes in heated casein: I. Preliminary study of solubility, gel filtration pattern and

amino acid pattern. *J. Food Technol.*, **9**, 301, (1975). II. Net protein utilisation, pepsin digestibility and available amino acid content. *J. Food Technol.*, **10**, 133.

16. Bjarnason, J. and Carpenter, K. J. (1970). Mechanisms of heat damage in proteins: I. Models with a cylated lysine units. *Brit. J. Nutr.*, **23**, 859.
17. Carpenter, K. J. (1973). Damage to lysine in food processing: Its measurement and its significance. *Nutr. Abstr. Rev.*, **43**, (16), 424.
18. Gennip, A. H. M. van (1976). *Protein enrichment of biscuits*, D. M. V. publ., Veghel, The Netherlands.
19. Widmann, G. (1975). Quantitative isothermal DTA-studies. *Thermochimica Acta*, **11**, 331–3.

19

Meat Products and Vegetables

P. Baardseth

*The Norwegian Food Research Institute, P.O. Box 50, N-1432
Ås-NLH, Norway*

INTRODUCTION

The purpose of many food-processing techniques is to slow down or prevent deleterious changes from occurring in food materials. These deleterious changes are often caused by contaminating micro-organisms, by chemical reactions among natural components of the food tissues, or by simple physical occurrences (e.g. dehydration).

The procedures used to prevent changes in foods caused by micro-organisms or chemico-physical reactions include removal of water; removal of other active components, such as oxygen or glucose; use of chemical additives; lowering of temperature; input of energy; and packaging. Most of the unit operations in food processing are based on combinations of two or more of these fundamental procedures.

In heat processing (input of energy) the desirable effects may be summarised as: favourable alteration of the characteristics of the product (browning reaction, textural changes, increased palatability, etc.); destruction of micro-organisms (sterilisation, pasteurisation); destruction of enzymes (peroxidase, ascorbic acid oxidase, thiaminase, etc.); improvement in availability of nutrients (gelatinisation of starches and increased digestibility of proteins); and destruction of undesirable food components (avidin in egg white, trypsin inhibitor in legumes). The undesirable effects of heat processing include changes in proteins and amino acids, carbohydrates, lipids, vitamins, minerals, and unfavourable flavour changes.[1] In other words, heat treatment causes desirable and undesirable effects on foods. The effects are complex and difficult to understand completely. In this paper I will not try to cover the whole field of meat products and vegetables, but rather talk about some aspects of it.

TENDERISATION OF MEAT

Industrial tenderisation of meat is a process which depends on temperature. Normal tenderisation is done by hanging a whole carcass or cuts of meat at 2–3 °C. This process takes a long time (3 weeks) and needs large storage space. The more modern vacuum method is no better in that context. The tenderisation time can be cut down by raising the temperature, but then it is difficult to control bacterial growth at the surface of the meat. On the other hand, if the temperature is higher than 45 °C, very few bacteria are able to grow, and the bacterial count decreases as the temperature is raised. At a temperature of 58–60 °C protein denaturation becomes extensive. By keeping the temperature between 45 and 60 °C, the meat would tenderise more rapidly and at the same time retain the raw character.

In a study done at the Norwegian Food Research Institute by Vold and co-workers[2] an attempt was made to look into this problem. The *m. semimembranosus* from 10 animals was tenderised by a conventional vacuum method and at 48 and 58 °C in 3, 7 and 12 hours. The quality changes were measured by chemical, physical, microbial and sensory methods.

The meat became more tender with storage time at both temperatures (48 and 58 °C), but not so tender as by the vacuum method. The water-binding capacity decreased with time, and faster at 58 °C than at 48 °C. The proteolytic activity and the bacterial count decreased with storage time. The meat became greyish on the surface at both temperatures, but more at 58 °C.

This study indicates that the storage temperature and time have great influence on the tenderness of the meat. Defects are also developed which are not tolerated in the meat industries. Further studies must be carried out.

THE FUNCTIONAL PROPERTIES OF PROTEINS

The functional properties of proteins are altered during thermal processing. But just what is happening at the molecular level is uncertain. One aspect of this problem has been closely investigated by Hägerdal and Martens.[3] The heat-induced denaturation of a well-known model protein, sperm whale myoglobin, was studied at neutral pH with water contents ranging from dry powder to thin solutions. The conformational transition was followed by a differential scanning calorimeter (DSC). This technique allows simultaneous determination of the transition temperature or denaturation temperature (T_m) as well as the heat of transition or enthalpy (ΔH_{app}). The

degree of irreversibility of the transition can also be determined by this
technique.

The variation in denaturation temperature with water content is shown in
Fig. 1. The curve is characterised by a broad minimum, with a minimum
transition temperature of 74 °C at about 50 % water content. There is also a
remarkable increase in T_m for decreasing water content. The highest T_m

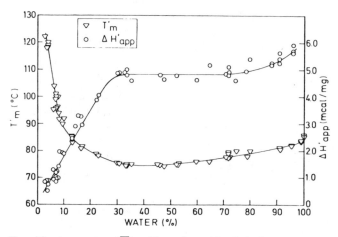

FIG. 1.   Transition temperature ($\triangledown$) and heat of transition ($\bigcirc$) of sperm whale myoglobin at
various water contents.

found was 122 °C at a water content of 3 %. No transition peak was observed
for water contents lower than 3 %, at maximum sensitivity of the
instrument. T_m also increased again from the minimum with increasing
dilution, to 86 °C for a water content of 99·6 %. Further dilution was limited
by the sensitivity of the instrument.

The curve shows that in a drying process for protein, the temperature
must never come above the T_m line, otherwise denaturation will occur. But
on the other hand, at low water contents the drying process may still be
speeded up by increasing the temperature to above 100 °C, thereby reducing
the drying time. Figure 1 also shows the apparent heat (ΔH_{app}) absorbed by
the protein during the transition process. The curve is characterised by a
steep linear increase in ΔH_{app} from 0·5 to 4·8 mcal/mg in the interval 3–30 %
water content. For water contents of more than 30 % there is a plateau
followed by a slow increase in the apparent transition heat from 4·8 to
5·8 mcal/mg at 96 % water. At water contents higher than 96 % the
determination becomes too uncertain to allow quantitative evaluation.

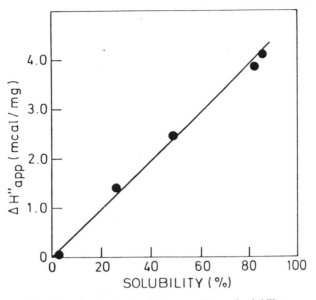

FIG. 2.   The relationship between ΔH_{app} and solubility.

Another experiment with the same protein shows a linear relationship between the resolubility and ΔH_{app} after heat treatment. Figure 2 indicates that the DSC gives a good measure of the degree of irreversible denaturation of the protein at various water contents.

Conclusions

The heating rate is shown to influence the final condition of the protein sample. At water contents below 30% only a certain fraction of the heat-treated protein sample underwent irreversible transitions. Above 30% water content the entire protein sample underwent irreversible transition.

A group at the Norwegian Food Research Institute is now working on some other aspects of the functional properties of proteins. They are trying to obtain basic knowledge about molecular protein–protein interactions, protein–water interactions and protein–lipid interactions by using various methods, and one of their aims is to be able to distinguish between, on the one hand, the thermal transition occurring within single protein molecules (denaturation, unfolding), and on the other hand the process of thermal aggregation of gelling.

Another project in the same group concerns the molecular transitions that take place in meat during a heating process. By increasing the

TABLE 1
Biological data on raw, cooked and fried sirloin

Measurement	Sirloin		
	raw	cooked	fried
TD	99·3	97·6	97·9
BV	79·0	78·3	75·3
NPU	78·4	76·4	73·7

temperature of small meat samples in the differential scanning calorimeter (DSC), at for instance 5 °C per minute, they can characterise the thermal denaturation of each of the different major proteins in the intact meat sample. Two distinct denaturation regions are found in mammalian meat tissue—one starting at 50 °C and another at 75 °C. (Tissue from birds and fish gives similar thermal denaturation patterns; the fish tissue gives lower transition temperatures.)

PROTEIN QUALITY

During heat treatment certain amino acids may interact with the glucose and/or ribose of the meat (Maillard reaction) and the nutritional values of these amino acids are lost when this occurs. Lysine, histidine, arginine and methionine seem to be particularly susceptible to degradation via this route.

Amino acids can also be destroyed by oxidation and by some modification of the linkage between the amino acids during heat

TABLE 2
Net protein utilisation (NPU) and true digestibility (TD) of four canned meats before and after sterilisation[7]

Canned meat	Sterilisation			
	Before		After	
	NPU	TD	NPU	TD
Goulash with potatoes	66	93	64	87
Spinach, pork and potatoes	68	91	60	88
Carrots, onions, beef and potatoes	72	92	62	90
Peas, minced meat and potatoes	55	93	53	90

treatment.[4] An experiment done by Eggum[5] (Table 1) illustrates some of these changes in the proteins of sirloin after heat treatment.

Biological experiments with canned meat and vegetable mixtures have also been done.[6,7] The results indicate good protein quality, but some reduction occurs due to the heat treatment and more than with meat alone (Tables 2 and 3).

TABLE 3
Net protein utilisation (NPU) and true digestibility (TD)
of canned baby food after sterilisation[6]

Canned baby food	NPU	TD
Veal and vegetables	60	92
Chicken and vegetables	62	91
Lamb and vegetables	56	92
Beef and vegetables	59	93
Liver and vegetables	59	90
Tuna fish and vegetables	58	91

A project just started at the Norwegian Food Research Institute will examine different methods of testing protein quality. We are also going to study how the protein quality in a model system changes with temperature and time, in the presence of reducing sugar, fat, different pH, and water activity, and to test the quality both with chemical and biological methods.

BLANCHING PRIOR TO FREEZING OF VEGETABLES

Blanching and thermal sterilisation are the most common types of heat processes used for plant tissues. Blanching is a mild heat treatment involving exposure of plant tissues to water or steam for a few minutes at about 100 °C and 1 atm pressure. Prior to freezing, dehydration, or irradiation, blanching is done primarily to inactivate the enzymes, whereas prior to heat sterilisation it is done for several reasons, the most important being to remove air from the tissues. During severe thermal processes, such as sterilisation or cooking, damage to the edible quality and nutritive value of plant tissues may result from (*a*) leakage of solutes into the surrounding medium; (*b*) mechanical loss of solids; (*c*) volatilisation; (*d*) oxidation; and (*e*) hydrolysis and other degradative reactions. The design of a thermal process will depend on the objectives desired. For example, slow or moderate heating allows activation of enzymes which can affect colour,

TABLE 4

The total lipid, fatty acids and relative peroxidase values in various fresh vegetables[8]

	Total lipid (%)	Fatty acids (%)						Relative peroxidase value
		Palmitic	Stearic	Oleic	Linoleic	Linolenic		
Carrot	0·4	18	1	6	69	6		1
Onion	0·1	23	2	23	45	7		<1
Swede turnip	0·1	17	2	18	25	38		200
Brussels sprout	0·4	22	2	7	20	49		170
Cauliflower	0·4	20	2	10	16	52		130

texture, flavour and nutritive value. This occurrence can be either desirable or undesirable, depending on the product.

An experiment done at the Norwegian Food Research Institute was designed as follows. Different vegetables were blanched at different times and temperatures, frozen and stored at −20 °C and −30 °C. Every third month samples were analysed both chemically and sensorially. The chemical analysis involved determination of peroxidase, total lipid content and fatty acid composition, as shown in Table 4.

The vegetables analysed contain small amounts of fat, less than 0·5 %, but the fat was highly unsaturated. The peroxidase level differs greatly from one vegetable to another.

TABLE 5
The blanching temperature and time for carrots and swede turnip before freezing

	Temperature (°C)	Time (min)
Carrots	90	3
(cubes)	90	1·5
	98	1·5
	80	10
Swede turnip	90	8
(cubes)	90	3
	95	3
	95	1
	80	10

The selected heat treatment (some examples are shown in Table 5) inactivates the peroxidase almost 100 % and no re-activation has been observed during storage. Neither has the total lipid content or the fatty acids changed.

The sensory analysis showed that the samples differed in texture, flavour and colour. The high-temperature short-time treatments are preferred even after nine months' storage.

STORAGE OF DEHYDRATED VEGETABLES

During storage, dehydrated vegetables are quite susceptible to changes which depend on the condition of storage (temperature, oxygen level, moisture content) as well as on the processing procedure and on the quality of the raw material.[9] Lipid oxidation is one of the major factors limiting the shelf life. The lipid content of vegetables is low, less than 1 % in dried material, but because the lipid contains mainly unsaturated fatty acids, oxidation occurs readily. Lipid oxidation results principally in undesirable flavour, which can make the vegetable unacceptable, as demonstrated in an experiment with dried swede turnip (Table 6).

TABLE 6

The unsaturated ratio and sensory evaluation of dried swede turnip during storage at 23 and 5°C, packed in paper bags coated with polyethylene[10]

Storage temperature (°C)	Weeks storage	Unsaturated ratio[a]	Sensory[b] evaluation
23	0	2·9	7·0
	4	2·7	6·9
	8	2·5	6·0
5	0	2·9	7·0
	5	1·9	2·5
	9	1·2	1·8

[a] (Linoleic acid + linolenic acid)/(palmitic acid + stearic acid).
[b] 10 is the best score.

To minimise the lipid oxidation in the dried vegetable, antioxidants can be added and/or the product packed in a material which excludes oxygen. The storage temperature is not so important when oxygen is excluded.[11]

REFERENCES

1. Lund, D. B. (1973). Effects of heat processing. *Food Technol.*, **27**, 16.
2. Vold, E., Martens, H., Russwurm, H., Jr, Tjaberg, T. and Høyem, T. (1973). Tenderisation of meat at 48°C and 58°C (Norwegian), unpublished data.
3. Hägerdal, B. and Martens, H. (1976). Influence of water content on the stability of myoglobin to heat treatment. *J. Food Sci.*, **41**, 933.
4. Bender, A. E. (1972). Processing damage to protein food. *J. Food Technol.*, **7**, 239.
5. Eggum, B. O. (1967). New results in amino-acid research on protein quality during heat processing (Danish). Kursus vedrørende Nyt inden for biokemi og fysiologi, Tune Landbrugsskole, 30 October–1 November.

6. Baardseth, P. (1972). Protein quality in Norwegian canned baby food (Norwegian). *Tidsskrift for Herm.-ind.*, **58**(10), 3.

7. Hellendorn, E. W., de Groot, A. P., van de Mijll Dekker, L. P., Slump, P. and Willems, J. J. L. (1971). Nutritive value of canned meals. *J. Amer. Dietetic Ass.*, **58**, 434.

8. Baardseth, P. (1975), unpublished data.

9. Sapers, G. M. (1975). Flavour and stability of dehydrated potato products. *J. Agric. Food Chem.*, **23**, 1027.

10. Baardseth, P. (1976). Quality changes in dried vegetable products (Norwegian). NINF information No. 5.

11. Sapers, G. M., Panasink, O., Talley, F. B. and Shaw, R. L. (1974). Flavour quality and stability of potato flakes: Effects of drying conditions, moisture content and packaging. *J. Food Sci.*, **39**, 555.

20

The Baking of Cereals

ARNE SCHULERUD
The Norwegian Cereal Institute, STI, Akersvn. 24c, Oslo-dept. Oslo 1, Norway

Thermal processing is much used in the cereals industry. In flour milling, gentle heat is applied to the wheat to facilitate penetration of added water in order to mellow the kernels and save mechanical energy in the milling. More severe heat treatment has been tried to suppress excess enzyme activity in damaged wheat and thus improve baking quality of the flour.

Several snack products are made in an extruder under high pressure and elevated temperatures. A partial gelatinisation of the starch takes place, and on leaving the cylinder's narrow orifice the product expands to the desired fluffiness. In the puff process, wheat, rice or oats are heated in a closed cylinder to 180 °C under pressure. When the lid is opened abruptly the kernels explode to a volume many times the original. Flavour also is formed, from very little in puffed rice to a very pronounced flavour in oats, probably because of the high contents of reactive, unsaturated fat in the latter.

A flour consists of about 70% starch in the form of granules, protein particles 8–16% and small amounts of fat, minerals and fibre, plus some moisture. Thiamin is the most important vitamin, α- and β-amylases are always present, further proteases, lipases etc.

When water is added to the flour, the protein particles swell to form the gluten matrix, in which the starch granules are embedded. Added yeast ferments present sugar to give carbon dioxide which raises the dough.

Temperature is essential to dough development. At low temperatures (below 25 °C) fermentation is slow, and so is the activity of the enzymes. Therefore, for 'weak' flours with a large enzyme content, low dough temperatures are recommended. With strong flours, giving a large quantity of elastic gluten, elevated temperatures (30 °C and more) are required to give enzymes and other substances the opportunity to degrade excess elasticity of the gluten.

This 'ripening' of the dough, however, can also be obtained by intense mechanical treatment of the dough, i.e. mixing. Thereby much heat is generated which must be considered on setting the temperature of the dough liquid.

During the dough rest, carbon dioxide is formed, but this is expelled during the dividing and moulding process. Therefore, the dough pieces are given another fermentation period, the so-called proof, to produce the necessary new gas bubbles to give the desired porosity before the loaves enter the baking oven. The heat, penetrating from the surface to the centre of the loaves, causes coagulation of the swollen gluten and gelatinisation of the starch. These two processes, which start at about 60 °C, serve to solidify the dough so that the bread gets its final shape.

Heat transfer takes place in 3 ways:

(a)   Conduction from the hot bottom of the oven.
(b)   Radiation from the sides and the ceiling.
(c)   Convection from the surrounding hot air.

Since heat penetration of dough is slow, baking must necessarily take some time (20–40 min). Therefore, high-frequency heating has been tried. Since heat is generated throughout the whole loaf at once, the baking, i.e. coagulation and gelatinisation, can be finished in one-third to one-quarter of the usual time. However, this system has had no appreciable success, largely due to technical difficulties. One of these is the fact that the energy tends to concentrate on the highest parts of the loaves, and unless they can be kept on exactly the same level uneven baking will result. It should be mentioned, however, that the Norwegian Central Institute for Industrial Research has developed a high-frequency oven for crispbread, where the thin and flat dough pieces are better suited to this method of baking. High-frequency baking gives no extra heat for crust formation and therefore such heat must be applied after the first baking. This can be done by infrared radiation, which accumulates on the surface and does not penetrate. Such infrared baking has been tried on thin objects like biscuits, but is not common.

The heat characteristics of the baking oven greatly influence the enzymes in the dough. Among the enzymes, α-amylases and their behaviour are most important. β-Amylase, which produces maltose for fermentation, is generally able to react but its activity is conditioned by the presence of α-amylase for splitting the amylopectin branches. Now, this enzyme may be present in too high or too low a concentration. The splitting effect is most active in the freshly gelatinised starch, and if it is too large, the starch can be so much degraded to dextrins and maltose that a sticky bread results. This can be overcome by lowering the pH of the dough, since cereal α-amylase has its optimal activity at pH 6·3 at 70 °C.

If the activity is too small, it can be enhanced by the addition of malt, fungal or bacterial α-amylase. While malt amylase is inactivated a little above 70 °C, fungal amylase is destroyed below the gelatinisation temperature and thus is safe against over-dosage. Bacterial amylase, on the other hand, may be resistant up to 100 °C and therefore can continue its work in the baked loaf. For this reason, such amylases are not commonly used.

The correct amount of α-amylase activity is desirable because a limited degradation of the gelatinised starch counteracts the staling of the bread.

Since bread dough contains 40–45 % moisture, and very little is lost during baking, temperature in the bread crumb will not exceed 98–9 °C. The surface of the loaf, however, will undergo an appreciable dehydration, and here temperature may rise to 150–60 °C. Thickness and colour of the crust will depend on the baking time. The colour, as well as flavour formation, to a great extent is influenced by the presence of sugars and available amino groups from the proteins in various types of Maillard reactions. Model experiments with sugars and different amino acids give very different aromas, and the question of synthetic bread aroma has not yet found a satisfactory solution.

A special product is the German pumpernickel type. Here a whole rye bread of sour dough is baked for 16–20 h at low temperatures in a well-closed oven to prevent evaporation. During this extremely long baking time appreciable quantities of sugars are formed, and these react with amino groups to form a dark, nearly black bread with a very special aroma. It should be noted, that protein value of this bread is markedly reduced because of the strong binding to the sugars.

Bread is a good source of thiamin. Since this vitamin is thermolabile, the loss during baking has considerable interest. Available literature on the subject seems rather confusing, partly due to analytical difficulties in older works, but also because the size and shape of the bread, pH and baking temperatures and times will influence the losses. In a recent Norwegian experiment, brown bread containing 30 % whole wheat produced in a commercial bakery, was baked for 38 and 45 min, respectively. After cooling, crumb and crust were separated, and thiamin was determined microbiologically both in the bread as a whole, and in crumb and crust respectively. The thiamin loss under baking was 25 % for the bread baked 38 min, and 29 % for the 45 min baking. Since the loss in the crumb was the same for both, namely 21 %, it was the crust which was responsible for the loss at progressive baking time, as could be expected.

The hot bread coming from the oven should be cooled before wrapping to avoid vapour condensation inside the plastic film. Normally, this will take an hour at least, and necessary racks or travelling belts require a lot of space in the bakery. A new patented process, developed by the Norwegian Cereals Institute, utilises the evaporation heat to cool the bread, which is set under vacuum right after baking. In this way, the bread can be sufficiently cooled in 20 sec.

In some cases (expeditions, military purposes etc.) it is desirable to preserve bread for weeks and months. Present-day plastic films fully prevent drying out, but chemical preservatives cannot sufficiently prevent mould growth. In such cases, heat sterilisation of the air-tight packages is effective.

For common bread, staling represents a serious problem. In the fresh state the crumb is soft and elastic, which is attributed to a special condition of the starch gel. This state is stable above 60 °C, but below this temperature the gel undergoes a retrogradation which makes the crumb hard and less elastic. Since staling proceeds most rapidly around the freezing point of water, hot or semi-hot storage overnight has been tried, but with limited success. However, when fresh bread is rapidly frozen, the fresh state is preserved, and on rapid thawing the bread still presents itself as fresh or nearly so. Several technical problems had to be solved before the deep-freezing of bread could be made commercial. Bread crumb contracts on freezing, and if the crust is too hard, it cannot yield to the contraction, and separates from the crumb. Limited baking time, or softening the crust by a certain absorption of moisture from the inner part of the loaf before freezing is therefore recommended. Since the air in the freezing cabinet is very dry, bread in cold storage should be protected from evaporation. These problems have been solved, and many bakeries today use freezing for rationalisation of their production. Private households also practise freezing of their bread. Norway has a very high percentage of home freezers, and the housewives have learned to make use of them.

Nevertheless, the desire for fresh bread in the shops is very strong, and since hours of work in bakeries are restricted by a special law, it has always been difficult to provide shops with sufficient fresh bread in the morning. Recently, the bakery workers have obtained an agreement not to work on Saturdays. Now, the Cereals Institute has developed a freeze–thawing system which has been installed in some large bakeries. Here, after a certain conditioning, the fresh bread enters into cabinets for rapid freezing. To avoid excess tensions in the loaves, freezing is interrupted when temperature in the centre of the loaf is $-15\,°C$. At a given time in the night heat elements are switched on, so that the bread can be brought out semi-hot and 'fresh' next morning. This system is well suited to meet the wishes for early deliveries, and of course it can also be used to deliver fresh bread on Saturdays. The other bakers, however, consider this to be unfair competition and are trying to get it stopped. If they succeed, it will not be the first time that technical progress has been stopped for political reasons.

21

Thermal Processing and the Technology of Wine

W. HEIMANN

Institut für Lebensmittelchemie der Universität Karlsruhe, Kaiserstrasse 12, D-7500 Karlsruhe, Federal Republic of Germany

During recent decades important new wine-growing regions have developed not only in Europe, but also in other parts of the world such as the USA, Argentina, Chile, Turkey and North Africa. As a consequence, the total wine production has increased considerably (see Table 1). Until recently, the traditional simple working procedures still prevailed in the manufacture of wine wherever this product was produced. These discontinuous manufacturing procedures are still being applied in small wineries. During recent years, however, large-scale manufacturing plants have come into existence in the Federal Republic of Germany, and also some small wineries have joined up into large co-operative units, requiring continuously operating, automatic manufacturing procedures, which have consequently been developed.

Since grapes represent a natural product which is highly susceptible to chemical, biochemical and microbiological influences, rapid processing is of utmost importance. Preservation and storage, in combination with such

TABLE 1
Fields of wine and wine-production in Europe and in the world

	Fields of wine Europe/World (ha × 10^{-6})		Production of wine Europe/World (hl × 10^{-6})	
1952	—	8·74	—	186·4
1956	6·72	8·82	163·6	217·6
1959	—	9·57	—	243·7
1971	—	9·87	222·0	284·0
1974	7·30	10·14	267·9	339·8

measures as cooling, freezing, drying or chemical treatment prior to processing, common practice in many other natural products, are not possible with grapes and are prohibited by law.

These circumstances mean that grapes must be processed immediately after harvesting by adequate and systematic technological processes. Wine technologists have met these requirements by developing new machines, equipment and several new procedures. Large quantities of grapes can be handled in the relatively short time that is available for harvest and processing (2–4 weeks) in modern, continuously operating, and frequently also automatic, production plants. Of particular importance in this respect is the application of heat, the effect of which may be successfully combined with chemical and enzymatic treatments.

MANUFACTURE OF WHITE WINES

The following sequence of operations is generally observed in the manufacture of white wines: the grapes are crushed (mash), the juice (must) is separated and subsequently fermented to wine. These operational steps may be carried out in different ways. It is important, however, that any delay is avoided, since the grapes may, at the harvest temperature, become rapidly infected with undesirable bacteria, wild yeasts and moulds.

To prevent this microbiological spoilage, the must or press juice is sterilised by sulfurous acid which is added in the form of $K_2S_2O_5$ (potassium metabisulfite) which is readily converted into available sulfurous acid when dissolved in acid solution. In some instances this compound may also be added already to the freshly harvested grapes prior to crushing. Sulfurous acid inhibits the growth of harmful micro-organisms and provides protection against oxidation which otherwise spoils the colour (brownish tint) and taste (flavour loss) of the finished wines. Since sulfurous acid is a specific inhibitor, inactivating polyphenolases, it also prevents enzymatic browning if added to the grape must. This explains why it is preferable and, indeed, almost essential that the fresh must is sterilised by sulfuring (a common practice in early Egypt, ancient Greece and Rome).

A new variant in wine-makers' techniques for the inactivation of the phenolases and elimination of micro-organisms is the application of heat; during the so-called HTST process the white-wine must is exposed to high temperatures for short times. In the largest German winery, all clarified musts are heated to 87 °C for about 2 min. After subsequent cooling to 15 °C an active culture of yeast is added to initiate fermentation.

TABLE 2

Influence of mash heating on pressing time

Mash	Pressing time (min)
Crushed—pressed immediately	85
Crushed—allowed to stand for 3 days	43
Stemmed—pressed immediately	83
Stemmed—heated to 80 °C for 2 min	50
Stemmed—heated to 45 °C for 2 h	52
Stemmed—heated to 45 °C for 2 h plus enzyme	34

One further advantage of the heat treatment of must is a favourable influence on the protein stability of the finished wines.

Heat treatment of the mash may be regarded as another promising way to make use of thermal processes in the manufacture of wine. Experiments have shown so far that the heat treatment of mash results in shorter pressing times. The yield of must obtained in this way is simultaneously increased. This yield increase is due to the fact that the grape pectins disintegrate more rapidly when heated, leading to a decrease in viscosity which facilitates pressing. The pectinolytic effect observed under heat may be enhanced by the addition of pectic enzymes (pectinase) (see Table 2). This term is used for and enzyme preparation consisting of polygalacturonase, pectin methylesterase and pectin transeliminase, which is obtained commercially from various fungi, including *Penicillia* and *Aspergilli*. It should be noted

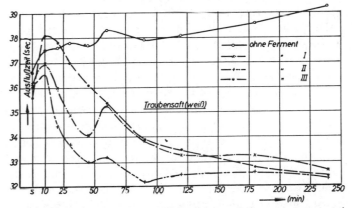

FIG. 1. The influence of temperature (45 °C) and pectolytic enzymes on pectins. Key: Ausflußzeit (sec) = Out-flow time (seconds); Traubensaft (weiß) = Grape juice (white); ohne Ferment = Without enzyme.

that a mash temperature of not less than 45 °C must be observed if pectinolytic enzymes are used, which have been found to have an activity optimum in the range 45–55 °C (Fig. 1).

The wines produced from heat-treated mash, are also characterised by enhanced protein stability.

MANUFACTURE OF RED WINES

Much greater advances have been recorded in the manufacture of red wines by heat treatment than in the manufacture of white wines in past years. In the manufacture of red wines, the main emphasis is laid on the development and preservation of the colour, i.e. the red-wine pigments which represent a particular and desirable quality characteristic. The colouring matter of red wine (anthocyans) is located in small cells of the microscopic range (cyanoplasts) of the grape skin and is surrounded by living plasma. The colour can dissolve only when the cell plasma has been destroyed. To accomplish this plasmolytic process, mechanical treatment alone is not sufficient. Several procedures are available for this purpose, including mash fermentation, mash heating and the addition of enzymes (see also Fig. 5).

Mash Fermentation

This procedure is based on the fact that the alcohol which is formed during the mash fermentation, denatures the cell and extracts the colouring matter from the grape skins. The sequence of operations in the traditional manufacture of red wine using mash fermentation is as follows: after crushing the red grapes and sulfuring, the mash is fermented *in toto* with admission of air over a period lasting 3–10 days. The alcohol-containing must, subsequently obtained by pressing, undergoes a secondary fermentation to red wine. It has to be noted that the extended duration of the mash fermentation in the presence of air involves a permanent danger of infection by micro-organisms. The development of colour varies for this reason (see Fig. 2) and is disturbed due to partial enzymatic and qualitative reduction to red-brown condensation products (melanoidins). This results in a noticeable reduction of quality of the red wine in colour and taste. Furthermore, more tannins are released through the alcohol from the hulls and kernels, so that harder and less smooth wines are obtained. These disadvantages of open fermentation are reduced by a series of modified processes, namely, mash fermentation in closed, enamelled vats away from the air—the mash is moved for the duration of fermentation by injection of previously compressed fermentation acid, so that colour and aroma are well developed. Among the novel methods involving closed mash fermentation, I would like to mention the Vinomat and Roto procedures, which have been approved methods of mash fermentation for red-wine production for ten years.

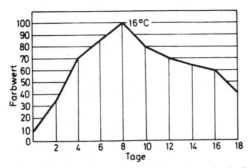

FIG. 2. Influence of the fermentation period on the extraction of red wine colouring during open fermentation (after Nessler). Key: Farbwert = Colour value; Tage = Days.

In spite of the unmistakable technical progress, these methods of mash fermentation are slowly being displaced in the larger factories for reasons connected with the sequence of operations. The application of heat alone (mash heating) or in combination with pectinolytic enzymes as 'thermal fermentation' is now in use more and more as methods for the production of grape must.

Mash Heating

The pigments of the red grapes may also be extracted from the skins by using heat to break down the cell walls of the grapes. The principle of mash heating lies in the rapid heating of the mash to 75–85 °C at which temperature the mash is allowed to stand for some minutes (HTST procedure). Experience gathered so far has led to the application of optimal temperatures and heating times in the indicated range for various varieties of grape.

For mash heating (HTST), several types of heating equipment are available (centrifugal heater, tube heater, plate heater), not forgetting the spiral or centrifugal heater of Rosenblad (Fig. 3) which is often used and which also serves the purpose of mash cooling in a countercurrent arrangement. In the mash-heating procedure, the release of colouring matter is excessively caused by heat. The increase in pigment yield (as

FIG. 3.   Rosenblad spiral heater.

extinction value) as a function of the temperature of the mash is shown in Fig. 4. Heating mash to temperatures of 75–85 °C for a few minutes ensures a good pigment yield. The must has a strong, full, ruby-red colour and yields a wine for which no correcting wines are required.

Under heat treatment, the polyphenol oxidases are inactivated, which prevents the pigments from becoming oxidised, and the wines do not turn brown any more. The inactivation of the polyphenolases is particularly important when mouldy grapes are among the harvested fruit. In these cases the mash has to be heated rapidly in order to ensure that the optimal range of activity of the polyphenolases (30–60 °C) is quickly passed.

Due to the heat shock, harmful micro-organisms are also eliminated (yeasts, bacteria); this, too, is essential mainly in mouldy grapes. Undesirable microbial reactions as, for instance, the formation of acetic acid (*Acetobacter*, *Xylinum*), are thus inhibited.

It was occasionally found that the musts contained more tannin after the HTST process, which may affect the sensorical quality of the finished wines.

Difficulties in clarifying and filtering the must were frequently reported to be a disadvantage of mash heating alone; native grape pectins which are still present due to incomplete disintegration are responsible for this problem. In these cases it is advantageous to use a thermal treatment using pectinolytic enzymes ('thermal fermentation') after the mash heating (HTST). The temperature optima of these enzymes (45–55 °C) should be observed.

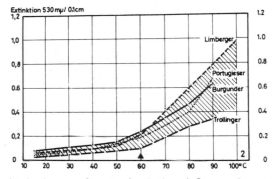

FIG. 4.   Mash heating in the manufacture of red wine—influence of temperature on the extraction of colouring substances. Key: Extinktion = Extinction coefficient.

Commercially available enzymes (pectolytic enzymes)—usually mixtures of pectases and pectinases—attack the native grape pectins at their ester and glycoside bonds. The grape pectins are finally decomposed to compounds of low molecular weight. Table 2 has shown already that better pressing results are obtained from the use of pectinolytic enzymes. Both enhanced pressability of the mash and higher must yield can also be achieved with red grapes by the addition of pectolytic enzymes. Clarifying of both the juice and the wine is facilitated. Also an increase in colour intensity of the juices was observed in many cases (Fig. 5). For this reason a combination of treatments (mash heating and thermal fermentation) is frequently used in large wineries in the manufacture of red wine. The mash is first heated to 75–85 °C for a few minutes (HTST), and subsequently, after cooling, the pectolytic enzymes are added, for which the mash is held at a temperature of 45–50–C for 1–2 h. This combined procedure provides the possibility of treating various red grape varieties under optimal conditions. Mash heating, either alone or in combination with enzymatic treatment, is being widely and successfully applied in the manufacture of red wines.

In this respect, production plants from Imeca–Sick (Germany, Fig. 6) and Julian (France, Fig. 7) have proved to be particularly suitable for some

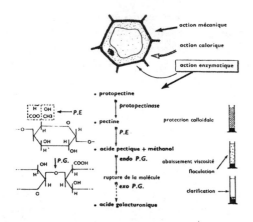

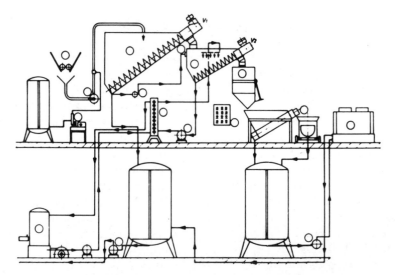

FIG. 5. Extraction of colouring substances and pectin separation from grape cells. Key: action mécanique = mechanical action; action calorique = thermal action; action enzymatique = enzyme action; acid pectique + méthanol = pectic acid + methanol; rupture de la molécule = break-up of molecule; acide galacturonique = galacturonic acid; protection colloidale = colloidal protection; abaissement viscosité = reduction of viscosity; floculation = flocculation.

FIG. 6. Automatic wine production plant manufactured by Imeca-Sick.

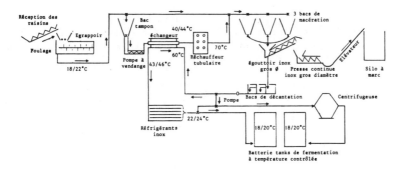

FIG. 7. Automatic wine production plant manufactured by Julian. Key: Reception des raisins = Incoming grapes; Egrappoir = Grape separator; Foulage = Soaking bay; Bac tampon = Makeup hopper; Pompe à vendange = Grape compression pump; échangeur = heat exchanger; Réchauffeur tubulaire = Tubular reheater; 3 bacs de macération = 3 maceration hoppers; égouttoir inox gros ø = large diameter stainless steel delivery pipe; Presse continue inox gros diamètre = Large diameter stainless steel continuous press; Elévateur = Elevator; Silo à marc = Silo for grape skins; Bacs de décantation = Decanting vats; Pompe = Pump; Réfrigérants inox = Stainless steel coolers; Centrifugeuse = Centrifuge; Batterie tanks de fermentation à température contrôlée = Fermentation tanks kept at controlled temperature.

years. Both plants operate continuously and automatically. They can be adjusted to the different requirements of the various types of red wine, including all gradations from pale pink to deep red. A combination of these procedures with mash thermal fermentation is also feasible.

BIBLIOGRAPHY

1. Troost, G. (1972). *Technology of Wine*, Eugen Ulmer, Stuttgart (German).
2. Vogt, E., Jacob, L., Lemperle, E. and Weiss, E. (1974). *The Wine*, Eugen Ulmer, Stuttgart (German).
3. Bujard, M. (1967). *Étude sur la vinification en rouge*, Wädenswil, Switzerland.
4. Haushofer, H. and Bayer, E. (1967). Results of investigations applying the new process 'Vinomat' for red wine preparation. *Mitteilungen Klosterneuburg* (Austria), **17**, 185–200.
5. Haushofer, H., Meier, W. and Bayer, E. (1969). New possibilities for avoiding of colour reduction in the red wine preparation. *Mitteilungen Klosterneuburg* (Austria), **19**, 344–60.
6. Haushofer, H., Meier, W. and Beyer, E. (1971). The new process of red wine preparation system 'Roto'. *Mitteilungen Klosterneuburg* (Austria), **21**, 389 97.
7. Heimann, W., Reintjes, H. J. and Schiele-Trauth, U. (1958). About the heat-(thermal)fermentation of mash for red wine and white wine preparation. (German.) *Weinberg und Keller*, 383.

8. Heimann, W., Wucherpfennig, K. and Reintjes, H. J. (1958). Mash heating and fermentation of different kinds of fruits etc. (German.) *Food*, 117.
9. Heimann, W., Wucherpfennig, K. and Strecker, G. (1961). The thermal fermentation of mash of grapes for production of wines. (German.) *Weinberg and Keller*, 175.
10. Klenk, E. and Maurer, R. (1964). About the application of heat in the production of red wine (German). *Dt. Weinbau-Kalender*, 136.
11. Klenk, E. and Maurer, R. (1967). About the German red wine and its preparation. (German). *Dt. Weinbau*, 824.
12. Klenk, E. and Maurer, R. (1969). The stabilising of red wines (German). *Dt. Weinbau- Jb.*, 169.
13. Koch, J., Troost, G. and Geiss, E. (1954). Heat and coldness by the cellar treatment of wines (German). *Dt. Weinbau*, 57.
14. Konlechner, H. and Haushofer, H. (1956–8). Results of experiments of red wine mash, heat fermentation. *Mitt. Klosterneuburg, Rebe und Wein* (Austria). **158** (1956); **78** (1957); **169** (1958).
15. Konlechner, H. and Haushofer, H. (1961). Thermal treatment: a new process of red wine preparation. *Mitt. Klosterneuburg, Rebe und Wein* (Austria). **148,** 180.
16. Lemperle, E., Trogus, H. and Frank, J. (1973). Investigations to the colour of red wines: III. Colour yields in the newer processes of the red wine preparation (German). *Die Weinwissenschaft*, **28,** 181–202.
17. Marteau, G. (1969). Some aspects of wine preparation in the red wine preparation by thermal treatment in comparison with the classic wine preparation (French). *Progr. agric. et vitic*, no. 12.
18. Mayer, K. and Vetsch, U. (1968). Investigations into thermal processes of mash of red wines. *Z. Obst- und Weinbau* (Switzerland), 559.
19. Peyer, E., Schneider, R. and Perret, P. (1970). Experiments by mash heating of blue Burgundy grapes in 1969. *Z. Obst- und Weinbau* (Switzerland), 256.
20. Prehoda, J. (1965–6). Red wine preparation by mash fermentation (German). *Borgazdasag*, **1965,** 106; *Mitt. Klosterneuburg, Rebe und Wein*, **1966,** 463.
21. Strobl, G. (1968). Reduction of mash juice and mash fermentation by the new Roto process (German). *Winzer*, 103.
22. Wucherpfennig, K., Troost, G. and Fetter, K. (1964). Investigations into mash thermal fermentation of white and red grapes (German). *Dt. Wein-Ztg.*, 688.
23. Schobinger, U., Schneider, R. and Dürr, P. (1975). Red wine preparation by mash heating: The longtime heating alternative to the high temperature short time heating. *Z. Obst- und Weinbau* (Switzerland), **111,** (84.), 503–14.
24. Schobinger, U., Koblet, W. and Schneider, R. (1973). SO_2 reduction in the red wine preparation by high temperature short time heating of the mash. *Zeitschrift für Obst- und Weinbau* (Switzerland), **109,** (82), 353–65.
25. Tanner, H. (1974). Wine analysis in the service of the consumer and the present situation of the wine technology. *Chimia* (Switzerland), **28,** no. 10.
26. Julian, P. and Sines, D. (1975). The equipment of modern cellars. 4th International Oenology Symposium, Valencia, Spain, May 1975, ed. by Lemperle, E. and Frank, J., pp. 7–15.
27. Symposium about mash heating of red grapes. *Z. Obst- und Weinbau* (Switzerland), **1967,** 508.

22

Nature of Bound Nicotinic Acid in Cereals and its Release by Thermal and Chemical Treatment

R. KOETZ AND H. NEUKOM

Department of Food Science, Swiss Federal Institute of Technology, Universitätstrasse 2, CH-8006, Zürich, Switzerland

Nicotinic acid and its amide (niacin) have important biochemical functions as constituents of a number of coenzymes, e.g. NAD^+ and $NADP^+$. Nicotinic acid is widely distributed in nature; the richest sources of the vitamin are wheat germ, yeast and liver. Nutritional deficiency of nicotinic acid, or nicotinamide in man results in pellagra, a disease characterised by dermatitis, diarrhoea and dementia.

EXISTENCE AND SIGNIFICANCE OF A BOUND FORM OF NICOTINIC ACID

Initial evidence of the existence of a bound form of nicotinic acid came from vitamin assays in cereals. These determinations showed that low values were obtained when the analysis was done without previous digestion with alkali. After alkali treatment the apparent content of nicotinic acid increased. These findings have been confirmed by various workers,[1-5] both by chemical and microbiological analysis. Depending on the extraction procedure of nicotinic acid from natural materials, the values obtained varied by as much as 500 %.[6]

Consequently there was considerable uncertainty regarding the true nicotinic acid content. While discrepancies in the values obtained, were first attributed to differences in extraction efficiency,[2] further studies pointed out the existence of a bound form: alkali treatment of aqueous extracts of cereals raised the nicotinic acid content to that found by direct extraction with alkali. This was explained by the presence of a water-soluble derivate or precursor, which on hydrolysis, liberates free nicotinic acid.[3] Studies on the stability of the bound form indicated a pronounced lability towards alkali, while strong acid is required to release all of the nicotinic acid.[4]

An estimation of the nicotinic acid in the bound form can be obtained before and after hydrolysis. The results showed that, in maize and wheat, the aleurone layer is the site of the greatest concentration of the nicotinic acid, and there is evidence that the bound form predominates, [7] whereas the nicotinic acid of the scutellum and the embryo is reported to be free.[3] Apart from maize and wheat, rice and barley also contain 85–90 % of their total nicotinic acid in the bound form.[8]

The occurrence of bound nicotinic acid is of considerable nutritional importance and explains why maize is a poor source of available nicotinic acid, in spite of its relatively high content of total nicotinic acid. Nicotinic acid deficiency in rats[9] and pigs[10] could however be cured by feeding with alkali-treated maize. Similar effects were observed with tortilla,[10,11] a traditional staple food consumed in Mexico. Tortilla is prepared by cooking maize with limewater and baking the resultant mash. This pretreatment also liberates nicotinic acid and possibly explains why pellagra is virtually unknown in Mexico and South America. Maize also has a low content of tryptophane, a precursor of nicotinic acid; therefore, any treatment which liberates free nicotinic acid from its bound form is expected to improve the nutritional qualities of maize, provided of course that the treatment does not lead to the destruction of other nutrients.

RELEASE OF NIACIN BY THERMAL AND CHEMICAL TREATMENT

Nicotinic acid is one of the most stable of the B vitamins, being unaffected by heat, light, oxygen, acid or alkali; losses could only occur by leaching. It has been shown that about 70 % of the nicotinic acid in commercial flour is present in the bound form. During baking the amount of bound nicotinic acid fell linearly with increasing pH, from 70 % at pH 6·8 to zero at pH 9·6. A partial release can also be obtained by lengthening the baking time.[12] Generally speaking, the amount liberated by heat treatment can vary considerably in different foods. Food preparation methods, including roasting, pressure-cooking followed by baking, and cooking in an acid medium, are reported to affect the availability of nicotinic acid. In roasted maize the free nicotinic acid increased about 30 times (from 0·03 to 0·88 mg/100 g), in peas by only 3 times, and in rice by only one-tenth. Pressure-cooking of maize or baking into chapaties also increased the free nicotinic acid content.[13] Another practice of using maize, namely roasting cobs of sweet maize followed by drying and boiling, is reported to yield significant amounts of nicotinamide, together with free nicotinic acid. The formation of nicotinamide, which was observed only with sweet maize, could be due to ammonolysis of bound nicotinic acid, the ammonia being liberated from heated proteins.[14] These observations show clearly that the niacin content can be raised by thermal processing of foods. This is in contrast to the losses in nutritional value, usually observed by the heating of foods. A similar increase in niacin

TABLE 1

Thiamine and niacin content in milled rice (Rexoro, lot C):
parboiled versus raw rice

Products	Thiamine		Niacin	
	Parboiled μg	Raw μg	Parboiled μg	Raw μg
Rough	3·40	3·40	54·12	54·12
White milled	2·50	0·50	32·17	16·40
Bran	6·82	21·05	233·52	269·21
Hulls	1·81	2·41	27·92	48·52

occurs during roasting coffee; in this case however the formation of additional nicotinic acid takes place by thermal decomposition of trigonelline.[15]

Another well-known hydrothermic treatment which improves food value is parboiling. Parboiling of rice (steeping in water, steaming and drying),which was carried out in Asia to make dehulling easier, is still carried out because it brings about a number of improvements in the quality of the products. During this process soluble vitamins and minerals migrate into the endosperm and improve the nutritional value of the products. People who consumed parboiled rice were rarely affected by beri-beri, a disease caused by lack of vitamin B_1. Determinations of thiamine content in milled rice showed a large decrease of the vitamin content in the bran after parboiling.[16] Similar determinations of nicotinic acid in rice bran gave only a slight reduction after parboiling[16] (see Table 1). This fact indicates that the diffusion of niacin towards the endosperm is inhibited, which is consistent with the existence of a bound form of high molecular weight. It is also an indication that heat treatment alone releases only a small amount of niacin.

NATURE OF THE LINKAGE

The nature of the bound form which is the cause of the unavailability of nicotinic acid has occupied several research groups. Early experiments consisted of extracting bound nicotinic acid in order to obtain a soluble material enriched in niacin. Two types of bound niacin were initially described: (a) a polypeptide with a molecular weight of 12 000–13 000, which was called niacinogen (wheat, rice, maize), and is believed to contain niacin attached to a peptide,[17] and (b) a carbohydrate–protein complex extracted from wheat bran, for which the name niacytin has been suggested.[18] A similar preparation of bound nicotinic acid was obtained from maize gluten.[19] Fractionation of niacytin into a number of macromolecules containing nicotinic acid gave no separation of peptide and carbohydrate material.[20] It is very likely therefore that niacytin is composed of various glycoprotein fractions, to which niacin is bound.

To clarify the nature of the linkage, Mason and Kodicek[21] hydrolysed niacytin with acid and tried to isolate fragments containing bound nicotinic acid. By this method, they isolated a fragment which contained both nicotinic acid and glucose. Experimental evidence points to an ester linkage between nicotinic acid and glucose (see Fig. 1). Based on the work of

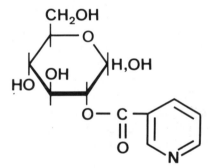

FIG. 1.   A possible model for the bound nicotinic acid 2-O-nicotinoyl-D-glucose

Mason and Kodicek,[21] we have also isolated a non-diffusible preparation containing nicotinic acid (NDM). Defatted wheat bran was extracted with 60 % ethanol and the extract dialysed and lyophilised. The composition of the NDM is shown in Table 2. Work on the composition of NDM and the exact nature of the linkage of bound nicotinic acid is still in progress. Preliminary experiments confirmed the assumption of Mason and Kodicek[21] that at least one type of linkage exists between glucose and

TABLE 2
Composition of the non-diffusible material
(%)

Protein		14·1
Carbohydrate (total)		71·1
Arabinose	22·1	
Xylose	8·3	
Mannose	3·8	
Galactose	14·8	
Glucose	22·1	
Ash		0·9
Nicotinic acid		1·0
H_2O		5·8
Unaccounted for (by difference)		7·1

nicotinic acid, as shown in Fig. 1. The exact location of the linkage and the structure of niacytin is still unknown.

REFERENCES

1. Oser, B. L., Melnick, D. and Siegel, L. (1941). Control of nicotinic acid in flour and bread. *Food Industries*, **13**, 66.
2. Snell, E. E. and Wright, L. D. (1941). A microbiological method for the determination of nicotinic acid. *J. Biol. Chem.*, **139**, 675–86.
3. Andrews, J. S., Boyd, H. M. and Gortner, W. A. (1942). Nicotinic acid content of cereals and cereal products. *Ind. Eng. Chem. (Anal. Ed.)*, **14**, 663–6.
4. Krehl, N. A. and Strong, F. M. (1944). Studies on the distribution, properties, and isolation of a naturally-occurring precursor of nicotinic acid. *J. Biol. Chem.*, **151**, 1–12.
5. Krehl, W. A., Elvehjem, C. A. and Strong, F. M. (1944). The biological activity of a precursor of nicotinic acid in cereal products. *J. Biol. Chem.*, **156**, 13–19.
6. Sweeney, J. P. and Parrish, P. P. (1954). Report on the extraction of nicotinic acid from naturally occurring materials. *J. Assoc. Offic. Analyt. Chem.*, **37**, 771–7.
7. Chaudhuri, D. K. and Kodicek, E. (1950). Purification of a precursor of nicotinic acid from wheat bran. *Nature*, **165**, 1022–3.
8. Grosh, H. P., Sarkar, P. K. and Guha, B. C. (1963). Distribution of the bound form of nicotinic acid in natural materials. *J. Nutr.*, **79**, 451–3.
9. Chaudhuri, D. K. and Kodicek, E. (1950). The biological activity for the rat of a bound form of nicotinic acid present in the bran. *Biochem. J.*, **47**, XXXIV.
10. Kodicek, E., Braude, R., Kon, S. K. and Mitchell, K. G. (1956). The effect of alkaline hydrolysis on the availability of its nicotinic acid to the pig. *Brit. J. Nutr.*, **10**, 51–67.

11. Laguna, J. and Carpenter, K. J. (1951). Row versus processed corn in niacin-deficient diets. *J. Nutr.*, **45**, 21–8.
12. Clegg, K. M. (1963). Bound nicotinic acid in dietary wheaten products. *Brit. J. Nutr.*, **17**, 325–9.
13. Rajalakshimi, R., Nanavaty, K. and Gumashta, A. (1964). Effect of cooking procedures on the free and total niacin content of certain foods. *J. Nutr. Diet. India*, **1**, 276–80.
14. Kodicek, E., Ashby, D. R., Muller, M. and Carpenter, K. J. (1974). The conversion of bound nicotinic acid to free nicotinamide on roasting sweet corn. *Proc. Nutr. Soc.*, **33**, 105A–106A.
15. Viani, R. and Horman, I. (1974).Thermal behaviour of trigonelline. *J. Food Sci.*, **6**, 1216–7.
16. Gariboldi, F. (1972). Parboiled rice. In *Rice, Chemistry and Technology*, Vol. IV, ed. by D. F. Houston, Monograph Series, St. Paul, Minnesota.
17. Das, M. L. and Guha, B. C. (1960). Isolation and characterisation of bound niacin (Niacionogen) in cereal grains. *J. Biol. Chem.*, **235**, 2971–6.
18. Kodicek, E. and Wilson, P. W. (1960). The isolation of niacytin, the bound form of nicotinic acid. *Biochem. J.*, **76**, 27p.
19. Christianson, D. D., Wall, J. S., Dimler, R. J. and Booth, A. N. (1968). Nutritionally unavailable niacin in corn: Isolation and biological activity. *J. Agric. Food Chem.*, **16**, 100–4.
20. Mason, J. B., Gibson, N. and Kodicek, E. (1973). The chemical nature of the bound nicotinic acid of wheat bran: studies of nicotinic acid-containing macromolecules. *Brit. J. Nutr.*, **30**, 297–311.
21. Mason, J. B. and Kodicek, E. (1973). The chemical nature of the bound nicotinic acid of wheat bran: Studies of partial hydrolysis products. *Cereal Chem.* **50**, 637–46.

23

The Influence of Heat on Meat Proteins, Studied by SDS Electrophoresis

K. HOFMANN

Bundesanstalt für Fleischforschung, D-8650 Kulmbach, Blaich 4, Federal Republic of Germany

INTRODUCTION

The investigation of the influence of heat on meat proteins had proved to be necessary for us in connection with another research programme, namely, the estimation of so-called foreign proteins which may be used as additives for meat products. In solving this problem the application of the polyacrylamide gel electrophoresis in presence of sodium dodecyl sulfate (SDS) seems to be very useful. The SDS electrophoresis allows identification and determination of the proteins of meat, soya, egg-white, milk and other proteins.[1-5] Proteins of different origin reveal characteristic patterns in the gel, so that foreign proteins in the meat can easily be detected. In presence of SDS the protein molecules are surrounded by a 'shell' of dodecyl sulfate anions. Because all protein molecules are charged uniquely negative, the only one property which now determines the mobility of the protein molecules or their subunits, in which they are split, is their molecular weight. Thus, SDS electrophoresis separates the polypeptide chains according to their molecular weight and enables a very simple measurement of the molecular weights of proteins.[2,6-11] This possibility is very valuable for the characterisation and identification of proteins in question.

EXPERIMENTAL PROCEDURES

In all cases flat gel slabs were used. The protein solutions were placed in small slots on top of the gels. In this way ten samples can be separated simultaneously side by side. The concentration of acrylamide in the gels was 8 %, using 1/30 cross-linker. A Tris-boric acid buffer pH 8·2 with 0·1 % SDS was used for preparing the gels, and 0·1 % mercaptoethanol was added. The protein samples were dissolved as a whole using the same buffer solution (details concerning the preparation of the samples are given later).

The usage of flat gels is of great advantage because for all samples on the same gel the conditions of electrophoresis and of the subsequent staining and destaining of the gels are equal, so their comparability is optimum. For staining the protein bands a solution of 0·2 % Coomassie brilliant blue R–250 in a mixture of acetic acid, ethanol and water was used. For optimum staining of the proteins in the gel it is very important to treat the gels with a trichloroacetic acid solution before staining, which fixes the proteins and additionally removes most of the SDS, which is probably important for proper staining of the proteins. The preparation of the gels and the equipment used for electrophoresis are described in detail by Hofmann and Penny.[2]

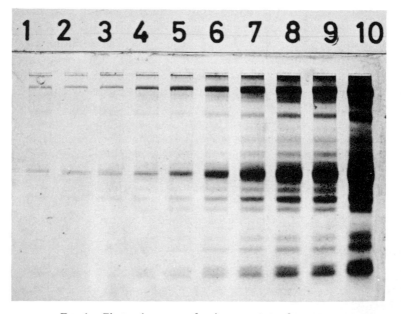

FIG. 1.   Electropherogram of various amounts of meat:

Run No.:	1	2	3	4	5	6	7	8	9	10
Meat (mg/ml)	0·5	1·5	2·5	5·0	10	20	30	40	50	100

QUANTITATIVE MEASUREMENT OF THE SEPARATED
PROTEINS

An important problem is the measurement of the amount of foreign proteins in meat products, or, alternatively, determination of the amount of meat protein itself. An electropherogram of a gel with various amounts of whole meat is shown in Fig. 1.

The two most distinct bands are those of actin (in the middle) and of myosin (on top of the gel), which represent two-thirds of the myofibrillar proteins. Beef—in contrast to pork—additionally exhibits a strong myoglobin band, which is seen as the lowest one in the gel. All the other bands are due to the minor fibrillar and the sarcoplasmic proteins. The amount of a protein in a gel is given by the area of the corresponding peak in the densitometric curve of the stained protein bands. Figure 2 demonstrates such a densitogram of whole meat, and underneath the facsimiles of the corresponding protein bands. As the composition of the myofibrillar

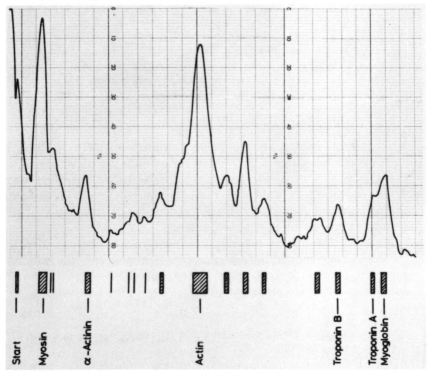

FIG. 2.   Densitometric curve of stained meat proteins in SDS polyacrylamide gel.

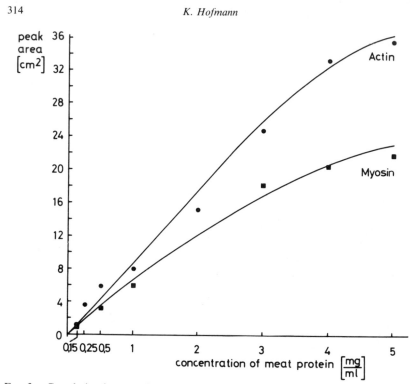

FIG. 3.   Correlation between the amount of meat protein and the peak areas of actin and myosin.

proteins is fairly constant it is possible to use one of the most distinct protein bands, such as myosin or actin, for quantitative measurement.

In Fig. 3 the peak areas corresponding to actin and myosin are correlated with the concentrations of the total protein in the solution of the sample. Under certain conditions, these curves may be used for determination of the meat protein content in an unknown sample.

However, these results are related to raw meat, and the question arises as to the way in which they are influenced by heat treatment or by the addition of salts. To answer these questions, extended experiments were carried out, the results of which are presented and discussed in the following.

EFFECT OF HEATING ON MEAT WITHOUT ADDED SALTS

The influence of heat on the electrophoretic behaviour of the meat proteins was studied by heating the meat samples up to various temperatures (100,

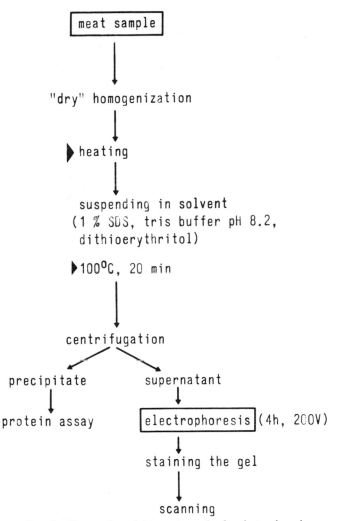

FIG. 4.   Preparation of the meat samples for electrophoresis.

110 and 120 °C) for two periods (15 and 30 min). The method of preparing
the meat samples is outlined in Fig. 4.

The meat is first homogenised in a 'Bühler-Homogenisator' (with rotating
knives) without addition of a liquid. We call it 'dry homogenisation'. Then
the samples are heated in screw-capped tubes or, at temperatures over
120 °C, in a small steel chamber (Wurzschmitt's bomb) using a thermo-
regulated paraffin bath. After heating, the samples are suspended and

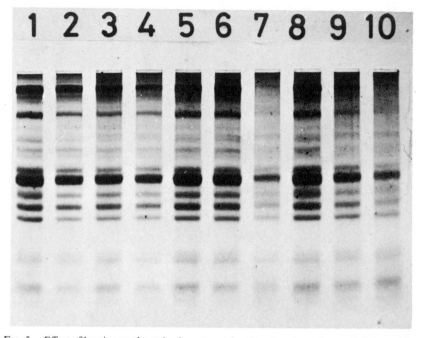

FIG. 5.   Effect of heating on the stained meat proteins. Run 1, unheated; runs 2–4, heated for 5, 15 and 30 min to 110 °C; runs 5–7 heated for 5, 15 and 30 min to 115 °C; runs 8–10 heated for 5, 15 and 30 min to 120 °C.

homogenised in a solvent containing SDS and dithioerythritol. The thiol reduces disulfide bonds which can be formed during heating. The function of SDS is twofold: SDS is necessary for the type of electrophoresis used and it enormously improves the solubility of the proteins. Repeated heating to 100 °C (water bath) accelerates the dissolution of the proteins and does not cause any significant additional protein changes. A small part of each sample remains insoluble (see Table 3) and is, therefore, centrifuged. The supernatant fluid which contains the dissolved proteins, is used for electrophoresis. The effect of heat on the proteins of meat (beef) is illustrated in Fig. 5. Comparison of the unheated sample (Fig. 5, run 1) with all the other heated samples (runs 2–10) reveals two different facts:

(a) Heat does not change the position of the protein bands in the gel, or in other words, the rate of migration. Roberts and Lawrie[11] reported that heating caused a loss of the electrophoretic mobility of meat proteins. However, this statement is clearly not valid for SDS electrophoresis. The disagreement may be explained by the fact that heat changes the number of protein charges[13] which are indeed important for the mobility in normal electrophoresis but not in the SDS technique. Furthermore, our results show that the size of the protein molecules is not irreversibly changed by heat. Although heating could have caused oxidation of sulfhydryl to intermolecular disulfide bridges, resulting in protein dimers, trimers and polymers, the formation of disulfide bonds is reversible by reduction with dithioerythritol, which is a constituent of the solvent for the proteins. Moreover, it is interesting to note that it was reported recently that proteins loaded with dodecyl sulfate and previously not reduced were reduced during electrophoretic separation.[14]

(*b*) Figure 5 shows that heating meat decreases the intensity of the stained protein bands, especially those of the myosin band. Actin obviously has a higher thermostability than myosin because of the higher molecular weight of the latter. This agrees with results of Kako[15] and Chrystall.[16] In earlier experiments we already found that a regular series of single proteins can be established according to the extent of the weakening of their stained bands induced by heat:[17] myosin (subunits, MW $\approx$ 200 000); serum albumin (68 000); G-actin (46 000); and myoglobin (17 000).

It seems to be a general rule that proteins of low molecular weight are

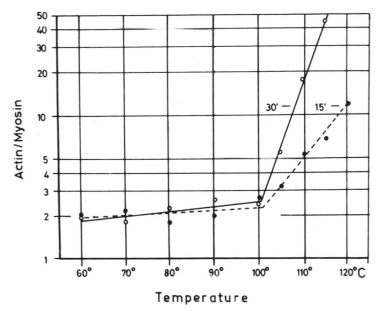

FIG. 6. Influence of heat on the relation of the peak areas of actin and myosin.

more resistant to heat than larger proteins. It should be remembered that
the band from actin is still very distinct and the band from myosin is still
visible in the samples heated to 120 °C. In contrast, it will be shown later that
at this temperature in the presence of various salts the intensities of the
myosin and the actin bands are decreased much more, in some cases even to
zero. As the myosin band is more weakened by the influence of heat than the
actin band, relation of the corresponding peak areas of actin to myosin in
the densitogram increases with increasing temperature and increasing time
of heating, particularly above 100 °C (Fig. 6). Above 100 °C the
actin/myosin ratio increases so drastically that in Fig. 6 a logarithmic scale
is used. It has to be realised that there is a sharp rise in the curves above
100 °C. The curves may be applied to estimate the extent of heat
treatment of an unknown meat sample. However, this possibility seems
to be limited to meat not mixed with any salt, because salts additionally
influence the intensity of the stained protein bands.

INFLUENCE OF SALTS ON THE STAINING OF MEAT PROTEINS

Sodium chloride, sodium diphosphate, sodium nitrite and potassium
nitrate are the salts mostly used in the preparation of meat products.
Hitherto it has not been known whether the presence of these salts has any
influence on the staining of the meat-protein bands in the gel. In

TABLE 1
*Effect[a] of salts mixed with meat on the staining of protein
bands in SDS–polyacrylamide gels*

Salt	Myosin	Actin	Myoglobin
NaCl	83	78	75
$Na_2P_2O_7$	133	93	135
$NaNO_2$[b]	17	29	52
KNO_3	83	88	62

[a] Relative amount of dye (Coomassie blue) bound to the
protein bands in comparison to samples without salt (%).
[b] $NaNO_2$ was added in mixture with NaCl ('Nitritpökelsalz').
The final concentration of the salts in the meat samples was as
follows: 3% NaCl, 0·3% $Na_2P_2O_7$, 150 ppm $NaNO_2$ (3%
Nitritpökelsalz) and 0·5% KNO_3.

investigations into the influence of heat on meat mixed with salts in amounts used in practice, the results shown in Table 1 were obtained.

The salts do indeed influence the intensity of the stained protein bands. NaCl and KNO_3 cause a moderate, and $NaNO_2$ a very drastic, decrease in the protein staining. In contrast, disphosphate increases the staining in the case of myosin and myoglobin. All these effects prove that the ions of the salts are bound at least in part to the proteins. Additionally, in the case of nitrite, a reaction with the dye-binding groups of the proteins may occur. The positive effect of phosphate on protein staining is possibly due to the bivalent character of the anion. The observed effects of the salts on protein staining in SDS gels should be taken into account in protein research.

EFFECT OF HEAT ON MEAT MIXED WITH DIFFERENT SALTS

Figures 7 to 10 demonstrate the effect of heat on myosin, actin and myoglobin from meat samples mixed with various salts. The time of heating was in all cases 30 min. Heating in the presence of NaCl has a strong negative effect on the intensity of the stained myosin band. The band disappears completely after heating to 100 °C. Actin is less sensitive.

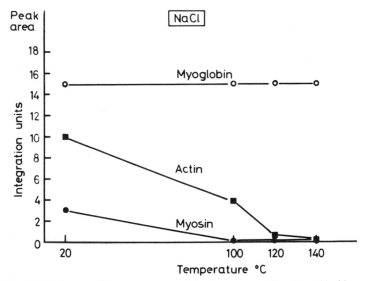

FIG. 7.   Influence of heat on meat proteins in presence of sodium chloride.

At 100 °C the decrease of the actin band amounts to about 60 %. After heating to 120 °C the actin band has mostly, and at 140 °C completely, disappeared. In contrast to myosin and actin, the band of myoglobin is extremely stable at each temperature used. This result shows again that the protein damage caused by heat decreases with decreasing molecular size.

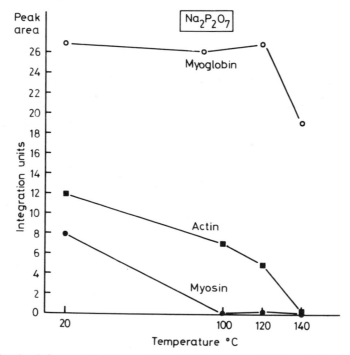

FIG. 8.   Influence of heat on meat proteins in presence of sodium diphosphate.

Comparison with the heated meat samples not mixed with salt reveals that in presence of NaCl the negative effect of heat on the intensity of the stained protein bands is apparently strengthened. The influence of the other salts (Figs. 8–10) is largely analogous to NaCl.

However, nitrite seems to have a particular effect, which can already be observed in the unheated samples: it decreases the apparent bands most. Once again, this effect may be due to the reactivity of nitrite. Very recently it was reported that appreciable amounts of nitrite can be bound to myosin with resulting modification of the protein.[18] In all cases myoglobin is the most stable band, its maximum decrease at 120 °C being only about 30 %,

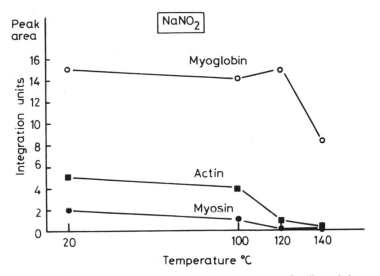

FIG. 9. Influence of heat on meat proteins in presence of sodium nitrite.

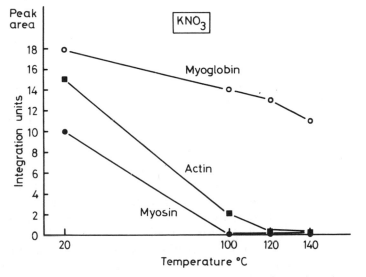

FIG. 10. Influence of heat on meat proteins in presence of potassium nitrate.

and at 140 °C about 50 %. An additional observation should be mentioned, which cannot be shown in the figures. Normally the background of the gels is more stained in the case of the heated samples than in the case of the unheated samples. However, the background of the samples mixed with salts is quite clear even in the case of heating to 140 °C. Either the salts inhibit the formation of substances causing the background staining, or they prevent the staining.

INVESTIGATION AND DISCUSSION OF THE CAUSES FOR THE HEAT-INDUCED WEAKENING OF THE PROTEIN BANDS

Obviously it cannot be assumed that the heat-induced decrease of the intensity of the protein bands stained with Coomassie blue is due to a decrease of the amount of protein in the gel. Instead, there are several possibilities which have to be taken into account. The various theoretical possibilities are shown together in Table 2.

TABLE 2

Possible reasons for the reduction of the intensity of the stained protein bands in polyacrylamide gels

---

(*a*) Decrease of the solubility of the heated proteins in the solvent (1 % SDS, boric acid Tris buffer, pH 8·2).

(*b*) Loss of the electrophoretic mobility of the heated and diluted proteins, leaving them in the slots of the gels.

(*c*) Decrease of the number of functional groups in the proteins responsible for the dye binding.

(*d*) Decrease of the amount of the single proteins as a result of thermal splitting of protein chains to form smaller fragments.

(*e*) Decrease of the amount of the single proteins as a result of the formation of products of higher molecular weight.

Further experiments were carried out in order to see which of the possibilities shown is correct, and these are now presented in turn.

(*a*) To discover whether the solubility of the heated proteins is reduced, the protein content of the insoluble residue was estimated and compared with the total protein content of the whole sample. The results, which are listed in Table 3, show that the solubility in SDS of even the most heated protein samples is rather high. The presence of the various salts has not had much influence. Heating to 100 °C decreases the solubility only slightly

(1–2 %). At 120 °C the averaged decrease is about 10 %, and at 140 °C about 30 %. It is evident that this relatively small decrease of protein solubility cannot explain the strong heat-induced decrease of the protein bands in the gel, as already demonstrated, but it can contribute to it only in part.

(*b*) A loss of electrophoretic mobility would be detectable by staining the startline, i.e. the immobilised protein in the slot of the gel. However, in the

TABLE 3

Effect^a of heat on the solubility of meat proteins in Tris buffer, pH 8·2, containing 1 % SDS

	Heat treatment			
Salt	Unheated	100 °C	120 °C	140 °C
—	99·0	98·1	88·5	70·8
NaCl	97·6	97·5	90·1	59·6
$Na_2P_2O_7$	98·8	98·1	92·6	62·2
$NaNO_2$^{b}	98·7	97·4	87·9	65·9
KNO_3	99·0	97·4	71·6	67·3

^{a} Solubility of total protein (%).
^{b} $NaNO_2$ in mixture with NaCl. The concentrations of salts used are the same as given in Table 1.

relevant investigations stained startlines could not be observed. Thus, heating does not cause a loss of the electrophoretic mobility of proteins soluble in SDS.

(*c*) To discover whether there is a decrease of the dye-binding of the heated proteins in the polyacrylamide gel the proteins were allowed to react with the SH reagents *p*-chloromercuribenzoate and silver nitrate labelled with the radioactive isotopes of mercury (^{203}Hg) and of silver (^{110m}Ag). The method of preparation of these meat samples is shown in Fig. 11.

One part of the samples was heated, another part was not heated. Both the raw and the heated samples were then treated with dithioerythritol for maximum reduction of disulfide groups—which may have formed during heating—to SH groups. Afterwards the SH proteins were allowed to react with nearly equimolar amounts of the labelled SH reagents and then prepared for electrophoresis as shown in Fig. 11. Finally, the radioactivity in the electrophoresis gel was measured by means of a scanner. The results, which have already been briefly reported,[19] show that the radioactivity of actin and myosin in the gel was not much less in the heated samples than in

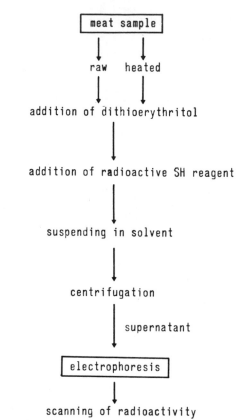

FIG. 11.   Preparation of meat samples labelled with radioactive SH reagents.

the unheated ones. This result was unexpected. It means that the protein
bands of the heated samples are still present in the gel, but their ability to be
stained with dye is reduced. Consequently it has to be assumed that heating
meat causes a loss of those functional groups of the proteins which are
responsible for the dye-binding. Coomassie brilliant blue, the dye used in
this experiment, possesses two sulfonic acid residues per molecule and
therefore may be bound by, among others, basic groups. As already
mentioned by Hamm elsewhere in this volume, the heating of meat proteins
results generally in a loss of dye-binding basic groups. Severe heat causes
damage to proteins in which the formation of 'isopeptides' by reaction of the
α-amino group of lysine is probably included. This may also partly explain
the described decrease in the intensity of the stained protein bands in the
gel.

(*d*) The thermal splitting of protein chains would result in the appearance of protein fragments of low molecular weight, causing additional bands in the gel. However, such bands could not be observed.

(*e*) The formation of proteins irreversibly aggregated during heating would also result in the appearance of additional bands which also could not be observed.

Finally, the fact that salts increase the effect of heating has still to be discussed generally. It is well known that salts used for meat products promote a loosening or even a dissociation of the protein complex actomysin, resulting in a swelling of the whole tissue. In this state the protein molecules might be more flexible and probably more reactive than in the fixed state. This hypothesis harmonises with the finding that myosin is more heat resistant in the myofibril when it is bound to actin.[15,20] It seems that in this way the presence of salts increases the reactivity of the proteins indirectly and therefore also the damage by heat.

The influence of salts which has been described is certainly not limited to the salts used in these experiments. Moreover it is to be expected that salts which usually occur in meat, such as lactate or phosphate, may also influence the effect of heat on the meat proteins. The content of these 'normal' salts in meat varies and depends, in addition, on several factors (e.g. degree of post-mortem changes).

All things considered, one can conclude that the quantitative determination of meat protein in meat products with the help of SDS electrophoresis, which was the starting-point of these investigations, becomes very complicated or even impossible. Since the protein bands of samples heated at high temperature are present in the gel, but are not stained at all or only weakly by Coomassie blue, another way of evaluating the amount of protein in the gel has to be found. Investigations in this field are in progress.

CONCLUSIONS

Heating meat decreases the intensity of the protein bands in SDS polyacrylamide gels stained with Coomassie blue. This decrease is only in part due to a decrease of the solubility of the meat proteins. The main cause is the weakening of the dye-binding of the proteins which is probably the result of heat-induced damage of the basic charged groups of the proteins. The weakening increases with increasing molecular weight of the meat proteins. Myosin is the most labile, myoglobin is the most stable. The presence of salts in the meat samples has an additional weakening effect on the protein staining. In principle, it is possible to determine the portion of meat protein in a meat product containing foreign proteins with the help of SDS electrophoresis. However, this is complicated by the influences described. Since the bands of the heated meat proteins are still detectable in the gel, further research is necessary for their quantitative determination.

ACKNOWLEDGEMENT

I wish to thank Mr E. Blüchel for his skilful technical assistance.

REFERENCES

1. Hofmann, K. and Penny, I. F. (1971). Identifizierung von Soja- und Fleischeiweiss mittels Dodecylsulfat-Polyacrylamidgel-Elektrophorese. *Fleischwirtschaft*, **51**, 577.
2. Hofmann, K. and Penny, I. F. (1973). Methode zur Identifizierung und quantitativen Bestimmung von Fleisch- und Fremdeiweiss mit Hilfe der SDS-Polyacrylamid-Elektrophorese auf Flachgelen. *Fleischwirtschaft*, **53**, 252.
3. Penny, I. F. and Hofmann, K. (1971). The detection of soya bean protein in meat products. *Proc. 17th Europ. Meeting Meat Res. Workers*, Bristol, p. 809.
4. Hofmann, K. (1973). Identifizierung und Bestimmung von Fleisch- und Fremdeiweiss mit Hilfe der Dodecylsulfat-Polyacrylamidgel-Elektrophorese. *Z. Anal. Chem.*, **267**, 355.
5. Mattey, M. E. (1974). *Food Res. Abstr. Circ.*, No. 518, citing Smith, P. R. (1974). 10th Anniversary Symposium: Meat and Meat Products, Institute of Food Science and Technology, Leatherhead, Surrey, UK.
6. Shapiro, A. L., Vinuela, E. and Maizel, J. V. (1967). Molecular weight estimation of polypeptide chains by electrophoresis in SDS-polyacrylamide gels. *Biochem. Biophys. Res. Commun.*, **28**, 815.
7. Weber, K. and Osborn, M. (1969). The reliability of molecular weight determinations by dodecyl–polyacrylamide gel electrophoresis. *J. Biol. Chem.*, **244**, 4406.
8. Weber, K. and Osborn, M. (1975). Proteins and sodium dodecyl sulfate: Molecular weight determination on polyacrylamide gels and related procedures. In *The Proteins*, 3rd ed., ed. by Neurath, H. Vol. 1, p. 179.
9. Dunker, A. K. and Rueckert, R. R. (1969). Observations on molecular weight determination on polyacrylamide gel. *J. Biol. Chem.*, **244**, 5074.
10. Scopes, R. K. and Penny, I. F. (1971). Sub-unit sizes of muscle proteins as determined by sodium dodecyl sulphate gel electrophoresis. *Biochim. Biophys. Acta*, **236**, 409.

11. Weber, K., Pringle, I. R. and Osborn, M. (1972). Measurement of molecular weights by electrophoresis on SDS-acrylamide gel. *Methods Enzymol.*, **26** (pt. C.), 3.
12. Roberts, P. C. B. and Lawrie, R. A. (1974). Effects of bovine l. dorsi muscle of conventional and microwave heating. *J. Food Technol.*, **9**, 345.
13. Hamm, R. (1966). Heating of muscle systems. In *The Physiology and Biochemistry of Muscle as a Food*, ed by Briskey, E. J., Cassens, R. G. and Trautmann, J. C., The University of Wisconsin Press.
14. Ruechel, R., Richter-Landsberg, C. and Neuhoff, V. (1975). Microelectrophoresis in continuous polyacrylamide gel gradients: IV. Effect of reducing agents on the electrophoresis of proteins in sodium dodecyl sulfate buffer systems. *Hoppe-Seyler's Z. Physiol. Chem.*, **356**, 1283.
15. Kako, Y. (1968). Studies on muscle proteins: II. Change of beef, pork and chicken protein during the meat products manufacturing processes. *Mem. Fac. Agric., Kagoshima University*, **6**, 1975.
16. Chrystall, B. B. (1971). Macroscopic, microscopic and physico-chemical studies of the influence of heating on muscle tissues and proteins. *Dissertation Abstr. Internat. Sect. B Sci. and Engng.*, **31**, 6050.
17. Hofmann, K. (1972). Identifizierung und Bestimmung von Eiweiss-stoffen mit Hilfe der Dodecylsulfat-Polyacrylamid-Gelelektrophose. Report at 'Kulmbacher Woche 1972', Bundesanstalt Kulmbach.
18. Woolford, G., Cassens, R. G., Greaser, M. L. and Sebranek, J. G. (1976). The fate of nitrite: Reaction with protein. *J. Food Sci.*, **41**, 585.
19. Hofmann, K., Hecht, H., Blüchel, E. and Egginger, R. (1973). Elektrophoretische Untersuchung der Hitzeschädigung von Muskeleiweiss. *Jahresbericht der Bundesanstalt für Fleischforschung*, Kulmbach, p. 68.
20. Penny, I. F. (1967). The effect of post-mortem conditions on the extractability and ATPase activity of myofibrillar proteins of rabbit muscle. *J. Food Technol.*, **2**, 325.

24

General Principles Involved in Measuring Specific Damage of Food Components During Thermal Processes

J. MAURON

*Nestlé Products Technical Assistance Co. Ltd, Research Department,
CH-1814 La Tour de Peilz, Switzerland*

INTRODUCTION

The general objective of food processing is to protect and preserve food, to destroy micro-organisms, to inactivate enzymes, to destroy inhibitors and toxic substances, to augment digestibility, to maintain or improve organoleptic properties and to produce more desirable physical properties and aesthetic characteristics. The attainment of these objectives is followed by the appropriate standard tests of food analysis, including acceptability by the consumer. However, pursuit of the primary objectives of processing often results in unwanted secondary effects, the so-called processing damage. As long as the latter reduces the acceptability of the food it is not very dangerous because the manufacturer will be compelled to eliminate it. The real danger lies in processing damage that reduces nutritive value without affecting consumer acceptance, perhaps even improving it. The detection of this type of damage is of increasing importance if we are to maintain a satisfactory standard of nutrition, since most of our food is now processed in one way or another. In principle, I see three possible approaches to the detection of processing damage to foodstuffs which leads to loss in nutritive value (Table 1).

The most obvious and classic method consists of determining the nutrient content before and after processing, and for a long time it has been the only approach. However, since you can only find out what you are looking for, this technique is not sufficient because it does not detect changes in nutrient availability, nor the possible formation of nutrient antagonists and of anti-nutritional or even toxic substances.

Another approach is to elucidate the mechanism of processing damage at the molecular level, studying structural modifications of the food

328

components that affect the availability of the various nutrients and the chemical reactions that lead to their deterioration.

The most important way of looking at processing damage is, however, from the biological point of view in which the actual biological consequences of processing are followed up in the living organism which eats that food. There are numerous interactions between these three

TABLE 1
Three approaches to determining processing damage

Food Analysis	Nutrient content before and after processing Special methods for *in vitro* availability
Study of Reaction Mechanisms	The browning reaction Severe heat treatment Alkali treatment Oxidation
Biological Evaluation	Animal bioassays for *in vivo* nutrient availability Nutritive value Classic toxicology Mutagenesis

approaches, as discussed below, leading constantly to improved and more relevant analytical techniques. Thus, present analytical methods for assessing processing damage contain many procedures that already result from the study of the precise chemical mechanisms involved or from the observed biological consequences.

CONVENTIONAL FOOD ANALYSIS

Nutrient Content
Conventional analyses of nutrients, such as carbohydrates, fats, amino acids, minerals, vitamins and trace elements and so on, before and after processing, will give basic information on nutrient losses due to the technical procedure envisaged. These conventional analytical methods are the standard techniques of food analysis and cannot be detailed here.[1] They essentially measure nutrient content. As a rule, the food must first be prepared for nutrient analysis. Thus, fats must be extracted, minerals concentrated by ashing, vitamins and amino acids liberated from complexes by acid or enzymic hydrolysis and so on. Although these

preparitive steps have been well standardised for different food materials, they can sometimes interfere with the detection of the processing damage. Indeed, processing may change the binding of a nutrient to macromolecules in the food and thus increase or decrease its availability to the mammalian organism without changing the nutrient content of the food because preparative steps such as acid hydrolysis completely obliterate the effect of processing, being themselves much stronger than the processing effect. For instance, the emphasis of the worker using microbiological assays for vitamin determination has been directed toward the development of extraction and hydrolysis procedures which tend to yield maximal figures for the material being assayed.[2] Certainly, autoclaving for 60 min at 15 psi and hydrolysing with 6N H_2SO_4 (recommended for biotin assays) have no counterpart in the animal organisms themselves. They measure total and not available vitamin content and are therefore unable to detect changes in vitamin availability due to processing. Enzyme hydrolysis is used instead of acid hydrolysis to liberate acid-labile vitamins from the food material, but once again the emphasis is on vitamin content rather than vitamin availability. Feeding trials with animals are the final criterion with which to test the biological potency of vitamins and the validity of the simpler, routine analytical tests should be checked periodically against the *in vivo* methods.

Processing may also transform a nutrient into another chemical form which is unavailable to the animal but still detectable by conventional analytical methods. A classic example is the transformation of lysine into a component that is completely unavailable *in vivo*[3] but still liberates 50 % of its lysine content during conventional acid hydrolysis. Obviously the usual analytical method for lysine, namely ion-exchange chromatography in the acid hydrolysate, grossly underestimates processing damage in this case. This is just one of the many examples to show that conventional nutrient analysis cannot be completely relied on for assessing processing damage; they must be checked and supplemented by other techniques that originate from a study of the reaction mechanisms and of the biological effects of processing. Some of these methods have already attained a satisfactory degree of accuracy.

Nutrient Availability

The availability of a given nutrient may be defined as the amount or percentage of that nutrient in the food which is actually utilised in the animal organism to fulfil its specific purpose, when it is the only limiting factor in the diet. Therefore, animal tests are obviously the reference points for a new method or a new material but they are too slow and expensive for a more or less routine quality control of a particular product or individual batches. For that we need shorter methods which can be carried through in fair numbers and, if possible, in a not too sophisticated laboratory. The development of relatively simple analytical tools to measure nutrient availability has been rather slow and many earlier hopes have faded away. Nevertheless, a few sound techniques have come forth and are still in the process of improvement.

For assessing vitamin availability, advantage can sometimes be taken of the fact that certain micro-organisms behave in a similar way to the animal as regards the utilisation of bound forms of the vitamin. For instance, in cereals nicotinic acid occurs mainly in a bound form that is nutritionally unavailable to certain bacteria such as *Lactobacillus casei* and to mammals, such as pigs and rats. Therefore, if we succeed in liberating the bound form from the food by a mild method that does not destroy the complex we can use the microbiological assay as an availability test. This has been done by Clegg[4] to study the availability of nicotinic acid in processed wheaten products. A weak acid extract of the test material was prepared containing both free and bound nicotinic acid, but the microbiological analysis measured only the free form which is available. A measured portion of the acid extract was treated with N NaOH to liberate the bound nicotinic acid, and assay of this solution gave the total nicotinic acid content of the test material. The difference between the two analyses was a measure of the bound nicotinic acid. Using this method the author was able to show that in baking wholemeal scones with increasing amounts of bicarbonate the percentage of available nicotinic acid increased with augmenting pH and heat treatment (Fig. 1) whereas total nicotinic acid remained unchanged. This is a good example of the use of a microbiological assay to determine vitamin availability.

Unfortunately, this principle cannot be applied generally as certain vitamins, such as B_6, for instance, are available to the animal also in the bound form, while the latter is unavailable to the micro-organism.[5] Since the liberation of vitamin B_6 from the multiple substances of a complex nature encountered in food is seldom completely successful, we are faced here with an interesting situation that vitamin B_6 content as measured by the conventional method may in fact underestimate vitamin B_6 availability for man. This should suffice to show that the determination of vitamin content, not to say availability, has become almost a 'science of its own' for each single vitamin, and can therefore only be handled properly by the specialist in the matter.

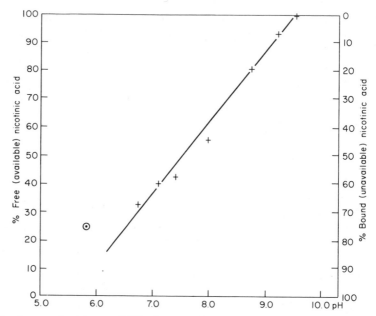

FIG. 1. Increased nutrient availability by processing. Baking wholemeal scones (×) at 265 °C for 10 min with NaHCO₃. Wholemeal flour (⊙).

At the present time a great deal of work is being carried out on amino acid availability. Originally, a somewhat similar approach was taken as with some vitamins, in the sense that microbiological assay for the amino acids was preceded by *in vitro* digestion with pepsin, trypsin, etc. The values obtained were usually lower than the corresponding availabilities for animals. Later, when chemical analysis of amino acids began to be used, more elaborate methods of *in vitro* digestion and subsequent dialysis were employed. When corrected for the necessarily incomplete digestion, the results showed good agreement with feeding tests in many foods such as, for instance, heated milk.[6] The technique is however too cumbersome to provide an economical means of quality control in practice.

An alternative is to carry out the microbiological assay with a micro-organism such as *Streptococcus zymogenes*, which is itself proteolytic.[7] The availability of several amino acids (Met, Iso, Lys, Leu, Val, Larg, His) could be determined in this way but most work has been done with methionine. Good correlations with corresponding animal assays on processed materials were found.[8] Although the method shows some weaknesses inherent in most microbiological assays, such as the need for maintaining cultural conditions so that absolute values remain constant from one assay to another, it can be recommended very much for determining available methionine in foodstuffs.

Let us now turn to chemical methods of measuring amino acid availability. The development of such techniques presupposes that at least something is known about the chemical mechanism of the damage. Since it has been found that all forms of thermal damage to proteins involve lysine side-chains, it is natural that measurement of the residual ε-amino groups has been the most studied approach for determining damage, the basic assumption being that, to be biologically available, lysine must have a free ε-amino group. Although this looks quite a straightforward proposition, the various methods developed to do just this have often surprisingly failed, giving significant values for lysine units involved in Maillard reactions.[9]

At present, two chemical methods appear to be theoretically sound and generally applicable to determine lysine availability in different foods: Carpenter's widely used reaction with FDNB (fluoro-dinitro-benzene)[10] and the guanidation reaction.[11] FDNB reacts with a free amino group of amino acid units in proteins and gives the corresponding dinitrophenyl (DNP) amino acid after acid hydrolysis. Yellow ε-DNP-lysine is then separated from the other DNP-amino acids and possible interfering compounds and measured colorimetically. The main technical problem in using the method is that with foods rich in carbohydrates a proportion of the DNP-lysine is destroyed during acid hydrolysis.[12] It has been shown in heated milk powder that 'FDNB-reactive lysine' and 'nutritionally available lysine' corresponded closely.[13] The agreement with enzymic digestion *in vitro* was equally good (Table 2).

Guanidation was proposed in 1964 as another general chemical method for measuring available lysine in processed food. It is based on the reaction with O-methylisourea. This transforms lysine molecules with a free ε-amino group into homoarginine, which is stable during acid hydrolysis. The guanidated sample is freed from the reagent by ultrafiltration and hydrolysis with 6N HCl. The homoarginine appears after arginine on the chromatogram. This method has been shown to be the most correct one with which to determine available lysine in all kinds of processed foods, since it gives 'true' values for all types of heat damage, as shown by Finot and Mauron[14] and Hurell and Carpenter.[15] The difficulty with this technique is its relative slowness (min–24 hr) and the slight incompleteness of the reaction (with some material only about 95 % of the available lysine reacts). These are, however, relatively minor drawbacks when one considers that it can be applied without correcting factors even in foodstuffs that are

TABLE 2
Results of analyses or assays for lysine (g/16gN) in cow's milk concentrated or dried in various ways[13]

Method of preparation	Total after acid hydrolysis	FDNB-reactive[a]	Digestion *in vitro* and dialysis	Rat growth assay
Freeze-dried	8·3	8·4	(8·3)[b]	8·4
Spray-dried	8·0	8·2	8·3	8·1
Evaporated	7·6	6·4	6·2	6·1
Roller-dried 2[c]	7·1	4·6	5·4	5·9
Roller-dried 4	6·8	3·8	4·5	4·0
Roller-dried 6	6·3	2·5	3·1	2·9
Roller-dried 8	6·1	1·9	2·3	2·0

[a] Determined by the Carpenter[10] procedure.
[b] The actual values were all lower than given here because of the limitations of the procedure. The values listed were calculated relative to the freeze-dried sample, for which the true digestibility was assumed to be 100%.
[c] The code numbers refer to samples prepared under increasingly severe conditions.

extremely rich in carbohydrates. The results we have obtained so far in our laboratory are promising (Fig. 2 and Table 3).

In our opinion the most efficient way to obtain accurate and rapid information on processing damage in protein foods using chemical methods only, is to prepare two hydrolysates, one of the food as such, the other of a guanidated sample and to run the corresponding chromatogram, one complete run and one of the basic amino acids only. The former measures amino acid content and gives some clues as to the type of

TABLE 3
Available lysine values by different methods g/100 g protein

Material	*In vitro* digestion	Carpenter[10]	Guanidation[11]
Lyophilised milk	8·3	8·4	8·2
Spray-dried milk	8·3	8·2	8·0
Roller-dried milk, 1	3·1	2·6	3·1
Roller-dried milk, 2	2·6	2·3	2·3
Beef meat	9·3	7·5	8·5
Zebu meat	8·5	7·9	7·6
Soya flour	6·3	6·2	6·0
Peanut flour	3·2	2·6	2·6
Sesame flour	—	2·4	2·2
Whole egg	6·7	—	6·8

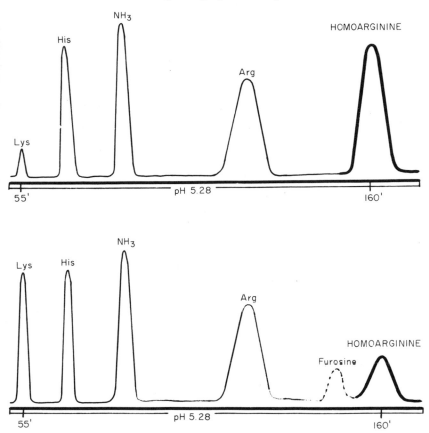

FIG. 2. Chromatography of guanidated, spray-dried (top) and roller-dried (bottom) milk.

processing damage (the so-called 'hot spots' we will consider below), the latter determines true lysine availability.

The disadvantage of all sound methods developed so far to determine nutrient availability is that they are relatively sophisticated and give reproducible and accurate results only in the hands of the specialist. There is a great need for a simpler, routine method that could be used for the quality control of individual batches of the same food. In this connection dye-binding methods and pepsin solubility tests as measures of heat damage to proteins should be mentioned. Their use is generally limited to the quality control for a single manufacturing plant because they lack general validity. For the application of these tests we refer to the specialised literature.[16]

THE MECHANISMS OF PROCESSING DAMAGE

The purely pragmatic analytical approach is obviously not sufficient since one wishes to anticipate the possible deleterious consequences of processing in order to prevent them. In order to be able to do this we need to have precise knowledge of the chemical reactions involved at the molecular level. Because of the great complexity of foodstuffs, progress in this field has been rather slow.

Fat

It is not fortuitous that the mechanism of fat deterioration has been most widely studied because fat stability plays a predominant role in food acceptance. Rancidity, namely the development of objectionable flavour and odour, may be caused by either hydrolytic or oxidative changes in fat. Hydrolytic rancidity involves chemical or enzymic hydrolysis of fats into free fatty acids and glycerol. Oxidative rancidity involves the addition of atmospheric oxygen in the presence of enzymes or certain catalysts. It is this second mechanism which leads to multiple losses in nutrient value. In addition to the loss in essential fatty acids, vitamins and proteins can be reduced in many ways by the secondary reactions of lipid oxidation. Figure 3 provides a simplified scheme of lipid oxidation.[17] We see that this involves

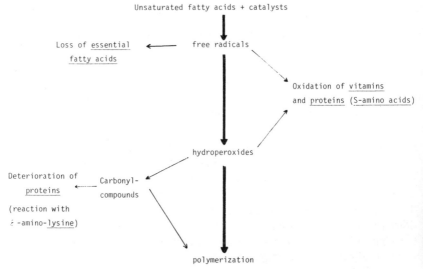

FIG.3.   Loss of nutrients by lipid oxidation.

a reaction between unsaturated fatty acids and catalysts to form free radicals; the latter combine with oxygen to give hydroperoxides that react in different ways. They can decompose or give rise to carbonyl compounds or polymerise. The free radicals and the hydroperoxides can interact with pigments, flavours, vitamins and proteins to give a variety of different products, and to inactivate nutrients by means of the oxidation of vitamins and sulphur-containing amino acids. The carbonyl compounds, on the other hand, can participate in non-enzymatic browning reactions, leading to the loss of protein value through inactivation of the ε-amino group of lysine.

Protein[18]

In the case of protein the situation is almost the opposite to that found with fats, in the sense that processing normally increases flavour and taste (browning reaction) whilst the nutritive value is often reduced.

We have already touched on two mechanisms that may deteriorate protein, namely oxidation and the browning reaction. Many other reactions occur during protein-food processing and cannot be all enumerated here. In order to simplify the matter we propose to distinguish, formally, four distinct types of processing damage to proteins (Table 4).[19]

(a) Mild reaction with carbonyl groups: this first type can be produced under relatively mild heating conditions in food that is rich in reducing sugars, such as milk. Its mechanism corresponds to the classic Maillard or non-enzymatic browning reaction. Lysine is almost exclusively made unavailable in this case.

(b) Severe conditions: the second kind occurs under much more severe heating conditions either in the presence or absence of sugars and leads to a fall in protein digestibility and in the availability of most amino acids in addition to that of lysine.

(c) Alkali treatment: the third type is prominent when protein is exposed to alkali treatment and results in a loss in cystine and lysine, in the first place, with the formation of lysinoalanine (LAL).

(d) Oxidation: the fourth kind comprises several oxidation reactions to which methionine and cystine are most sensitive.

We may now rapidly discuss these four types of processing damage.

The Maillard reaction has already been thoroughly described by Carpenter elsewhere in this volume. A simplified scheme of the reaction is given in Fig. 4.[20] We have been especially interested in the browning reaction as it occurs during milk processing. We found that in heated and

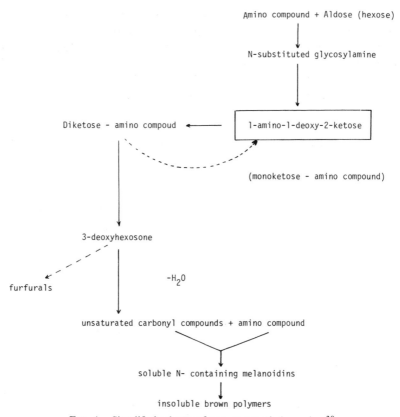

FIG. 4.   Simplified scheme of non-enzymatic browning.[20]

even severely over-heated milk most, if not all, of the deteriorated lysine is present in the form of the Amadori rearrangement product, namely the 1-amino-1-deoxy-2-ketose (deoxy-lactulosyl lysine). This corresponds to the first relatively stable intermediate compound of the Maillard reaction. We synthesised[21] the corresponding compound by reacting α-N-protected lysine with glucose and investigated the properties of the pure ε-N-deoxyfructosyl lysine (ε-DFL). On acid hydrolysis it yields furosine (20%) and pyridosine (10%). These two new amino acid derivatives are easily detectable in the chromatogram of the basic amino acids. Furosine appears after arginine in the ordinary run, whereas pyridosine is separated ahead of lysine.[22] The quantities and proportions of furosine and pyridosine found in an acid hydrolysate of a given food are constant (Fig. 5) and are proportional to the amount of lysine bound in the food as 1-amino-1-deoxy-2-ketose, but the recovery is different from that with free ε-DLF. Whenever furosine is found in a chromatogram it indicates that part of the lysine appearing on the same chromatogram is unavailable in the food and that, therefore, processing damage has occurred that needs further investigation. In the case of processed milk, furosine content can also be

TABLE 4
The main types of processing damage to protein

(*a*) Mild heat treatment + reducing sugar (carbonyl-groups); early Maillard

(*b*) Severe heat treatment
    Without sugar
    With sugar; late Maillard
      Reducing
      Non-reducing
    With oxidised lipids (carbonyl groups)

(*c*) Alkali treatment
    Carbonates
    Ammonia
    Strong bases

(*d*) Oxidations
    Molecular O_2
    Hydroperoxides ⎫
    Free radicals   ⎬ Lipid oxidation
    H_2O_2
    Photo-oxidation

used for a quantitative determination of the amount of lysine made unavailable during processing, as shown recently by Finot and Bujard.[23] The only thing that is needed is the chromatogram of the basic amino acids. Lysine deterioration can then be calculated from the lysine (x) and furosine (y) content according to the formula:[23]

$$\% \text{ lysine deteriorated (or blocked)} = \frac{3 \cdot 1 y \times 100}{x + 2y}$$

Another possibility for detecting lysine deterioration in a reaction of this first type is to stabilise the carbonyl-amino derivatives of lysine by hydrogenation and determine lysine afterwards in an acid hydrolysate. In this case lysine content corresponds to available lysine value.[15] However, this hydrogenation procedure is not valid for the detection of heat damage of type (*b*), especially when pure proteins are heated. These two examples

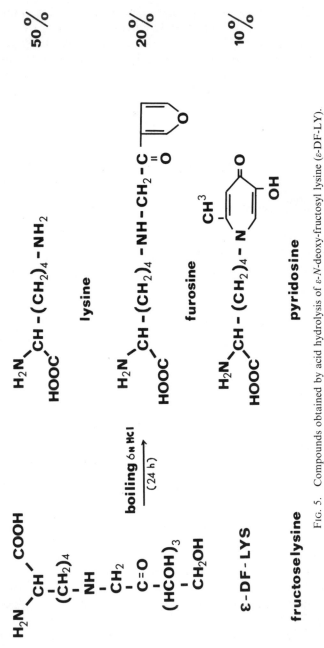

Fig. 5. Compounds obtained by acid hydrolysis of ε-N-deoxy-fructosyl lysine (ε-DF-LY).

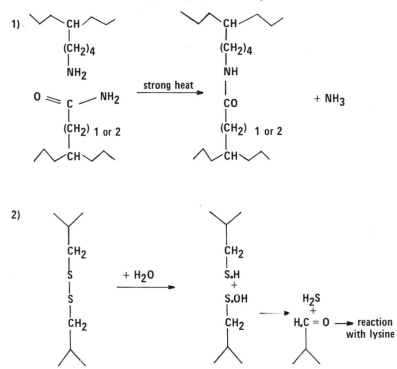

FIG. 6.   Reactions of pure proteins that have been severely heated.

show that a study of the reaction mechanism can lead to new analytical tools.

Two mechanisms are involved in heat damage of type (*b*), one when the protein is heated alone or with very little sugar, the other when the protein is intensely heated in the presence of a significant, but still relatively limited, amount of sugar. The former mechanism has been elucidated by Bjarnason and Carpenter.[24] It involves the formation of internal peptide links between the ε-amino group of lysine and the amide group of glutamic and aspartic acids and probably also with decomposed cystine residues (Fig. 6). These peptide links are split by acid hydrolysis, therefore leaving no trace of the amino acids in the chromatogram. Only the destruction of cystine remains as a sign of the extensive damage. The FDNB procedure, however, will show the decrease of FDNB-reactive lysine.

Severe heat treatment in the presence of some sugar in the system forms profuse cross-linkages between peptide chains of the protein through sugar degradation products and as a result indigestible peptides appear in the

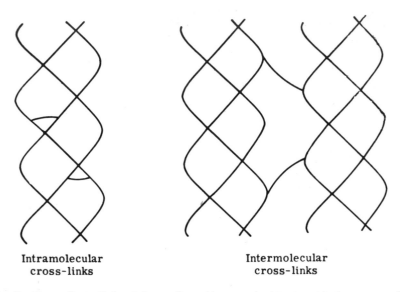

Intramolecular   Intermolecular
cross-links   cross-links

Fig. 7.   Types of cross-linkages that are formed by severe heat treatment in the presence of a sugar.

formation of dehydroalanine

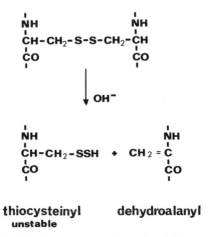

thiocysteinyl   dehydroalanyl
unstable

Fig. 8.   Alkaline treatment of proteins: initial reaction.

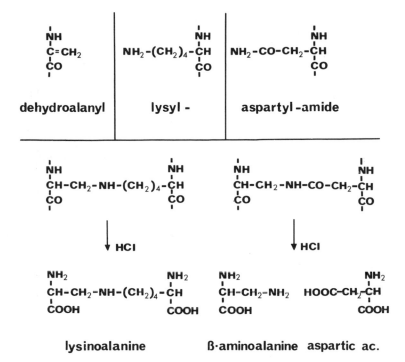

FIG. 9. Alkaline treatment of proteins: cross-linkage reactions.

faeces so that amino acids other than just lysine are lost.[25] On acid hydrolysis these cross-links are broken and most amino acids recovered, so that conventional amino acid analysis barely takes notice of this kind of damage, although it might be very important (Fig. 7).

Processing damage of the type (*c*), namely by alkali treatment is, on the contrary, easily detected by conventional amino acid analysis. Indeed, lysinoalanine (LAL), the main compound formed, comes off the column just before lysine and there is a concomitant drop in cystine and, to a lesser extent, in lysine content.[26,27] Another amino acid, ornithino-alanine, that is formed under even milder conditions (boiling with carbonate), emerges just before LAL. Yet another compound formed with ammonia or NaOH is β-aminoalanine.[29] The probable mechanism which leads to the presence of these amino acids in the acid hydrolysates of alkali-treated protein is represented in the Figs. 8 and 9. The reaction appears to start with a cleavage of cystine or serine by a β-elimination mechanism to form dehydroalanine and thiocysteine. The latter is unstable and

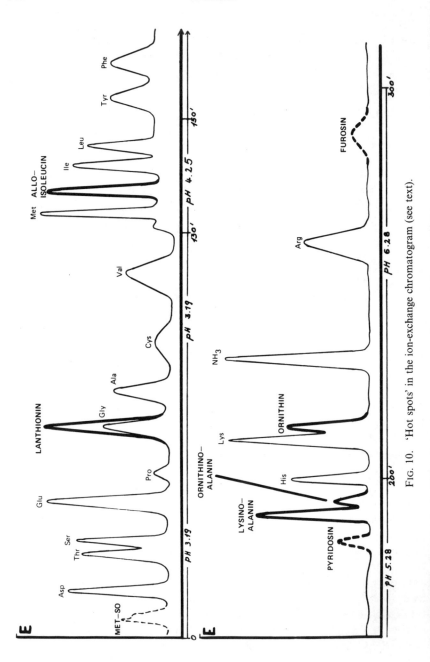

Fig. 10. 'Hot spots' in the ion-exchange chromatogram (see text).

decomposes to other sulphur-containing derivatives.[30] Dehydroalanine reacts with lysyl, ornithyl or amide side-chains of the protein to form new cross-links and on acid hydrolysis LAL, ornithinoalanine and β-aminoalanine are recovered. Although extensive damage of this type only occurs with alkali treatment of the protein, traces of LAL can be found in processed food that never sustained any alkali treatment (Table 5).[31] The mechanism of its formation under these conditions is not yet elucidated. Loss of amino acid availability is not measurable in these foodstuffs.

The final type of damage, by direct oxidation of the protein, has already been mentioned when fat oxidation was discussed. Other oxidative agents may be involved and under special laboratory conditions many different reactions could be detected. However, in normal food processing the only important damage derives from the loss of cystine and the oxidation of methionine to methionine sulphoxide.[17] The latter, being partially unstable on acid hydrolysis, has to be detected in an alkaline hydrolysate.[32] Under proper conditions it appears just ahead of aspartic acid.

In conclusion, we have plotted in Fig.10 the 'hot spots' in the ion-exchange chromatogram of an acid hydrolysate of processed protein food,

TABLE 5

LAL in home-cooked foods, food ingredients and commercial food preparations

Food	Cooking conditions	LAL (μg/g protein)
Egg-white	Fresh	none
Egg-white	Boiled 3 min	140
Egg-white	Boiled 10 min	270
Egg-white	Boiled 30 min	370
Egg-white	Pan-fried 10 min at 150°C	350
Egg-white	Pan-fried 30 min at 150°C	1 100
Milk, infant formula	Commercial sample	330
Milk, evaporated	Commercial sample	590
Milk, condensed		360
Sodium caseinate		6 900
Sodium caseinate		1 000
Whipping agent 1		6 500
Whipping agent 2		50 000
Soya protein isolate		0–370
Ovalbumin	1 h, 120°C, pH 2·0	150
Ovalbumin	1 h, 120°C, pH 2·0	250
Ovalbumin	1 h, 120°C, pH 4·6	310
Ovalbumin	1 h, 120°C, pH 6·0	570

indicating the 'markers' for the different types of processing damage. Whenever an unusual peak shows up in the vicinity of the indicated position, processing damage can be suspected and a more thorough investigation should be performed.

THE BIOLOGICAL APPROACH

Nutritive Value

The final answer as to the nutritional consequences of processing damage to food can obviously only be given by *in vivo* experiments. In this area, as in the analytical field, the techniques for evaluation of the potency of the various nutrients in the food involve the standard methods, comprising the measure of growth response; PER; balance-sheet studies; evaluation of tissue saturation including enzymic activity; excretion of nutrients; metabolic studies etc. This huge field cannot be covered in this paper. Only a few remarks will be made as regards the growth response of the animal.

The classic bioassay procedure which measures the animal growth response as a function of graded amount of the missing nutrient in an otherwise complete diet is theoretically the soundest method for determining the biological availability of essential nutrients such as vitamins and indispensable amino acids. (Figure 11 shows the rat growth response for lysine.) However, to measure every essential nutrient in a processed food in this way would be an overwhelming task and quite unrealistic. Generally, preliminary analytical work and some knowledge about the heat damage involved will guide the investigator so that only certain groups of nutrients will have to be tested for. Thus, for instance, one will devise an assay for protein quality instead of testing for every single essential amino acid.

In many cases a simple growth test can give valuable information when the comparison is made with the material before and after processing. The material will have to be incorporated into an appropriate standard diet whose composition depends on the group of nutrient one is looking for. If one is testing for protein quality, the standard diet will provide all essential nutrients, except protein. Such an assay can already give useful indications. If no change in growth response is observed with the material before and after processing, one can conclude that the limiting nutrient has not been affected and that no other nutrient has been made unavailable to the point of becoming growth-limiting. Such a result also shows that no overt deleterious compound has been formed.

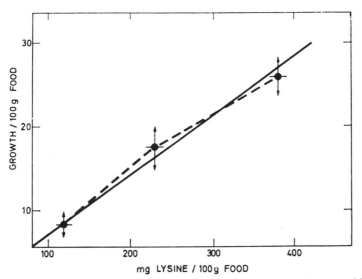

FIG. 11.   Rat growth response for lysine. Average standard curve from 18 assays with basal diet and two lysine supplements: linear extrapolation and curvilinear approximation.

If growth is increased after processing, it will indicate that, very probably, digestibility has been augmented. On the contrary, growth reduction would denote, either that nutrients have been damaged or made unavailable or that noxious compounds have been formed or that digestibility has been reduced. Adding now nutrients, single or in combination, to the diet and observing the effect on growth will give a clue as regards the nutrients that have been damaged. If these various additions have no effect it must be concluded that digestibility or general nutritional availability of the material has been reduced or that toxic compounds have been formed.

As an example of the usefulness of a very simple growth test, Fig. 12 shows its application to detect protein damage of type (a) in heated milk.[33] Spray-dried and scorched roller-dried milk were fed alternatively to rats with lysine and methionine supplementation. During the first period (3 weeks) the rats grew well with the spray-dried milk but very little with the roller-dried powder. Methionine had no effect on the roller-dried milk, showing that it is not damaged in the first place; however, it increased growth with spray-dried milk since the S-amino acids are limiting in milk. In the second period (3 weeks) lysine-supplemented roller-dried powder gave the same growth response as the spray-dried milk before, and an effect of methionine could now be seen. The animals, after switching from spray-dried to roller-dried powder in the second period, reduced their growth immediately and

J. Mauron

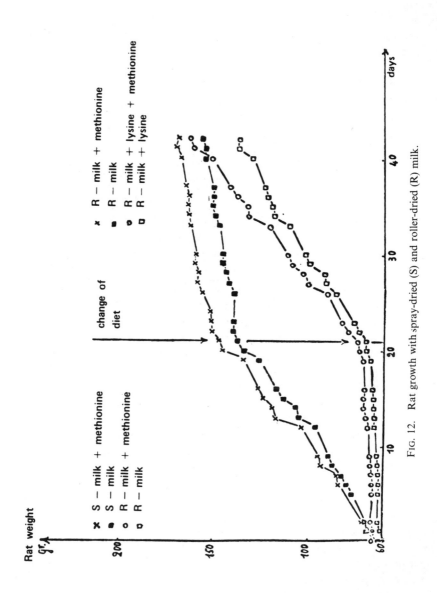

FIG. 12. Rat growth with spray-dried (S) and roller-dried (R) milk.

methionine remained without effect. This simple growth test shows that in the scorched roller-dried milk lysine is damaged in the first place and that no other essential amino acid is involved, with the possible exception of the S-amino acids. Nothing can be said about possible slight damage to the latter because it would be completely masked by the overwhelming damage to lysine and demonstrates one of the limitations of the procedure. Another conclusion can be drawn from the assay, namely, that over-heated milk does not contain any antinutritional or overtly toxic substance.

Let us now make the assumption that growth could not be fully restored by the addition of amino acids, either singly or in combination. This is actually the case with heat damage of type (*b*). Nitrogen balance-sheet tests may help in this situation since they will disclose the presence of indigestible peptides in the faeces, as shown by Ford[34] for instance. If this test is also negative a 'toxic' effect might be suspected. A simple way to test the latter hypothesis is to assay the suspected material as supplement to a well-balanced diet. If no growth depression is observed an acute toxic effect is very unlikely. In this case the original observation might be due to the intestinal absorption of amino acid derivatives which are not metabolised and whose quantities in the urine might be too small to be detected in the nitrogen balance-sheet. At this point, more elaborate metabolic studies must be performed to look for the excretion of unusual metabolites, in urine, for instance.

Let us now briefly survey, as an example, the nutritional consequences of the various types of protein food processing. We have already seen that the main lysine derivative formed in heat damage of the first type is the ε-N-deoxy-ketosyl lysine. It was therefore logical to attribute the growth failure with over-heated milk to the biological unavailability of this compound. This is, indeed, the case as we could show with a growth bioassay on rats using pure ε-N-deoxyfructosyl lysine (fructose-lysine).[3] The latter was found to be partly absorbed from the gut and excreted in the urine. Protein-bound fructose-lysine, however, was mainly retained in the intestine where it was partly degraded by micro-organisms (Fig.13).[35] We could also show that fructose-lysine, added to a complete diet, had no growth-depressing effect[36] which demonstrates that it has no antinutritional or 'toxic' properties. We can now recapitulate the steps in the elucidation of the heat damage in this example:

(*a*) Detection of heat damage by growth failure in rats.
(*b*) Identification of the nutrient involved by amino acid supplementation.

(c) Elucidation of the mechanism of the damage to the nutrient by chemical investigation.
(d) Synthesis of the damaged nutrient.
(e) Bioassay in rats with the damaged nutrient in a diet deprived of the nutrient.
(f) Metabolic studies.
(g) Growth assay with the damaged nutrient in a complete diet.

Such a scheme cannot be generalised but it shows that, even in a relatively straightforward case, the work involved is quite considerable.

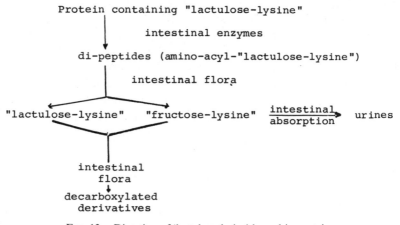

FIG. 13.   Digestion of 'lactulose–lysine' bound in proteins.

The biological consequences of severe heat treatment [type (b)] are less well elucidated. In a general way it can be stated that the loss of nutritive value can be ascribed almost entirely to a fall in the digestibility of the protein and to a further, rather uniform, reduction in the availability of all the amino acids. Actual amino acid destruction plays an insignificant role, except for cystine. We have seen that during intense heating new cross-linkages involving lysine are formed within the protein, either direct to dicarboxylic amino acid units, or through 'bridges'. It is supposed that these cross-links hinder the access of peptidases to the several adjacent peptide bounds leading to biological inactivation of the amino acids contained in that fragment. The fact that lysine in ε-(γ-L-glutamyl)-L-lysine is fully available to the rat;[3,37,38] does not impair this hypothesis of steric hindrance, because Finot *et al.*[39] have shown that the ε-peptide linkage is not broken by digestive enzymes in the gut but only in the kidney.

As regards the nutritional consequences of alkali treatment they are rather clearcut since it leads to the destruction of cystine and lysine. Only after severe treatment is the general amino acid availability also reduced through the formation of cross-links.

The traces of LAL formed by heat treatment alone do not result in measurable losses in amino acids.

The methionine sulphoxide formed in protein oxidation is partly available to the organism. In spite of some conflicting results it can be stated that free methionine sulphoxide is a rather satisfactory methionine source, whereas peptide-bound sulphoxide is considered less effective than methionine but still to a considerable extent after slow *in vivo* reduction to methionine.[32]

Safety Aspects

Some deleterious effects observed on feeding severely heat-processed foods may not be due to reduction in nutrient availability but to the formation of antinutritional and even toxic substances.

Frying oils have been extensively studied in this respect. For extensive reviews see those of Lang[40] and Potteau.[41] So-called over-heated or 'abused fats' which have been experimentally over-heated in the presence of air or oxygen (intensive bubbling) impair the health of experimental animals and the harm is proportional to the degree of over-heating.[40] Because of this finding some doubts about the wholesomeness of normally heated frying fats were cast, and extensive lifespan studies in rats with used frying fats were performed. In contrast to the results obtained with the experimentally over-heated fats, these chronic studies revealed no undesirable effects of any kind, as long as the fats employed corresponded to those obtained by so-called 'good manufacturing practice'.

This example clearly demonstrates a particular aspect of the toxicology of processed food. Since overt toxic effects are not to be expected the investigator has essentially two means at hand to observe what is nevertheless a phenomenon, either by exaggerating experimentally the processing damage or by performing a chronic toxicity test with the normally processed food fed in relatively high amounts. The first approach is much more rapid and can give interesting information but, strictly speaking, the results cannot be interpreted in terms of the properly processed food in either sense, positive or negative. The second approach is extremely time-consuming and expensive but it is the only relevant way of detecting the possible harmful effects of a processed food (Fig 14). The dilemma we are faced with is, therefore, that we cannot possibly test chronically in animals all processed material we would like to assay, so that we are still left with a choice to be made on the basis of short-term experiments. At present, I see no real solution to this dilemma.

An example to show our relative ignorance as regards the possible toxic effects of the most common processed food is again the browning reaction

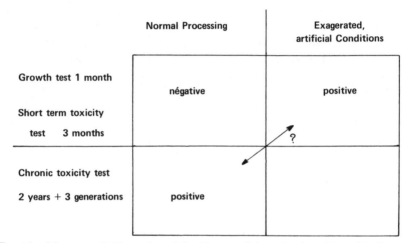

FIG. 14.   Diagrammatic illustration of the dilemma of the evaluation of harmful effects of processing.

that is performed daily at home and in industry. Thus, Adrian[12] has pointed out that the poor performance of animals given heat-damaged materials could be due, in part, to Maillard compounds having antinutritional or even toxic characteristics. This is based on his observation that the so-called 'pre-melanoidins' namely the soluble brown fraction obtained by autoclaving glycine with glucose, when added to the diet of rats could either stimulate appetite and growth or at higher levels depress them. He also observed that pre-melanoidins added in high amounts (20% of the protein) to a 10% casein diet reduced digestibility and N-retention. More important still, the pre-melanoidins (2 g N/kg) reduced the number of off-spring and created a state comparable to that of malnutrition. Chichester,[43] using a browned, synthetic amino acid diet found that the polysome pattern in the liver of the animals fed the mixture became disassembled, even when the mixture had been supplemented with all amino acids.

What are the consequences of these observations in respect to the heat damage to protein? They would tend to indicate that these are factors beyond those of merely reducing the availability of amino acids which have not been investigated. However, it should be borne in mind that all these observations were made with browning mixtures of free amino acids and glucose and not with proteins. It is doubtful, therefore, whether they have much relevance to the problem of heating proteins because the Maillard reaction products of a free amino acid and of a protein behave very differently in the digestive tract, as Finot could show using fructose-lysine.[35] We are currently performing assays with protein-bound pre-melanoidins in order to see whether the findings of Adrian and Chichester have any relevance in respect to heat-damaged proteins. Certainly this is a field which needs a lot of additional investigation.

There has been considerable controversy as regards possible toxic effects of protein damage of type (*c*) through the formation of LAL. Indeed, free LAL had been found to give cytomegaly of the proximal tubule of the kidney in the rat.[44] Protein-bound LAL was also found to be nephrotoxic in their study but not by De Groot and Slump.[45] Subsequent detailed studies by De Groot *et al.*[46] confirmed the kidney changes in rats fed free lysino-alanine but not in rats fed alkali-treated protein containing high levels of bound lysino-alanine. Negative results obtained in six other species suggest that LAL-induced cytomegaly is a phenomenon specific to the rat species.

In this context I would like to mention here some histopathological observations made in our laboratory in Orbe with sources of proteins that had never sustained any alkaline treatment. Three products were tested: a soya milk, a crude single-cell protein (SCP) from yeast and unpurified bacterial SCP. These materials were incorporated into a laboratory chow replacing 15, 25 and 50 % of the original protein. A control in which 20 % of the protein was replaced with casein was also run. Bi- and multi-nucleated epithelial cells were found in the tubuli of the corticomedullari junction and outer medulla of the kidney. Bi-nucleated cells, not present with laboratory chow, appeared with casein, increased with soya and were definitely stimulated with SCP (Figs. 15, 16). Multi-nucleated cells appeared with SCP only, although still to a moderate degree.[52] Although these histopathological kidney changes are probably strain-related, the progressive increase, first in bi-nucleated, then in multi-nucleated cells when one switches from laboratory chow to casein, to soya, to yeast and bacteria indicates that these diets progressively increase and perturb mitotic activity. The degree attained even with the last diet is still moderate and not alarming but

J. Mauron

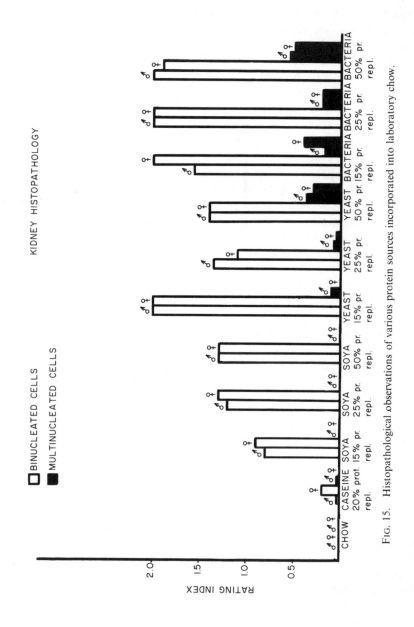

FIG. 15. Histopathological observations of various protein sources incorporated into laboratory chow.

cannot, on the other hand, be neglected. The tubular epithelium of the medulla of the kidney is a very sensitive target organ that appears to be extremely useful to histopathologists for detecting potentially harmful effects of food ingredients at a very low concentration.

This is just an example of how histopathology can be a useful tool in nutrition research and should not be left exclusively in the hands of the classic toxicologists who are used to working with pure chemical

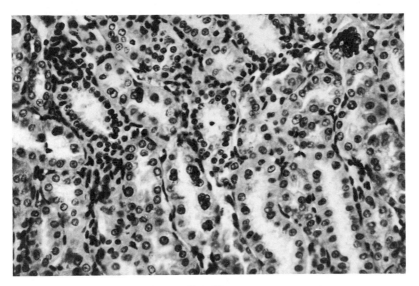

FIG. 16.

substances. Food technologists and nutritionists are generally somewhat afraid of toxicology and thus miss a lot of valuable information that the morphological sciences can give.

Finally, we cannot neglect any longer some newer aspects of toxicology, which are closely connected with the deterioration of our environment. It has been recognised that chemical mutagens are relatively widespread in our industrialised society and the whole field of testing for mutagens is therefore in rapid development. The need for simple, routine procedures was felt and such a test was actually developed by Ames.[47] It makes use of the special strains of histidine-less *Salmonella* mutants and determines the frequency of reversions to the wild type as a measure of mutagenicity. It gives positive results with most known mutagens and is considered suitable to test pure chemical substances and food additives. It has been further improved by pre-incubation with mammalian microsomes. According to our limited experience the Ames test is not well adapted to complex foodstuffs, because the presence of histidine and other growth factors or inhibitors leads to false positive responses. Another reason why we would not like to advocate this simple bacterial test for food is that its relevance to the mammalian organism is still a matter of controversy.[48] We may mention that we found pure ε-DLF not to be mutagenic in this test.

Chemical mutagenesis in mammals and man is now a field of intense research activity that cannot be discussed here.[49] The methodology for detecting mutations in mammals is generally very time-consuming and there is a need for a high degree of expertise. A method we can recommend that is relatively simple, is the dominant lethal test in mice or rats.[50] The material to be tested is fed to the males only, who are then mated weekly to females for 8 weeks. This experimental design allows the identification of the various stages of the sperm cells. The induction of dominant lethal mutations leads to the death of the fertilised eggs before or after implantation. The number of living embryos per female in the test group compared to the number of living embryos per female in the control group gives the percentage of dominant lethal mutations.

At this point one may wonder, of course, what these mutagenesis tests have to do with food processing. Well, in general probably very little, but the possibility can never be totally excluded that under certain processing conditions and in the presence of certain ingredients harmless food components may become mutagenic substances. Therefore, whenever there are well-founded indications in this direction from analogous situations, it is recommended that mutagenesis tests are included in the evaluation of heat damage to food.

I have tried to give you a broad view of the complex nature of processing damage to food and to describe some of the methodology involved in its evaluation. I am afraid that I was not able to delineate very well the great principles because the field appears to be in constant evolution. Certainly, there is a tendency to include more and more biological tests of increasing complexity in this field, showing also that in food science Weinberg's era of 'Big Biology' has arrived.

REFERENCES

1. Pomeranz, Y. and Meloan, C. E. (1971). *Food Analysis*, Avi Publ. Co., Westport, Conn.

2. Freed, M., Assoc. Vitamin Chemists (1966). *Methods of Vitamin Assay*, Interscience, New York.
3. Mauron, J. (1970). The chemical behaviour of proteins during food preparation and its biological impact (French). *Internat. Z. für Vitaminforschg.*, **40**, 209.
4. Clegg, M. K. (1963). Bound nicotinic acid in dietary wheaten products. *Brit. J. Nutr.*, **17**, 325–9.
5. Storwick, C. A. and Peters, J. M. (1964). Methods for the determination of vitamin B in biological materials. In *Vitamins and Hormones*, **22**, 833–54.
6. Mauron, J. (1970). Nutritional evaluation of proteins by enzymatic methods. In *Evaluation of Novel Protein Products*, ed. by A. E. Bender, R. Kihlberg, B. Lofquist and L. Munck, Pergamon Press, Oxford, England.
7. Ford, J. E. (1962). A microbiological method for assessing the nutritional value of proteins: 2. The measurement of available methionine, leucine, isoleucine, arginine, histidine, tryptophan and valine. *Brit. J. Nutr.*, **16**, 409–25.
8. Miller, E. L., Carpenter, K. J. and Milner, C. K. (1965). Availability of sulphur amino acids in protein foods: 3. Chemical and nutritional changes in heated cod muscle. *Brit. J. Nutr.*, **19**, 547–64.
9. Hurrell, R. M. and Carpenter, K. J. (1973). Contrasting results for the reactive lysine content of heat-damaged materials. *Proc. Nutr. Soc.*, **32**, 54A.
10. Carpenter, K. J. (1960). Estimation of available lysine in foods. *Biochem. J.*, **77**, 604–10.
11. Mauron, J. and Bujard, E. (1964). Guanidation, an alternative approach to the determination of available lysine in foods. In *Nutrition, Proc. 6th Intern. Congress*, ed. by C. F. Mills and R. Passmore, Livingstone, Edinburgh.
12. Handwerck, V., Bujard, E. and Mauron, J. (1966). The reduction of aromatic nitro groups during acid hydrolysis in the presence of carbohydrates. *Helv. Chim. Acta*, **49**, 419–23.
13. Mottu, F. and Mauron, J. (1967). The differential determination of lysine in heated milk: II, Comparison of the *in vitro* methods with the biological evaluation. *J. Sci. Food Agric.*, **18**, 57–62.
14. Finot, P. A. and Mauron, J. (1972). The inactivation of lysine by the Maillard reaction (French). *Helv. Chim. Acta*, **55**, 1153.
15. Hurrell, R. F. and Carpenter, K. J. (1974). Contrasting results for the reactive lysine content in heat-damaged material. *Proc. Nutr. Soc.*, **33**, 13A.
16. Carpenter, K. J. (1974). Chemical and microbial assays for the evaluation of processed protein foods. In *Nutrients in Processed Foods: Proteins*, ed. by P. L. White and D. C. Fletcher, Publishing Sciences Group, Acton, Mass.
17. Tannenbaum, S. (1974). Industrial Processing. In *Nutrients in Processed Foods: Proteins*, ed. by P. L. White and D. C. Fletcher, Publishing Sciences Group, Acton, Mass.
18. Mauron, J. (1972). Influence of industrial and household handling on food protein quality. In *International Encyclopaedia of Food and Nutrition, Vol. II, Protein and Amino Acid Functions*, ed. by E. J. Bigwood, Pergamon Press, Oxford.
19. Mauron, J. (1975). Nutritional evaluation of processed proteins (Germany). *Deutsche Lebensm, Rundschau*, **71**, 27–35.
20. Burton, H. S. and McWeeny (1964). Non-enzymic browning: Routes to the

production of melanoidins from aldoses and amino compounds. *Chem. and Ind.*, **1964**, 462–3.

21. Finot, P. A. and Mauron, J. (1969). The inactivation of lysine by the Maillard reaction (French). *Helv. Chim. Acta*, **52**, 1488–95.
22. Finot, P. A., Viani, R., Bricout, J. and Mauron, J. (1969). Detection and identification of pyridosine, a second lysine derivative obtained upon acid hydrolysis of heated milk. *Experientia*, **25**, 134.
23. Finot, P. A. and Bujard, E. (1976). Unpublished results.
24. Bjarnason, J. and Carpenter, K. J. (1970). Mechanisms of heat damage in proteins. *Brit. J. Nutr.*, **24**, 313–29.
25. Carpenter, K. J. (1973). Damage to lysine in food processing: Its measurement and its significance. *Nutr. Abstr. Rev.*, **43**, 424–51.
26. De Groot, A. P. and Slump, P. (1969). Effects of severe alkali treatment of proteins on amino acid compositions and nutritive value. *J. Nutr.*, **98**, 45.
27. Woodward, J. C. and Short, D. D. (1973). Toxicity of alkali-treated soya protein in rats. *J. Nutr.*, **103**, 569.
28. Ziegler, K., Melchert, I. and Lürken, C. (1967). N, δ-(2-amino-2-carboxyethyl)-orthinine, a new amino acid from alkali-treated proteins. *Nature*, **214**, 404.
29. Asquith, R. S., Booth, A. K. and Skinner, J. D. (1969). The formation of basic amino acids on treatment of proteins with alkali. *Biochim. Biophys. Acta*, **181**, 164–70.
30. Catsimpoolas, N. and Wood, J. L. (1964). The reaction of cyanide with bovine serum albumin. *J. Biol. Chem.*, **239**, 4132.
31. Sternberg, M., Kim, C. Y. and Schwende, F. J. (1975). Lysinoalanine: Presence in foods and ingredients. *Science*, **190**, 992–4.
32. Cuq, J. L., Provansal, M., Fuilleux, F. and Cheftel, C. (1973). Oxidation of methionine residues of casein by hydrogen peroxide. *J. Food Sci.*, **38**, 11–13.
33. Mauron, J. (1956). The deterioration of lysine during certain heat treatments of milk and its effect on growth (French). *Internat. Z. für Vitaminforschg.*, **27**, 85–96.
34. Ford, J. E. (1973). Some effects of processing on nutritive value. In *Proteins in Human Nutrition*, ed. by J. W. G. Porter and B. A. Rolls, Academic Press, London.
35. Finot, P. A. (1973). Non-enzymic browning. In *Proteins in Human Nutrition*, ed. by J. W. G. Porter and B. A. Rolls, Academic Press, London.
36. Mottu, F. and Mauron, J. (1970). Unpublished results.
37. Waibel, P. E. and Carpenter, K. J. (1972). Mechanisms of heat damage in proteins: 3. Studies with ε-(γ-L-glutamyl)-L-lysine. *Brit. J. Nutr.* **37**, 509.
38. Finot, P. A. and Mauron, J. (1977). *In vivo* lysine availability from N-substituted lysines. *Brit. J. Nutr.*, in press.
39. Finot, P. A., Mottu, F. and Mauron, J. (1977). Syntheses and *in vitro* enzymic hydrolysis of N-substituted L-lysines. *Brit. J. Nutr.*, in press.
40. Lang, K. K. (1973). The physiological effects of heated fats, especially frying fats (German). *Fette, Seifen, Anstrichmittel*, **75**, 73–6.
41. Potteau, B. (1973). Heated oils: IV. Nutritional evaluation of the quality of frying fats (French). *Rev. Fr. Corps Gras*, **20**, 471–80.
42. Adrian, J. (1974). Nutritional and physiological consequences of the Maillard reaction. *World Rev. Nutrition Dietetics*, **19**, 71–122.

43. Chichester, C. O. (1974). Protein and amino acid—interactions with other macronutrients. In *Nutrients in Processed Foods: Proteins*, ed. by P. L. White and D. C. Fletcher, Publishing Sciences Group, Acton, Mass.
44. Woodward, J. C. and Short, D. D. (1973). Toxicity of alkali-treated soya protein in rats. *J. Nutr.*, **103**, 569.
45. De Groot, A. P. and Slump, P. (1969). Effects of severe alkali treatment of proteins on amino acid composition and nutritive value. *J. Nutr.*, **98**, 45.
46. De Groot, A. P., Slump, P., Feron, V. J. and van Beck, L. (1976). Effects of alkali-treated proteins. Feeding studies with free and protein-bound lysinoalanine. *J. Nutr.*, in press.
47. Ames, B. N., McCann, J., Yamasaki, E. (1975). Methods for detecting carcinogens and mutagens with the *Salmonella*/mammalian-microsome mutagenicity test. *Mut. Res.*, **31**, 347–64.
48. McCann, J. and Ames, B. N. (1976). Detection of carcinogens as mutagens in the *Salmonella*/microsome test: Assay of 300 chemicals *Discussion. Proc. Nat. Acad. Sci. USA*, **73**, 950–4.
49. Vogel, F. and Röhrborn, G., eds. (1970). *Chemical Mutagenesis in Mammals and Man*, Springer Verlag, Berlin.
50. Röhrborn, G. (1970). The activity of alkalating agents: I. Sensitive mutable stages in spermatogenesis and oogenesis. In *Chemical Mutagenesis in Mammals and Man*, ed. by F. Vogel and G. Röhrborn, Springer Verlag, Berlin.
51. Weinberg, A. M. (1966). Prospects for Big Biology. *The News* (Oak Ridge National Laboratory, Tennessee) Dec. 16.
52. Lugginbühl, H., Bexter, A. and Würzner, H. P. (1972). Unpublished results.

25

Biological Changes—A Consensus

A. E. BENDER

Department of Nutrition, Queen Elizabeth College, University of London, Campden Hill, London W8 7AM, England

The perspective of the observer influences the conclusions he draws as to the effects of processing on foodstuffs and may even alter his objectives. For example, the food manufacturer has a major interest in small changes in appearance, flavour and texture that may make or mar his sales; the consumer is often interested in attempting to make comparisons—frequently impossible to evaluate—between manufactured and home-produced dishes; the nutritionist has his special interests which very often differ from the preceding two.

Although it has long been a cliché to say that people choose dishes that they like to eat and not for their nutritional content, nevertheless the ultimate reason for eating is to provide energy and nutrients so any discussion of the effects of processing on the physical, chemical and biological properties of foods can suitably conclude with a consideration of nutritional changes. There is a vast literature on the subject, recently swollen through the introduction of nutritional labelling in the United States, together, perhaps, with a belated sense of responsibility by some manufacturers. To avoid unjustified conclusions this literature must be considered against a total background which can be summarised under ten headings.

PERSPECTIVE

Inevitability of Nutritional Losses

Many processes, particularly those that involve precooking, water treatment and certainly those that involve removal of unwanted parts of the food (such as in the milling of cereals or the extraction of foodstuffs from their original sources) must inevitably result in losses however carefully they are carried out.

Beneficial Effects

Quite apart from the separate topic of nutritional enrichment, some foods improve in nutritional value through processing, e.g. destruction of toxins and increased protein quality of legumes, and the liberation of bound niacin in cereals.

Limitation of Damage

It is possible to point at the same time to the relatively unimportant losses that take place in most processes and to individual instances of serious losses. However, most nutrients are relatively stable and when losses do occur on a serious scale they are often restricted to the two labile nutrients, vitamin C and thiamin. Much ill-informed condemnation of the food-processing industry as a whole has been based on losses of vitamin C.

Processing versus Home Cooking

Manufacturing losses, where they do occur, are often in place of, rather than in addition to, those that inevitably accompany cooking in the home. For example, canned foods have already been cooked and may be consumed without further preparation or simply reheated, and many dried and frozen foods that have been blanched need a shorter final cooking time than the fresh food. This is exemplified in Table 1, which shows little difference after final cooking between the vitamin C contents of fresh frozen and freeze-dried garden peas. These required boiling for 10 min, 3·5 min and 2 min, respectively.

The air-dried and canned peas shown in the same table, with a lower vitamin C content than the other three products, differ since canning caused the greater loss during processing while the air-dried peas suffered their main loss during cooking.[2] In similar fashion Abrams[3] showed that fresh Brussels sprouts required 20 min cooking and lost 36 % vitamin C while the frozen product required only 10 min cooking and lost a total of 41 % of the vitamin.

Dietary Importance of the Food

Criticism is sometimes levelled at losses that are of no practical importance in the diet as a whole. While the food scientist may devote much effort to measuring even small changes that take place in specified foods during specified processes the nutritionist must consider the diet as a whole.

TABLE 1

Vitamin C content of garden peas after cooking (mg/100 g)[1,2]

Process	Fresh		Frozen	Freeze-dried	Air-dried	Canned
	Variety 1	Variety 2				
Time boiled (min)	10	10	3·5	2	15	Brought to boil
Volume of water/(Volume of food)	1·2	1·2	1	12·3	23·6	—
Vitamin C (mg/100 g)	16·4	18·5	14·0	15·8	11·3	9·2

Peas: percentage loss of vitamin C after the stages indicated[1]

Fresh		Frozen		Canned		Air-dried		Freeze-dried	
		Blanching	25	Blanching	30	Blanching	25	Blanching	25
		Freezing	25	Canning	37	Drying	55	Drying	30
		Thawing	29						
Cooking	56	Cooking	61	Heating	64	Cooking	75	Cooking	65

Losses from foods that supply only a small proportion of the nutrient in question are not of any practical significance. There can be a marked difference between this observation and the consumers' opinion. The salesman must be able to show that his new product is at least as good, nutritionally, as the one he is attempting to replace. It does not help sales to show that the differences are nutritionally insignificant. For example, there is no shortage of protein foods in western diets so that if meat substitutes were slightly lower in protein quantity or quality than meat this would be of no significance in the eyes of the nutritionist but could be a major detriment to the acceptance of the product.

Vulnerable Groups

Regard must be paid to sections of the community who may be relying on a limited number of manufactured foods as a source of their nutrients. Babies may be obtaining all their nutrients from a single manufactured product—it was the introduction of a new heat process of infant milk food that first showed that vitamin B_6 was essential to babies. Elderly people often consume a limited range of foods so that changes of little importance in the average diet may be of considerable importance to their nutritional status. Moreover, it must always be borne in mind that average figures for food consumption and nutrient intake cover a very wide range and include individuals living on very few foods indeed, and possibly suffering mild degrees of malnutrition.

Significance of Losses

In Western communities, where food processing is carried out to a considerable extent, affluence and the variety of foods available virtually rule out nutritional deficiencies so whatever losses occur are not very important. However, changes in methods and in eating habits and the introduction of new foods emphasise the need for continual vigilance. It is in the developing countries where there is total reliance on a limited number of foods that the main problems occur. Often it is the traditional rather than the modern methods of food processing that cause the problems. For example, it has little significance in Europe whether rice is milled or fortified but the milling of rice has led to large scale thiamin-deficiency over the centuries in the Far East. Although losses of vitamin C in food processing are considerable there are few signs of even sub-clinical or covert malnutrition in Western Europe but traditional air-drying of fruits and vegetables leads to scurvy in winter months in Eastern Turkey.

In the present context it is the introduction of modern techniques into developing communities that may create problems. Replacement of traditional foods by processed ones may improve or worsen nutritional status. Since malnutrition is already a problem, processing losses that are not important in the industrialised communities may be of great importance in developing countries.

Basis of Comparisons

Incorrect conclusions have often been drawn from making wrong comparisons. For example, processed food is often compared with the raw food which would normally undergo a cooking process before being eaten (see Table 1). Similarly there can be enormous variations in the properties, including nutrient content, of foods of different varieties, produced under different conditions.

Losses versus Advantages

While there are often avoidable losses of nutrients through poor control of processing, there are also inevitable losses even under the best conditions and this is presumably a price worth paying for the advantages provided. Pasteurisation of milk provides safety at the expense of some vitamins; the convenience of ready-peeled potatoes preserved with sulphur dioxide is at the expense of some of the thiamin; the flavour of heated protein foods is at the expense of a small part of protein quality; the preservation of comminuted meat products in Great Britain with sulphur dioxide, which is both a safety and an economic advantage, is at the expense of the thiamin.

No Choice

Finally, whatever the nutritional losses involved in food processing or even textural and flavour changes, it is not always a choice between the attractiveness and high nutrient content of fresh unprocessed food on the one hand and the lower quality (if it is lower) of the processed food on the other, but a choice between heated, preserved, canned or dried foods on the one hand and no food at all on the other. For example the garden pea is available in the fresh form in Great Britain for about 2 months in the year, so for 10 months the comparison is not between processed and fresh but between processed and none at all.

DRAWING CONCLUSIONS

There are many variable and apparently contradictory findings in the literature, often due to varying materials, differing experimental conditions and, less excusably, to insufficient information provided by authors. Although published papers are supposed to provide full experimental details there are many reports that lack adequate information of the nature of the raw materials, specific details of processing and evidence of the reliability of analytical method used. Methods of assay are sometimes open to question. For example, the indophenol method of estimating vitamin C is still used by some workers on the grounds that most of the vitamin is present as ascorbic acid, but evidence is not provided that this is still true after processing and storage.

Large increases in the vitamin A (carotene) content of vegetables have been reported after heat treatment. While it is possible that there is enzymic destruction of carotenoids during extraction from the raw foods, so reducing the amount determined, Nutting *et al.* suggested[4] that the greater part of the difference was due to low extraction from raw material compared with the heat-treated food. A second reason for drawing incorrect conclusions from earlier work is that the different carotenoids vary considerably in their biological potency and only total carotenoids were measured. The total amount may not be affected by processing, leading to the assumption that there has been no loss, but that isomerisation into less potent forms of the vitamin does occur and was not realised in the earlier work.[5] The bulk of the carotenoids in fresh vegetables consists of the all-*trans* isomers and heat converts part of these, the extent of isomerisation depending on the severity of treatment, into *neo* isomers of lower biological potency. Hence canning can reduce the vitamin A content of green vegetables by 15–20 % and that of yellow vegetables by 30–35 %. Similarly, earlier reports of the effects of processing on folate are invalid because most assays prior to about 1970 were unreliable.

Reports of changes in protein quality can depend on the method of measurement. Chemical determination of amino-acid composition is no guide to the nutritive value of a protein if any part of the amino acids is biologically unavailable, since the preliminary acid hydrolysis liberates all the amino acids. An extreme example is provided by the two samples of meat canned and stored for 136 and 110 years, respectively, that showed the same amino-acid composition as fresh meat but had net protein utilisation of 29 and 37 respectively, compared with 75 for that of fresh meat.[6] Furthermore, even biological assay can be misleading unless damage is sustained to the limiting amino acid. For example, preparations of canned evaporated milk that are brownish in colour have obviously suffered loss of available lysine through the Maillard reaction. Since milk protein contains a relative surplus of lysine and is limited by the sulphur-containing amino acids it could sustain some loss of available lysine without detection in most of the usual bio-assays. While such a loss, since the lysine was in surplus, may be of no significance when considering the quality of the milk protein by itself, it could be significant when milk so damaged is consumed with cereal in such proportion that lysine is limiting in the mixture.

An obvious problem arises with non-homogeneous materials from the impossibility of analysing the same sample before and after treatment or of comparing the effects of two processes on the same sample. For example, Zarnegar and Bender[7] examined the loss of vitamin C from potatoes through mechanical damage inflicted in machine peeling. A set of 12 potatoes was selected for analysis of the raw material and another 12 subjected to the required treatment. Overall there were four different experimental treatments each with its set of 12 raw potatoes as controls. The problem is to decide whether a statistically significant difference between control and treated samples has any real meaning since the sets of untreated controls showed differences in vitamin C content as follows: (mean mg/100 g $\pm$ SD with range in parentheses) (1), $12\cdot4 \pm 1\cdot8$ ($9\cdot0$–$15\cdot5$); (2), $12\cdot8 \pm 1\cdot8$ ($9\cdot7$–$16\cdot1$); (3), $12\cdot6 \pm 1\cdot3$ ($8\cdot6$–$14\cdot0$); (4), $11\cdot7 \pm 1\cdot7$ ($8\cdot3$–$14\cdot6$). In one instance the treatment resulted in a statistically significant loss compared with its control, the difference however was less than the difference between the two control sets of samples (1) and (4) above.

Similarly, Miller et al.[8] carried out a number of trials on roller-dried, instantised and canned pinto beans using three separate sets of samples. The range of variation between the three control samples was 13 % for vitamin B_6, 18 % for niacin, 59 % for thiamin and 100 % for folate—considerably greater than many values reported in the literature for processing losses. Two papers from the same laboratory further illustrate some of the difficulties in drawing conclusions. The authors reported that there was no loss of iron from French beans and drumstick (*Moringei olifera*) in their 1972 paper but a loss of 10–14 % in the same method of cooking in their 1974 paper. A loss of 60 % of the calcium was reported from French beans during pressure cooking in the earlier paper but no loss in the later paper.

Some differences, well substantiated by experimental evidence, may be so small as to have little or no nutritional importance. For example Wing and Alexander[9] showed that chicken cooked for $1\cdot5$ min in a microwave oven retained 91 % of vitamin B_6 in the meat and $1\cdot5$ % in the drippings—total retention $92\cdot5$ %. This compared with 83 % retained when conventional cooked for 45 min with $5\cdot4$ % in the drippings—total retention $88\cdot4$ %.

(Miller *et al.*[8] found the coefficient of variation of vitamin B_6 analysis to be $\pm 9\%$.)

What is much less excusable is the absence of evidence in many reports of the reliability of the assays, an absence of sufficient evidence from which to apply any test of statistical significance, and failure to draw correct conclusions from the results provided. For example, a paper published in 1952 reported that storage of lamb at $-18\,^{\circ}\mathrm{C}$ for 6 months resulted in 250 % increase in niacin and the authors suggested that it was being synthesised during the storage period. Subsequent reviewers charitably stated that 'the authors clearly had analytical difficulties'.

Vitamin assays are far from precise and few workers would claim accuracy better than $\pm 10\%$ for any of the B vitamins and probably no better than 20–30 % for many of them, yet without providing any evidence of reproducibility one group of workers reported losses from rice stored for 2·5 years as 29·40 % thiamin, 5·44 % riboflavin and 3·77 % niacin! In a similar way there are tables of single values of amino-acid composition of protein foods without any information as to their reliability, and figures given for methionine without, at the same time (since it is the combined sulphur-containing amino acids that are of interest) giving any value for cystine.

An example of incorrect conclusions drawn from experimental results (without evidence of their reliability) was that reporting the losses of thiamin and riboflavin from pork stored at $-1\,^{\circ}\mathrm{C}$. The values at the times stated were as follows for thiamin:

Time (weeks)	8	16	24	32	40	48
Losses (%)	19	26	33	24	29	21

Thus the loss after 48 weeks was the same as that after 8 weeks yet the authors concluded that losses increased to 33 % after 24 weeks storage. Similarly for riboflavin:

Time (weeks)	8	16	24	32	40	48
Losses (%)	19	20	19	20	17	14

There would appear, on the evidence presented, to be no loss after the first 8 weeks yet the authors concluded that the loss increased steadily up to 20 % of the initial amount after 32 weeks.

PREDICTIONS AND MATHEMATICAL MODELS

The food processor and particularly his marketing division would like to be able to forecast the stability of the various nutrients during the process in

A. E. Bender

TABLE 2
Vitamin C in green beans^a after canning at two temperatures (mg/100 g)[18]

	Ascorbic acid			Dehydroascorbic acid		
	Fresh	Drained beans	Brine	Fresh	Drained beans	Brine
8 min at 124°C						
Cultivar 1	14·0	0	0·9	7·0	2	0·5
Cultivar 9	11·0	5·0	4·0	5·6	3·4	1·0
Mean of 15 cultivars	17·0 ± 5·0	2·5 ± 1·6	3·7 ± 2·2	6·5 ± 2·0	2·8 ± 1·2	1·9 ± 1·1
Loss		84%	77%		54%	67%
25 min at 116°C						
Cultivar 1	14·0	3·9	3·8	7·0	3·1	2·8
Cultivar 9	11·0	3·4	2·1	5·6	2·1	2·1
Mean of 15 cultivars	17·0 ± 5·0	3·4 ± 1·6	4·1 ± 2·3	6·5 ± 2·8	2·6 ± 1·3	1·8 ± 1·0
Loss		79%	75%		56%	71%

^a Figures rounded off.

question and during subsequent storage. Evidence can be drawn from the published literature and also from model (simple) systems studied in the laboratory. Results from the latter, however, may not always be applicable to the far more complex system existing in foods. For example, pure pteroyl glutamate is stable at temperatures of 230 °C for 60 min but is partly destroyed by much less severe treatment in many foods where it is present in combined form.[10,11]

There appears to be no change in the nutritive value of meat subjected to canning and normal cooking procedures but when it is heated in the presence of other materials there is a fall in nutritional value.[12]

The rates of destruction of vitamin C have been shown by some authors to follow a first-order reaction but there appear to be different mechanisms operating in different foods. Freed *et al.*[13] produced a nomogram for the rates of destruction of ascorbic acid under a variety of conditions, and Wanninger[14] also produced a mathematical model taking account of temperature, water and oxygen. Farrer[15] carried out similar calculations for the loss of thiamin; and Kramer[16] dealt with storage of dry and canned foods. However, Labuza[17] pointed out that there are several anomalies in the published reports of vitamin C destruction which could be explained by different mechanisms. In addition to oxidation, losses of ascorbic acid in some foods run parallel to the extent of non-enzymic browning. In this reaction the activation energy increases with decreasing moisture while the loss from enriched wheat flour and a corn–soya–milk mixture decreased with decreasing moisture. In orange juice crystals losses were the same in air and in vacuum and there were losses even below the monolayer of moisture. Labuza suggested that the mechanism of ascorbic acid destruction may differ at different moisture contents—possibly oxidation being the main cause of destruction at low moisture and browning at high moisture content.

The impossibility of producing a mathematical model in the present state of knowledge is illustrated by the observations of Marchesini, Majorino *et al.*[18] (Table 2). They showed that even among a range of 15 cultivars of green beans grown under identical conditions the rate of loss of both ascorbic and dehydroascorbic acid differed enormously. Under the same canning conditions (8 min at 124 °C) one cultivar lost all the ascorbic acid (AA) and three-quarters of the dehydroascorbic acid (DHA), while another lost only one-quarter of the AA and a quarter of the DHA. At a lower temperature (25 min at 116 °C) the latter cultivar lost more AA than the former cultivar but less DHA. To complicate any predictions of vitamin C loss the same authors showed that the average losses were the same at 124 °C and at 116 °C. Yet another difference was that the amount of vitamin leached into the brine in the cans differed among the cultivars.

CONCLUSIONS

Since the role of each food in the diet must be taken into account it is not possible to generalise about the extent of the losses that take place during the processing of food. The one generalisation that is possible is that few manufacturers take account of nutrition in their exercise of quality control. Since each food, each process and the variation within each factory can produce different results it is not possible at present to predict nutritional changes with any degree of reliability.

REFERENCES

1. Mapson, L. W. (1956). Effect of processing on the vitamin content of foods. *Brit. Med. Bull.*, **12**, 73–7.
2. Robertson, J. and Sissons, D. J. (1966). The effects of maturity, processing, storage in the pod and cooking on the vitamin C content of fresh peas. *Nutrition*, Vol. XX, **1**, 21–7.
3. Abrams, C. I. (1975). The ascorbic acid content of quick frozen Brussels sprouts. *J. Food Technol.*, **10**, 203–13.
4. Nutting, M. D., Neumann, H. J. and Wagner, J. R. (1970). Effect of processing variables on the stability of beta-carotene and xanthophylls of dehydrated parsley. *J. Sci. Food Agric.*, **21**, 197–202.
5. Sweeney, J. P. and Marsh, A. C. (1971). Effect of processing on provitamin A in vegetables. *J. Amer. Dietetic Assoc.*, **59**, 238–43.
6. Bender, A. E. (1962). Loss of nutritive value of proteins through processing and storage. *Proc. 1st Int. Congress Food Science and Technology*, **3**, 449–55.
7. Zarnegar, L. and Bender, A. E. (1971). The stability of vitamin C in machine-peeled potatoes. *Proc. Nutr. Soc.*, **30**, 94A.
8. Miller, C. F., Guadagni, D. G. and Kon, S. (1973). Vitamin retention in bean products: Cooked canned and instant bean products. *J. Food Sci.*, **38**, 493–5.
9. Wing, R. W. and Alexander, J. C. (1972). Effect of microwave heating on vitamin B_6 retention in chicken. *J. Amer. Dietetic Assoc.*, **61**, 661–4.
10. Colman, N., Green, R. and Metz, J. (1975). Prevention of folic acid deficiency by food fortification: II. Absorption of folic acid from fortified staple foods. *Amer. J. Clin. Nutr.*, **28**, 459–64.
11. de Ritter, E. (1976). Stability characteristics of vitamins in processed foods. *Food Technol.*, **30**, 48–54.

12. Hellendoorn, E. W., de Groot, A. P., van der MijllDekker, L. P., Slump, P. and Willems, J. J. L. (1971). Effect of heat sterilisation and prolonged storage. *J. Amer. Dietetic Assoc.*, **58**, 434–41.
13. Freed, M., Brenner, S. and Wodicka, V. O. (1949). Prediction of thiamin and ascorbic acid stability in stored canned foods. *Food Technol.*, **3**, 148–51.
14. Wanninger, L. A. (1972). Mathematical model predicts stability of ascorbic acid in food products. *Food Technol.*, **26**, 42–5.
15. Farrer, K. T. H. (1950). The thermal loss of vitamin B_1 on storage of foodstuffs. *Austral. J. Exp. Biol. Med.*, **28**, 254.
16. Kramer, A. (1974). Storage retention of nutrients. *Food Technol.*, **28**, 50–60.
17. Labuza, T. P. (1972). Nutrient losses during drying and storage of dehydrated foods. *CRC Critical Reviews Food Technol.*, **3**, 217–40.
18. Marchesini, A., Majorino, G., Montuori, F. and Cagna, D. (1975). Changes in the ascorbic and dehydroascorbic acid content of fresh and canned beans. *J. Food Sci.*, **40**, 665–8.
19. Thomas, M. H. and Calloway, D. H. (1961). Nutritional values of dehydrated foods. *J. Amer. Dietetic Assoc.*, **39**, 105–16.

26

Perspectives on the Conference—Pointers for the Future

A. M. Pearson

Department of Food Science and Human Nutrition, Michigan State University, East Lansing, Michigan 48824, USA

The conference was appropriately opened by Professor Eric von Sydow, who traced the history of IUFOST and reviewed its functions, one of the principal ones being that of sponsoring symposia on foods such as the one that brings us together on this occasion.

Dr Tore Høyem then reviewed the design and purpose of the symposium and aptly pointed out that the theme is all-inclusive since any temperature affects foods, influencing not only their physical and chemical properties but also taste, aroma and nutritive value. He concluded that the food industry could ill afford to allow foods to spoil or to throw away any components, and that proper use of thermal processing could play an important role in this regard.

Director Oskar Kvåle then welcomed to Oslo those attending and traced the history and expansion of the Norwegian Food Research Institute. Of particular interest were the new facilities of the Institute built at a cost of 27·5 million Norwegian Kroner, which we were permitted to tour. I can assure you, Director Kvåle, that we envy you with your well-designed and well-equipped laboratories, with your surplus of space, but especially your budget of 7·7 million Norwegian Kroner. Our envy, however, is only a wish that we may all do as well, and we wish you continued success in your endeavours to establish an outstanding Food Research Institute—you are well on your way!

A most impressive lecture by Professor Skulberg concluded the opening session. The importance of not only producing but in preserving food was highlighted by Professor Skulberg as he indicated that half of the world's population is underfed and that the 20 % rich have 20 times more than the remainder. He observed that increased efficiency is not the starting point for aiding the underdeveloped countries, but must come through the transfer of knowledge and the practice of scientific principles. He pointed out that Norway possesses great natural resources, namely fisheries, agriculture and pure water, and I might add, honest and hard-working people. These resources are a great asset, not only to Norway but to the entire world. Professor Skulberg then emphasised some areas that need attention: (1) improving the viability of the cod larva; (2) more extensive use of grass and other roughages for feeding livestock with increased availability of grains for man; (3) more efficient use of the fishery supplies in providing high quality food for man; (4) better and safer manufacturing methods for preserving foods; (5) improved food hygiene practices; (6) better packaging for foods; and lastly, (7) to restore public confidence in the food industry through honest practices and concern for consumers.

At this time, I should like to express the thanks of all the participants to IUFOST for sponsoring this timely and important symposium, especially to Professor von Sydow for his support and encouragement. Special thanks are extended to Director Kvåle, to Dr Høyem and to all of the members of the staff of the Norwegian Food Research Institute who have worked so untiringly to make the arrangements for this symposium near-perfect. I should also like to express the gratitude of all concerned for the excellent tour of the Norwegian Food Research Laboratory and to the staff of Stabburet A/S for the fine tour of their operation and also for the delicious lunch.

TIME–TEMPERATURE RELATIONSHIPS FOR NORMAL PROCESSING OF FOODS

The first technical session dealt with time–temperature relationships for different thermal processes, covering a number of diverse topics on some problems and solutions in food processing. Professor Thijssen and his colleague, Dr Kerkhop, began by pointing out the effect of water content and temperature of processing control, the rates of the processes involved in food preservation, regardless of whether it is process kinetics, physical properties or chemical and physical equilibria. If the effects of temperature and moisture content are known quantitatively, every operation can be optimised. Bearing these principles in mind, the authors then derived a series of equations by which they could predict the parameters affected by water content and temperature. Such equations could be used to predict any adverse effects, such as chemical changes or nutrient losses. The process of optimisation in spray-drying of a liquid food containing a heat-labile aroma was utilised as an example of these principles by showing how the temperature and moisture content of the droplets change during spray-drying. Finally, the authors demonstrated how a new calculation procedure could be used to optimise quality retention in the sterilisation of canned foods.

In the discussion following this interesting presentation, the question was raised as to the influence of lack of homogeneity of canned foods upon process optimisation. This would appear to be an area worthy of further investigation by those with a mathematical bent to solving problems. It was also wisely pointed out in the discussion that in the end process optimisation must be tested under practical conditions. Nevertheless, application of the mathematical approach to process optimisation provides a ready tool that can be applied in avoiding costly trials on parameters unlikely to yield any information of value.

Professor Hallström presented a very informative discussion on heat preservation involving liquid foods in continuous flow pasteurisation or UHT. He pointed out that thermal preservation involves heating to a certain temperature, holding the product at this temperature for a certain period of time followed by cooling to a desired storage temperature. The maximum temperature and time of holding at that temperature are chosen to destroy certain bacteria likely to be present in the food. He then discussed the techniques available for heating continuous flow systems, which included the indirect heat-exchangers (tubular, plate or scraped surface types), the direct heat-exchangers (steam injection and steam infusion), the electrical methods (resistance, inductance or dielectric heating) and heating by friction. All of these methods are being used by the food industry with the possible exception of the electrical methods. Professor Hallström then demonstrated how mathematical kinetic models could be developed for biological, chemical and physical changes in foods using the time–temperature formula for the process so that different processing systems can be compared and evaluated. He showed how these models could be utilised in calculating the lethal effects on bacteria, for enzymatic changes and even for sensory changes in foods.

The discussion following this paper concentrated on applications to processes by different types of heat-exchangers, such as in reclamation of the flavour constituents from orange peel as an ingredient for soft drinks or deodorisation of soya beans by steam injection. Processes utilising a combination of more than one type of heat-exchanger to achieve efficient heating were also considered. An understanding of the various heat-exchangers and their best uses should lead to better procedures for heating. Finally, Professor Hallström pointed out that UHT is a promising new development but has not been widely used by the food industry.

Dr Hermans then discussed heat preservation involving liquids and solid foods in batch autoclaves or in hydrostatic sterilisers. He first pointed out that the initial steps in preparation of the foods are essentially the same until the food is hermetically sealed and transported to the steriliser. At this point, the batch-type autoclaves make the process strictly a batch operation, whereas the hydrostatic sterilisers become a continuous process. He pointed out that there are advantages and disadvantages in both the batch and continuous processes. In both types of sterilisers, agitation speeds up the rate of heat penetration. In batch autoclaves this is achieved by rotating the basket and in continuous sterilisers by rotating the carrier. Dr Hermans concluded that the radius and the revolutions per minute affect the rate of heat transfer. Since batch operations have different radius for containers at different locations (centre versus outside), there is a compromise in choosing the processing condition for batch operations, whereas the hydrostatic steriliser is constant in its radius and all are processed the same. Dr Hermans pointed out that glass jars with ventilating caps and plastic containers offer special problems due to their inability to withstand vigorous agitation. Plastic pouches due to their shape and size have advantages in rate of heat penetration and rapidity of achieving sterilising temperatures, but they suffer from more frequent leaks, require care in handling and are difficult to scale up to large production throughput.

This particularly timely paper discussed a variety of topics including batch and hydrostatic sterilisation, speeding up processing by agitation and use of plastic and flexible packages and glass jars with self-ventilating caps. Questions were raised concerning the policies of different countries in acceptance of the flexible packages and suggested they are meeting with some success but are not universally acceptable. It was finally suggested that special processes may require specially designed packages, which current technology is and will continue to make available.

Professor Leniger discussed the important topic of concentration by evaporation and aroma recovery. Although it was pointed out that there are several procedures for concentration by evaporation, including among others freeze concentration and membrane processes, Dr Leniger concluded that heating was the best method for concentration. Processing conditions and the influence of initial and final concentration on the quality was considered. The cost, selection and design of equipment relative to efficiency and quality of the final product were discussed. The influence of the flow properties of the product in the evaporator and the heat transfer were carefully outlined and considered. The author also considered process optimisation and concluded that expensive evaporation methods are important only when consumers are willing to pay for added quality. Only in special instances does it appear to be advisable to attempt to recover aroma volatiles—only when the improvement in quality will pay for the added costs of recovery.

The discussion following this paper suggested that the other methods of concentrating foods should receive some attention. Freeze concentration and perhaps even membrane concentration procedures may be as efficient as concentration by evaporation since costs of fuel have increased so dramatically. Aroma concentration is worthy of consideration as Professor Leniger so wisely stated 'only when the consumer will pay for the added quality'. Systems for recovering aroma volatiles need to be developed that are simpler and easier to use with a resultant reduction in costs.

The final paper of the first session by Dr Dagerskog represented a scientific approach to understanding the time–temperature relationships in industrial cooking and in frying. He pointed out that one can calculate the combined influence of time and temperature on rheological, chemical and the sensory properties of foods, which can be expressed as a C- (cook) value and Z_c-values (the temperature rise in °C for a 10-fold increase in reaction rate for chemical and sensory changes). For example, a C-value of 6 equivalent 100 °C minutes was shown to apply to small cylinders of white potatoes and a Z_c of about 17 °C for the three varieties tested. Using computer stimulations for the varying processing variables (thickness, starting temperature, cooking temperature programme and cooking medium) it was possible to optimise temperature distributions and the related time–temperature dependent quality changes. The simulations closely approximated the situation for actual cooking conditions. Unconventional processes, such as microwave cooking, and infrared heating, were also studied. Frying of foods was considered, especially the use of double-sided contact frying versus single-sided frying, which had distinct advantages in forms of lower temperatures, faster cooking and less shrinkage.

In the discussion it was pointed out that the cooking process is not understood from the standpoint of what is actually happening. The use of scientific procedures for studying cooking and the changes in foods is only in its infancy and studies on combined methods of cooking are needed to complete an analysis of what occurs during cooking. As sound scientific methods are applied in analysis of the effects of cooking methods upon the physical, chemical and organoleptic changes occurring during heating, cooking will emerge from an art to that of science, and only then we will be able to fully explain the phenomena involved.

PHYSICAL, CHEMICAL AND BIOLOGICAL CHANGES RELATED TO DIFFERENT TIME–TEMPERATURE COMBINATIONS

The first presentation of the second session by Professor Hamm covered the changes in the muscle proteins during the heating of meat. It was first shown that there are a great many proteins in meat, namely, the myofibrillar proteins—actin, myosin and more especially the combined protein actomyosin—the connective tissue proteins and the sarcoplasmic proteins. The emphasis was placed on actin, myosin and actomyosin in view of their quantitative and qualitative importance. Professor Hamm, by means of charts, demonstrated that the myofibrillar proteins showed marked changes in two phases, the first being between 30 and 50 °C and involving unstable cross-linkages between the unfolded proteins. The second phase occurred between 60 and 90 °F, which seems to be due to denaturation and shrinkage of the connective tissue, collagen, and/or of formation of new cross-linkages in the coagulated actomyosin system. The changes in tenderness, rigidity and water-holding capacity during heating of meat clearly take place in these two stages. In addition, heating above 80 °C causes a marked increase in H_2S release, probably from the SH groups of the myofibrillar proteins. Finally, it was pointed out that coagulation starts at about 40 °C, so is probably not due to disulphide cross-linkages which do not increase until the temperature reaches 70 °C or greater. Time did not permit more details on other protein changes influencing the physical, chemical and organoleptic properties associated with cooking of meat. Professor Hamm thus provided us with the basis for a fuller understanding of the effects of heat on meat.

This excellent paper elicited considerable discussion on such questions as to how one can measure the degree of doneness in meat, which could be a treatise on physical changes alone, or what is the role of oxidation during cooking? Another interesting question dealt with the effect of heat and rate of heating on the stability of meat emulsions, which is an important topic to meat researchers and processors.

Professor Morton gave a scholarly presentation—in New Zealand English—after commending those using English so beautifully even though it was not their native tongue. His treatise dealt with changes in fat during heating. Although he carefully detailed some possible effects of structure upon fat stability during heating, his principal contribution may have been to point out how gel permeation chromatography, liquid–liquid chromatography and gradient elution chromatography can be utilised for isolating breakdown products from heating of fats and oils. He reviewed the pertinent literature on long-term feeding studies with over-heated fats, and concluded that recent data suggest there is no evidence of their being harmful. Nevertheless, he suggested this matter should not be put to rest until the evidence becomes unequivocal. He pointed out that snack foods absorb considerable quantities of frying oil, which tends to be more highly oxidised than that of the fat itself. Finally, Professor Morton pointed out that countries in the northern hemisphere should concentrate on producing edible oils in order to become self-sufficient.

A lively discussion ensued following this paper on how proteins, carbohydrates and fats may interact on heating, which is the usual situation in natural food systems, yet it was admitted that little is known on this important subject. Other questions of a practical nature included how consumers can effectively evaluate whether an oil is still sound and the differences which occur during the cooking of fish and meat in oils. Finally, it should be pointed out that time did not permit the discussion of such timely topics as warmed-over flavours due to oxidation of meat, poultry and fish or the possible role of malonaldehyde—a common product of oxidation—as a carcinogen.

The heating of carbohydrates was the topic of Dr Birch, especially as it is related to chemical, physical and biological changes. He first stated that marked hydrogen bonding is a characteristic of carbohydrates that greatly influences their physical properties. He pointed out that the polyhydroxy nature of carbohydrates is not only responsible for hydrogen bonding *per se*, but also gives rise to their sweetness, high water solubility, high melting and boiling points and their crystallisation or gelatinisation properties, as well as the lack of odour associated with their low vapour pressure. Dr Birch showed how the cyclic structure of carbohydrate molecules makes them undergo conformational changes at different temperatures and is the basic cause for their reactivity in foods, especially because of their relatively large number of asymmetric centres. Probably the most interesting part of the report (at least to this author) dealt with increased calorific values for a mixture of sucrose and citric acid and the data showing that raw potato starch in the diet of the albino rat led to a tremendous increase in the size of the caecum and was lethal in about 10 days. However, heated potato starch in the diet would completely alleviate the toxicity. Finally, it was shown that carbohydrates increased the performance of tested human subjects on IQ tests.

During the discussion it was indicated that heating of meat causes little loss in biological value, even when canned, but the addition of vegetables caused a marked decrease in the quality of the meat protein. No explanation was forthcoming so that this interesting observation needs further investigation. A related observation suggested that breaking of meat products may also cause increased amino acid losses during heating. The mechanism for the increased calorific value of heated mixtures of carbohydrates and citric acid needs to be elucidated further as does the role of carbohydrates in improving the intelligence quotient of man. Finally the effects of heating raw potato starch to eliminate toxicity in the white rat needs further elucidation. This paper certainly indicates that the effects of heating on carbohydrates should be a fruitful field for some time to come.

Dr Hurrell gave a timely discussion on the Maillard reaction in foods, which was co-authored by his colleague, Dr Carpenter. The paper traced the original evidence for the damage to high quality proteins on heating in the presence of glucose and demonstrated that lysine supplementation restored the original PER value. With the exception of milk and eggs, few protein foods have sufficient levels of reducing sugars to cause problems. In production of dried egg proteins, the use of glucose oxidase or yeast fermentation is utilised to remove the reducing sugars. Starches are not damaging to protein foods during processing unless yeast or some other source of amylase is present to break down the starch to sugars. Sucrose in foods has been shown to gradually invert to form glucose and fructose, and under severe conditions may cause some damage to foods containing proteins. Dilute acids increase hydrolysis, whereas dilute alkalis inhibit the hydrolysis of sucrose. The authors pointed out that excessive heat and low moisture contents tend to accelerate the carbonyl–amino acid reaction. They concluded that the Maillard reaction is not a serious nutritional problem in the Western world, where excess protein consumption is common, with the possible exception of baby foods or perhaps milk products and dried eggs, but even in these products undesirable flavours and colours are probably self-limiting.

Questions were raised concerning the advantages to be gained from fermentation of the sugars in heat processed meat meals, and which amino acids and reducing sugars are involved in browning of meat products. The relative importance of oxidation as a contributing cause to the Maillard reaction was also raised without resolution as was the importance of gelatinisation during the browning of bread. Although the desirable aspects of browning in baking of bread and cooking of meat were mentioned, these changes would warrant more detailed discussion, and perhaps, research.

Vitamin losses during thermal processing were discussed by Dr Benterud, who concluded that nicotinic acid, riboflavin and tocopheryl esters are extremely thermostable, whereas thiamine is particularly labile with losses of up to 20–70 % being reported in meat and fish products. Although lesser losses of thiamine have been reported during canning of fruits and vegetables, high-acid foods such as fruits and tomatoes retain more of the original vitamin. Dr Benterud discussed all of the other vitamins, which were intermediate in stability between thiamine and vitamin C and the highly stable vitamins previously mentioned. In general, vitamin destruction is influenced by pH, peroxide content, traces of metals and oxidation due to exposure to air but decreased by antioxidants and protein complexes. The variable nature of individual vitamins under different conditions, which was so carefully covered by the author will be omitted due to time limitations, except to state that the UHT treatment appeared to result in better retention of most thermolabile vitamins, such as thiamine, vitamin C, vitamin B_6 and folic acid.

Questions of importance raised during the discussion included the relative stability of added vitamins versus the naturally occurring vitamins and the influence of encapsulation as a means of stabilising added vitamins. Obviously, the latter would be effective only when protected from water solubilisation or in dry foods.

Dr Svensson next discussed the inactivation of enzymes during thermal processing. He first pointed out that blanching by heat to inactivate the enzyme is a standard practice for fresh vegetables before freezing or drying. Although most of the enzymes causing deterioration in vegetables have been identified, inactivation of peroxidase activity is commonly used as a measure of the adequacy of heat treatment due to its relatively high resistance to thermal destruction. The author clearly indicated that reduction of moisture content or an increase in heating either alone or in combination together increased the effectiveness of inactivation. He finally concluded that the effects of the whole spectrum of degrading enzymes and their properties with their influence on the physical, chemical and organoleptic changes should be studied in relation to heat inactivation—only when this is accomplished will the best procedures be known for individual food products.

A most interesting discussion followed in which one of the participants pointed out that enzymatic problems could occur, even at moisture contents below 10 %. It was also indicated that low storage temperatures inhibit enzymatic changes by reducing the activity of the enzymes, yet there may still be problems depending on the properties of surviving enzymes. A further question concerned the heat inactivation of enzymes in meat as a means of avoiding problems in flavour during high temperature storage of irradiated meats, which have improved flavours and odours following heating. The area of enzyme inactivation in foods was barely touched. Furthermore, enzymes in the food industry—desirable and undesirable effects—would seem to warrant a symposium on its own. Since our discussion here has dealt solely with the undesirable effects, the broader topic could be most meaningful, as enzymes can be both useful and can cause problems.

Dr Ledl discussed the role of thermal degradation in meat products by concentrating on the effects of heating on aroma and flavour. He pointed out that there are two basic procedures for studying flavour and aroma: (1) isolation, identification and evaluation of the contribution of volatile compounds, usually by GLC–MS procedures; and (2) systematic evaluation of the products of heated model systems. It was shown that precursor components of meat are water soluble, low molecular weight compounds that are often unstable on heating or in the isolated condition. Dr Ledl briefly and succinctly covered the patent literature showing precursors of 'meat-like' flavours and aromas. He then pointed out the essential role of a number of sulphur compounds, beginning with such simple compounds as H_2S up through a number of sulphur-containing cyclic compounds that have odours or flavours reminiscent of meat. The possibility of combining two or more simple compounds to produce the ring-structures from H_2S, carbonyls, pyrazines and other compounds was briefly mentioned.

One of the important considerations in development of meat flavours and aromas is the wide variety of substances identified in volatiles from heating meat and the possible interaction to form components resembling meat flavours and aromas. The faint hydrogen sulphide odour of cooking meat is well known and desirable at low concentrations, yet becomes extremely offensive at higher levels. The combination of H_2S with carbonyls and other substances to produce cyclic sulphur compounds with meat-like flavours and aromas has recently been reported to have more complicated interactions. To date, it has been impossible to rule out the possibility that several compounds in combination together may in mixtures give the desirable aromas and flavours of meat. The question on the possible toxicology of contributing compounds could also be most important to the meat industry, and if they are in fact toxic need to be recognised and blocked before formation. The meat flavour system appears to be one of the least understood and most complicated food flavour and aroma systems and should receive added emphasis. In our laboratory we are attempting to predict the combination of simple components that most nearly approach the odour and flavour of meat by actual panel tests and a complicated statistical procedure using computers to predict the best combination and level of compounds.

The final paper of this session dealt with the effect of temperature and time on emulsion stability and was presented by Dr Rydhag. In an elegant series of graphic illustrations the stability of emulsions was shown to be dependent on the angle between the solid and the two liquid surfaces, with the most stable emulsions occurring when the contact angle with solid at the interface is close to 90 °C. To achieve this it is usually necessary to add surfactants which modify the hydrophilic–lipophilic balance at the solid particle surface. Dr Rydhag then demonstrated how surfactants associate in the emulsion to give ordered structures, i.e. to form liquid crystals of lamellar and hexagonal types. The liquid crystalline structure was shown to influence emulsion behaviour. Finally, Dr Rydhag pointed out that the properties of emulsions are influenced by temperature and its effects upon the liquid crystalline structure.

In the discussion that followed Dr Rydhag said that data are available showing that proteins may function as surfactants in emulsions, thus, food proteins could be useful in emulsion systems. It was also shown that slow cooling tends to stabilise emulsions and has distinct advantages. Meat emulsions were briefly discussed but not being true emulsions it was not clear whether or not the same principles would govern their stability. Food emulsion systems is a topic where there is great need of further research and discussion.

The third session of the symposium dealt with current problems in applications of thermal processing by the food industry. The first paper by Dr Morvich of a manuscript prepared by Skramstad covered the topic of application of heat as a means of preserving fish products. The history of development of safe processes for producing canned fish in Norway was traced. It was shown that there are three distinct phases, namely: (1) coming-up time; (2) holding time; and (3) cooling time—each of which influences the process. Heating of sardines and other fish containing the intact bones was found to improve the organoleptic properties, which in part is associated with softening of the bones, and in exception to most heat processes improves both safety and the organoleptic properties. The minimum F_0 value for canned sardines was found to be 6·0 minutes at 112 °C, giving a 60 min processing time. In general, fish products were rich in vitamin D and B_{12}, but low in vitamin C and thiamine. One question following this interesting presentation dealt with the possibility of using STHT processing for fish products. Dr Morvich pointed out the process is difficult to use, but best results occur with low-temperature-long-time processing. This is probably even more true when one considers the advantages of the usual process in improving palatability. It was clearly indicated that controlled rates of cooling are essential to get the necessary heating during processing but also because too rapid cooling can cause reinfection. An area of considerable interest dealt with reasons fresh fish produce higher quality meat balls than frozen fish. The differences in rates of cooling for the water and the can contents also elicited considerable discussion.

The second paper by Dr Søbstad described the production of fish-meal in which the fish is first heated to boiling to partly coagulate the proteins and release the lipids. Some of the low molecular weight proteins (20–25 % up to 45 % of the protein) are liberated with the lipids and are carried away in the soluble fraction. After removal of the lipids, the stick water is concentrated by evaporation. The heat hydrolyses the proteins in the stick water, to a variable degree, which then increases the viscosity making concentration difficult. The stick fluid and the residue (pressed material) are added back together before drying. Dr Søbstad indicated that the best fish-meal is produced by using 80 % pressed cake and 20 % stick fluid concentrate (30 % moisture). During heating, oxidation of the lipids may occur and lead to cyclic structures having dark colour, poor amino-acid availability and decreased digestibility.

Questions raised dealt with relationship of price and nutritive quality and measurement of protein quality. Some important questions were raised on the vital topic as to how to measure protein quality and the use of different species, especially of monogastric and young ruminants. It was pointed out that heat processing causes sulphoxide formation and could be just as badly damaged by freeze-drying. These two excellent discussions on processing of fish along with the nutritional problems raised by Professor Skulberg would suggest that a symposium dealing with fish production and utilisation would be an extremely important topic for further consideration by IUFOST.

Professor Samuelson, with his enthusiasm and skilful use of visual aids gave a masterful discussion on pasteurisation and sterilisation of dairy products that imbued the audience with his own love for the subject. He began by lucidly outlining the differences and reasons for both pasteurisation and sterilisation, followed by a brief outline of the processes involved. He pointed out the reasons and results of each step in both processes. Since it was so clearly explained by Professor Samuelson, I will simply refer you to his text for fear that I may miss something.

The discussion indicated that selection of pasteurisation temperatures is a compromise between complete destruction of the micro-organisms and the enzymes produced and the degradative quality changes brought about by more severe heating. It was also brought out that the nutritive value of milk that is repasteurised or even reheated can destroy the folic acid, which is protected during the first heating process by vitamin C. This could be a most important nutrition question, especially for babies where reheating of bottles is common. It was also brought out that this problem could be compounded in countries permitting repasteurisation. The possibility of pasteurisation on farm or even going beyond to concentration to decrease transportation costs seems worthy of consideration by researchers and manufacturers.

Dr Schut next discussed evaporation and drying of dairy products, indicating that the processes must be considered from four standpoints: (1) nutritional changes caused by the processes; (2) changes in the sensory properties; (3) technological behaviour; and (4) the effect on price. The importance of these processes can vary depending upon the population for which they are intended. Heating during evaporation and the drying process can cause browning and may continue even during storage. Finally, Dr Schut presented data on the differential thermal analysis (DTA) showing the curves differed with different amounts of heating, and suggested that DTA be investigated as a possible tool for monitoring changes in milk during heating.

The discussion following concentrated on two topics: (1) the significance of the DTA data, which are not fully explored at this time; and (2) the possible use of whey proteins in soft drinks. It was clearly pointed out by Dr Schut and Professor Samuelson that whey proteins are rich in the high quality serum proteins—β-lactoglobulin and albumin—and could offer distinct advantages to any group needing high-quality protein. At present the evidence is not clear on this subject and data are not available—only Dr Bender's 'meaningless mean' which suggests the average person in the Western world would have little to gain but says nothing about smaller segments of the population. Certainly efforts should be made to use this high-quality protein food—whey. It seems likely that this so-called by-product could become a valued item in commerce, now that we are concerned with environmental factors. Nutritional research to determine where whey proteins could best be used should be encouraged.

Dr Baardseth, one of our hosts of this conference, clearly outlined some of the research on meats and vegetable products at the Norwegian Food Research Institute, clearly showing how temperatures between 45–58 °C could be used to tenderise meat. The product suffered some from surface browning but otherwise the process seemed quite helpful. Using model systems of whale myoglobin, she showed that proteins having a high water content could stand high temperatures without adverse effects, however, at lower water contents the damage increased during heating. This suggests that high temperatures may be useful in the initial phases of drying but should be reduced as the moisture content is lowered. Finally, she pointed out an interesting study in which dehydrated vegetables stored at 5 °C surprisingly enough suffered more organoleptic damage than those stored at 23 °C.

The discussion following covered such important topics as to why the difference between the acceptability and oxidation of the dehydrated products stored at 5 and 23 °C. Many possible factors, including cycling of the coolers, the moisture content, previous history and oxygen uptake were suggested as possible causes. Certainly analyses for lipases would seem to be crucial. The DSC method of following changes in products during heating was suggested and seems worthy of consideration by all. Finally, it is not clear why reducing sugars cause a decrease in biological value of certain products (combined meats and vegetables) but seems to be related to destruction of methionine.

Dr Schulerad gave an excellent discussion on cereals as related to the effects of baking, by first reviewing the steps in bread making which included mixing, fermentation, moulding and proofing, pointing out the significance of each step. He pointed out temperature is important in each step including development, fermentation and baking. During baking, temperature control is crucial to coagulate the gluten, gelatinise the starch, setting the texture, develop the crust and develop the flavour and aroma. The activity of α-amylase is essential to both the changes in the proteins and in development of flavour and aroma. Rapid cooling is desirable before wrapping and a new vacuum process for accomplishing this step was described. Staling results from a thermolabile change in the starch structure but can be prevented by rapid freezing followed by rapid thawing before use. A completely new process to freeze and thaw bread to deliver fresh bread regardless of work hours was described by Dr Schulerad.

The discussion following dealt with the contributions of the body of the bread to flavour, although the crust contains the greatest concentration of flavour constituents. The effect of the type of flour and its composition markedly influences the bread-making process. It was indicated that canned bread at low pH (below 5·0) could be a problem. The role of oxidising and reducing agents and of emulsifiers was briefly considered. A question relative to the problem of phytate in binding calcium, and more recently iron, was considered, the latter may be a particularly serious problem. After this discussion one was left with the feeling that much of bread making is an art and that more complete discussion coupled with increased research could change the process to a more scientific basis. Certainly further programmes to discuss the problems and processes could be useful.

Dr Koetz, in collaboration with his co-worker, Dr Neukom, discussed the release of bound nicotinic acid by chemical and thermal processes. It was shown that either acid or alkali treatment resulted in an increase in the availability of bound nicotinic acid in maize and other cereal crops. It was also shown that heating and drying of sweet corn increased the available niacin. Studies showing that parboiling of rice improved the nutritional value by release of thiamine and niacin was reported. Elegant studies on the structure of bound niacin were also reported.

In the discussion it was pointed out that the exact structure of bound niacin is not known, although it appears to be a mixture of glycoproteins and not be the simple compound of niacytin as originally proposed by Kodicek. Studies on cereals and their processing to provide maximum nutritive value of the B vitamins, and particularly by release of bound niacin, could be very important to segments of the population consuming large proportions of their diet in the form of cereals.

The final paper of this session was presented by Dr Hoffmann and covered the influence of heating of the meat proteins upon their SDS-gel electrophoresis patterns. In this interesting study, decreases were observed in the intensity of the staining patterns of the myofibrillar proteins on heating, although heating had no effect upon the distance the various proteins migrated. The effects of heating were more marked at temperatures above 100 °C. The addition of various salts also caused a marked decline in the intensity of staining. Dr Hoffmann concluded that the decreased staining of these protein bands by Coomassie blue can be attributed to damaging of the functional groups responsible for dye-binding.

Questions dealt with the possible use of multivariate computer design for identification of foreign proteins that may be added to meat as extenders. This approach certainly seems to be a useful one but does not overcome the problem of the decrease in the intensity of staining. Peaks using DSC showed that patterns were quite different for different species and suggested that still another method may be useful for identification of foreign proteins. Studies of these kinds should be encouraged and should ultimately lead to methods of measuring not only the kind of added proteins in meat products but also to their quantitation.

Dr Mauron opened the last session by elegantly discussing the principles involved in measuring specific damage of food components during thermal processing. It was first pointed out that processing is designed to preserve foods but that side-effects are often undesirable. There are three ways to approach measurement of processing damage: (1) food analysis by either chemical means or *in vitro* availability; (2) study of reaction mechanisms, i.e. browning, heat treatment, alkali treatment or oxidation; and (3) biological evaluation using animal bio-assays, determination of nutritive value; classic toxicology and mutagenesis. The advantages and disadvantages of the various procedures for measuring amino acids and certain B-complex vitamins were discussed. Dr Mauron then discussed the main types of damage to protein, namely: (1) mild heat damage; (2) severe heat damage; (3) alkali damage; and (4) oxidation. He pointed out the mechanisms and markers for damaged proteins, such as methionine losses, sulphoxide and lysinoalanine formation.

Discussion dealt with problems in biological testing, specifically in regard to the Ames test, which could be dangerous if used for food testing or perhaps even for individual ingredients. The test organism is very sensitive, but has no mechanism for repair and is abnormal in other ways.

Professor Bender, in reaching a consensus, neatly sidestepped the issue and thereby left the discussion to me. His lucid discussion of the background to studies of nutritional changes in food processing and some examples of misinformation and correct information has put this symposium in perspective. Procedures for properly designing experiments to evaluate results and to correctly interpret them is one area in which researchers should be extremely careful, and as Professor Bender pointed out, it is only from this standpoint we can reach a consensus. Only when correct information becomes available can we correctly assess the losses of nutrients in food processing.

Let me say that we can reach a consensus in saying that this symposium has provided much important information on food processing and has pointed out some topics that need further research and discussion. If we have been successful in doing this, I submit that this symposium has served a vital purpose. Now I would suggest that we should each make use of the interchange of ideas that has taken place here.

Again may we thank IUFOST and our hosts of the Norwegian Food Research Institute especially to Director Kvåle and Dr Høyem and their colleagues for their efforts in making this symposium a success.

Index